Fourth Edition

The Canadian North

Issues and Challenges

Robert M. Bone

OXFORD
UNIVERSITY PRESS

OXFORD
UNIVERSITY PRESS

Oxford University Press is a department of the University of Oxford.
It furthers the University's objective of excellence in research, scholarship,
and education by publishing worldwide. Oxford is a registered trade mark of
Oxford University Press in the UK and in certain other countries.

Published in Canada by
Oxford University Press
8 Sampson Mews, Suite 204,
Don Mills, Ontario M3C 0H5 Canada

www.oupcanada.com

Copyright © Oxford University Press Canada 2012

Library and Archives Canada Cataloguing in Publication

Bone, Robert M.
The Canadian north : issues and challenges / Robert M. Bone. — 4th ed.
Includes index.

ISBN 978-0-19-544503-9

1. Canada, Northern—Geography—Textbooks. 2. Canada, Northern—Economic conditions—
Textbooks. 3. Natural resources—Canada, Northern—Textbooks. 4. Native peoples—Canada,
Northern—Textbooks. I. Title.

HC117.N5B65 2012   917.19   C2011-908059-1

Cover image: The Canadian Press / Jonathan Hayward

Oxford University Press is committed to our environment. This book is
printed on paper which has been certified by the Forest Stewardship Council®.

Printed and bound in Canada.

1 2 3 4 — 15 14 13 12

Contents

List of Figures

List of Tables

List of Vignettes

Preface

The North is a fascinating region of Canada. I hope that the students reading this book will not only agree, but also become familiar with its unique character and challenges. Over the last 50 years, profound changes have taken place in the Canadian North, including the emergence of the resource industry as the dominant factor in the economy, the rise of Aboriginal political power, and the threat to the northern environment and its peoples by industrial and urban pollution. New challenges have emerged, including global warming and Arctic sovereignty, as well as new opportunities, such as the resurgence of scientific activities spurred by the International Polar Year in 2007–8.

Megaprojects define the nature of the resource industry and the scope of impact on the environment and Aboriginal peoples. These projects are not distributed evenly across the North, but are concentrated in the Subarctic. The hidden costs of resource development, especially its effect on the land, are now recognized in the much more stringent environmental impact assessment procedure. However, massive cleanup challenges remain due to abandoned mine sites that contain toxic and radioactive wastes.

As we move forward in the twenty-first century, Aboriginal peoples have ceased to be marginal players in the process of economic development and have become major players through the land claims process. Aboriginal participation in megaprojects— whether as partners in the proposed Mackenzie Valley pipeline project, as owners of vast northern property as a result of land claim agreements, or as active participants in environmental impact assessments—has changed the way of doing business in the North. Still, Aboriginal workers hold relatively few jobs in the northern economy. This low rate of participation is a complex subject worthy of a much fuller discussion than found in this book. Nevertheless, geography and culture play a role. Most Native settlements, for example, are both small in size and remote in location. As a result, they are ill-suited for Western-style economic development, thereby stranding their residents from participating in the market economy. Aboriginal residents of Native settlements live in familiar places with strong ties to their former traditional land-based economy, but such communities remain isolated from Canadian society. This isolation is a double-edged sword. On the one hand, isolation limits contact with the larger Canadian society and thus ensures a continuing of the past lifestyle, insofar as that remains a possibility. On the other hand, isolation limits opportunities found in the rest of Canadian society and, beyond, in the international community.

An economy based on huge projects to extract resources from the environment is not part of traditional Aboriginal culture, which stresses collectivism and respect for the environment. Yet, the flexible nature of Aboriginal culture is coming to grips with resource development in different ways. Comprehensive land claim agreements illustrate this flexibility by allowing a foot in the market economy through Aboriginal corporations and another foot in their traditional world based on the land and wildlife. Perhaps we can consider such institutional developments as 'hybrid' institutions that blend the old with the new.

The readers of this book will note that resource development is perceived differently by northerners and southerners. The first group sees the North as a homeland. They look at resource development from three perspectives:

- What benefits do northerners receive?
- What effect does it have on the environment, especially wildlife?
- What impact will it have on local cultures?

While southerners have become more aware and accepting of northern perspectives, they tend to regard the North as a 'developing' frontier with resource companies leading the way.

The first edition of this book was published in 1992; the second in 2003; and the third in 2009. What is remarkable to me is the magnitude of change that has taken place since the first edition appeared; even more astounding is the prospect for greater change in the future. Social change is one element, especially the advances made by the Aboriginal community where they have gained control over some of the economic and environmental levers affecting their lands. Political change is another element, with the emergence of Nunavut and Nunavik, despite a recent political stumble in the latter when a final agreement failed in a ratification vote among the Nunavummiut. Economic change is triggered by the growing world demand and rising prices for resources, especially energy. Finally, back in 1992, few people would have thought that Arctic sovereignty, climate change, and the race for the seabed of the Arctic Ocean would take centre stage. For that reason, a new chapter on the geopolitics of the Arctic has been added to this new edition.

Like the previous editions, the fourth edition is built on close relations with geographers and other northern scholars, plus the support of the staff at Oxford University Press, particularly Patricia Simoes, the developmental editor, and Richard Tallman, the copy editor. Over the years, Richard has become familiar with my writing and we work well together in moving the manuscript through its final stages. The anonymous readers selected by Oxford University Press, who made a number of constructive suggestions and brought a number of issues into focus, deserve special mention. One recommendation was to merge the chapters on resource development and megaprojects, which has been done in this edition. Finally, a special note of appreciation goes to my wife, Karen, for her support and willingness to shoulder so many of my family responsibilities while I disappeared for long periods of time into my 'writing room'.

Robert M. Bone
July 2011

1

Northern Perceptions

The North is a unique region within Canada. Canadians have several images of the North. While most Canadians have never visited the North, this region holds a special place in their minds and hearts. In many ways, the challenges and issues facing northerners symbolize the nation's sense of purpose and suggest the course of its future development. Yet, what exactly is the North? For most, the North represents the colder lands of North America where winters are extremely long and dark, where permafrost abounds, and where few people live. For Aboriginal peoples, the North is their homeland where comprehensive land claims provide them with more control over their destiny within Canada. The North is also the place where global warming is predicted to have a greater impact than elsewhere in Canada. One possible consequence could lead to summer shipping through the Northwest Passage, thus providing a shorter route between Asia and Europe. Global warming, the Northwest Passage, and Arctic sovereignty are discussed more fully in Chapter 2 and Chapter 8.

Canadians also perceive of this cold and diverse environment in two ways: either as a resource frontier or as an Aboriginal homeland. For the indigenous peoples who first occupied these lands many thousands of years ago, the Canadian North forms a series of Native homelands. Others accept the frontier version of the North, which implies building southern-like communities and developing a resource economy to serve the needs of the North American and global economies. While the vision of the frontier dominates the overall direction of northern development, the homeland theme surfaces in areas where Native people comprise the majority of the population, where they hold Aboriginal rights to the land, and where they have established a form of self-government. While the economic and political forces shaping the North of tomorrow swing back and forth between these two visions, the process of accommodation is softening both positions. The fundamental question is: Will the North of tomorrow replicate the course of development found in southern Canada or will it break away from that template? Already, there are signs of a different course. Nunavut, for example, represents a political divergence from the other two territories and the 10 provinces, and it represents an Inuit response to the northern homeland question. Another example is the formation of the Aboriginal Pipeline Group, which can acquire a one-third ownership in the proposed $7.8 billion Mackenzie Gas Project natural gas pipeline, reflecting both a desire on the part of northern Aboriginal peoples to participate in resource development and recognition by corporate Canada—in this case, four major oil and gas companies[1]—that Aboriginal participation is necessary for business in the North (see Vignette 1.1). Yet, the high hopes for this bold project have dimmed along with the dreams of the Aboriginal Pipeline Group (see Chapter 5 for more details).

Vignette 1.1 **What Is Development?**

In Western culture, the concept of development describes a process of long-term change in society. In this sense, development is linked to **modernization theory**. Driven by economic growth flowing out of industrialization, development is concerned with economic, social, and political factors shaping society. It includes not only the notion of economic growth but also social and political changes that involve people's health, education, housing, security, civil rights, and other social characteristics. Seen in this light, development is a normative concept involving Western values, goals, and beliefs. It follows that development in the Canadian North occurs within the market economy and is supported by the Canadian political system. Unfortunately, this concept of development views traditional values and ways as 'hindrances' to economic and social progress. For that reason, until the late twentieth century, the place of Aboriginal peoples within the Canadian version of development was often ignored or dismissed. Since then, the legal validation of Aboriginal rights, the establishment of modern land claim negotiation processes, and the participation of Aboriginal organizations in the market economy mark the 'necessary' inclusion of Aboriginal peoples within the Western development process. Still, participation in the market economy depends on having the capital or equivalent resources. The Aboriginal Pipeline Group, for example, has obtained the option for a one-third ownership in the proposed Mackenzie Valley gas pipeline. Participation takes different forms but was greatly encouraged by comprehensive land claim agreements that provide the capital for Aboriginal regional corporations such as the Inuvialuit Regional Corporation and the Makivik Corporation. Impact benefit agreements, on the other hand, represent another avenue of participation and ensure that the Aboriginal people living in areas designated as having **Aboriginal title** receive negotiated benefits from resource companies.

Defining the North

The North is both easy and difficult to define. In a narrow sense, the North refers to the three territories. But such a definition ignores the natural conditions found on both sides of the sixtieth parallel. In this text, the geographic definition and spatial extent of the North is defined by its two natural zones or biomes, the Arctic and Subarctic (Figure 1.1).

The Arctic and Subarctic extend over a vast area—nearly 80 per cent of Canada—that includes the three territories and reaches into seven provinces (only the three Maritime provinces do not extend into the Subarctic). As well, the extent of the natural vegetation making up these biomes is clearly indicated on the landscape. For example, the southern boundary of the North is marked by the places where the boreal or northern coniferous forest of the Subarctic gives way to other natural vegetation zones such as the grasslands of the Canadian Prairies. At the same time, the North's southern edge closely corresponds to the southern limit of permafrost, indicating the interrelationship of climate, natural vegetation, and permafrost. The Arctic exists in the three territories and in four of the seven provinces (Quebec, Newfoundland and Labrador, Ontario, and Manitoba), while the Subarctic occurs in all seven provinces with a northern landscape (British Columbia, Alberta, Saskatchewan, Manitoba, Ontario, Quebec, and Newfoundland and Labrador) and two territories (Northwest Territories and Yukon). Each biome is described in detail in Chapter 2.

Yet, often understated in such a natural definition of the North are the extremely varied biophysical diversity, cultural differences, and historic events found in this polar

Figure 1.1   The Canadian North

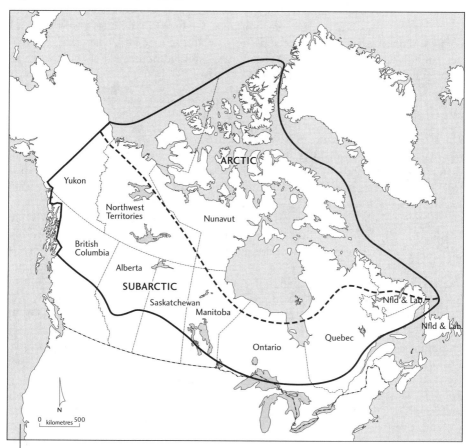

The Canadian North consists of: the Arctic and Subarctic. The southern boundary of the Canadian North corresponds with the southern limit of the boreal forest. From a political perspective, the North consists of the three territories (Yukon, Northwest Territories, and Nunavut) and the northern areas of seven provinces (British Columbia, Alberta, Saskatchewan, Manitoba, Ontario, Quebec, and Newfoundland and Labrador).

realm—all of which provide the Canadian North with its unique character and thus distinguish it from other parts of the **Circumpolar World**. Any definition and understanding of the North must account for the history and prehistory, the natural resources and physical environment, and the peoples of this vast region. Aboriginal peoples arrived in Canada's North thousands of years before the Vikings reached Greenland in the late tenth century and then Newfoundland around the year 1000. The value of the region's mineral resources were the focus of the Klondike gold rush in the late nineteenth century, which gave substance to the North as a northern Eldorado.[2] Other examples focus on the North's cold environment as a frozen land where permafrost is commonly found, where ice covers the Arctic Ocean for most of the year, and where winter nights are long and cold. And, finally, an appreciation of the North must include a homeland definition where Aboriginal peoples, born and raised there, have a special, deeper commitment to that place. Louis-Edmond Hamelin (1979: 9) described this feeling as 'a trait as deeply anchored as a European's attachment to the site of his hamlet or his valley'. In this single statement, Hamelin has captured the geographer's notion of a sense of place and the parallel idea of regional consciousness.[3] In this text, these three images of the North—as a northern resource Eldorado, as a cold environment, and as an Aboriginal homeland—reoccur in the following chapters. The richness and diversity of its physical geography are captured in the next chapter, while its cultural/historical geography is the focus of Chapter 3.

A Different Region

The North is a different region from the rest of Canada. As part of the Circumpolar World, it has more similarities with other high-latitude regions, such as Greenland, than it does with southern Canada. Northern peoples, too, have experienced the historic encroachment of nation-states, thus placing this region and its peoples within larger political settings, such as Canada, Denmark, and Russia.

But what about the differences between the North and the rest of Canada? To begin with, the North has the coldest environment in Canada, with permafrost—permanently frozen ground—abounding, the arctic ice pack covering much of the Arctic Ocean, and the flashing of the aurora borealis across its winter skies. Second, the North is by far the largest region in Canada. The North accounts for 76 per cent of Canada's geographic area. A third revelation is that the North is almost equally divided between the Territorial North and the provincial norths (Table 1.1). While most Canadians realize that the three territories occupy a vast area of Canada, the sheer size of the northern area found in the seven provinces is surprising to many people. Of all the provinces, Quebec has the largest 'northern' area, which makes up 81 per cent of that province. Two provinces—Newfoundland and Labrador and Manitoba—are next, with 74 per cent of their territory classified as 'northern'. Northern lands make up 65 per cent of Ontario and 50 per cent of Saskatchewan. Alberta and British Columbia trail behind, with 47 and 40 per cent respectively.

A fourth fact marking the North as a different region from the rest of the country is its small population and its strong Aboriginal composition. Lying beyond Canada's ecumene, this thinly populated landscape contains fewer than 5 per cent of Canada's population—less than 1.5 million people. Of the two northern biomes, the Arctic

Table 1.1 Geographic Size of the Canadian North, by Province/Territory

Province or Territory	Total Area (000 km²)	Northern Area (000 km²)	North (%)	Canada (%)
Newfoundland/Labrador	405	300	74	3.0
Alberta	662	310	47	3.1
Saskatchewan	651	325	50	3.3
British Columbia	945	375	40	3.8
Manitoba	648	480	74	4.8
Ontario	1,076	700	65	7.0
Quebec	1,542	1,250	81	12.5
Provinces	*5,929*	*3,740*	*63*	*37.5*
Yukon	482	482	100	4.2
Northwest Territories	1,346	1,346	100	13.5
Nunavut	2,093	2,093	100	21.1
Territorial North	*3,921*	*3,921*	*100*	*38.8*
Canada	*9,985*	*7,661*	*76*	*76.3*

Source: Statistics Canada (2005).

contains fewer than 100,000 people, making its population density of 0.01 persons per km² one of the lowest in the world. Furthermore, Aboriginal peoples (Inuit) form over 80 per cent of those living in the Arctic; and for much of the Subarctic the Aboriginal population represents a majority. This demographic fact and the associated cultural change taking place are at the heart of critical economic and political issues and challenges (see Vignette 1.2).

Lastly, climate change is affecting the North's physical geography and those changes have cultural, economic, and political implications. Climate change—most noticeably in the form of higher summer temperatures—is taking place more rapidly in the North than in other regions of Canada. Does this imply that physical geography is no longer a stable or fixed feature of the northern landscape? What are the long-term implications for the Arctic and Subarctic biomes, the Northwest Passage, and Aboriginal peoples? These topics, along with climate change and global warming, are discussed more fully in forthcoming chapters. Other features are unique to the North, such as the sense of isolation felt by newcomers to the North, are discussed below.

Sense of Isolation?

One of the defining features of the North is the different perceptions between those born and raised in the North and those who have come to the North for economic opportunities. Those who originate in the North, mainly Aboriginal peoples, have a strong **sense of place**. In simple terms, the North is their home. On the other hand, newcomers frequently have a hard time adjusting to the northern lifestyle, which is typified by small communities with limited urban amenities and with long travel

---

Vignette 1.2   Culture Change among Aboriginal Peoples

The culture of a society provides its members a frame of reference for interpreting and responding to events and ideas. What happens when that frame of reference is lost? An argument could be made that the pace of change for Aboriginal cultures in northern Canada has been overpowering and that their frames of reference are being replaced by the dominant Canadian frame of reference. One indicator is the demise of Aboriginal languages and the adoption of English—despite the efforts of the Aboriginal Peoples Television Network, founded in 1999, the ubiquity of the English language in remote northern communities has been a fact of life for the past generation.

An important question is: Are the cultural changes taking place in northern Canada a matter of adaptation or domination? Residential schools were clearly an attempt at assimilation and domination, while comprehensive land claim agreements are a means for Aboriginal peoples to adapt to a changed reality. However, do the resulting co-management boards flowing out of comprehensive land claims provide more power to Aboriginal participants or do they simply draw Aboriginal peoples more deeply into Canadian society? The 2001 Paix des Braves Agreement between the Quebec Cree and the Quebec government, whereby the province is enabled to complete the final phase of the James Bay Hydroelectric Project and the Cree receive $3.6 billion over 50 years, as well as full responsibility for their own welfare and economic development (see Chapter 7), provides some insights into this complex question. Nonetheless, what is given up in such agreements—often, the last remnants of a traditional lifestyle; a way of being and thinking and relating to others and the land; and a large part of a land base that has been central to cultural and spiritual identity for hundreds or thousands of years—is not easily equated with a sum of money, however large.

---

connections to their places of origin and previous residence, often major cities in southern Canada. While some adjust, many leave within five years. These transplanted Canadians living in remote northern centres, such as Aklavik, are affected by a sense of isolation from the rest of Canada (Figure 1.2). This psychological barrier can affect newly arrived residents, who live in communities where air transportation represents the only means of reaching southern Canada. Such communities are found in the three territories and, with the exception of British Columbia, in all seven provinces with northern areas.

This sense of isolation has a negative effect on recruiting skilled workers and professional people, and accounts for the high job turnover and out-migration. Companies and governments often employ incentives to attract workers to live in remote centres. Northerners, for example, receive a special income tax deduction—Northern Residents Deduction—for living in the North, but this deduction varies by tax region. In the more northerly (less accessible) tax region the deduction is pegged at 100 per cent, while in the less northerly (somewhat more accessible) tax region the deduction falls to 50 per cent. By reducing the amount of taxable income, northerners are able to reduce their personal income tax. Because the geographic definition of these two regions determines the amount of the Northern Residents Deduction and the size of northern living allowances for federal employees, Statistics Canada has joined in the search for a working definition of the North (McNiven and Puderer, 2000). A 'tax' definition of the North is complicated by two factors: (1) improvements in the national transportation system that create greater accessibility to northern communities; and (2) population increases of northern urban centres that lead to the availability of more public and private services. This dynamic element to northern boundaries is discussed further in this chapter under the subsection 'Nordicity'.

Isolation also takes the form of a limited transportation network. The most striking limitation is the heavy reliance on air transportation. All communities are served by air transportation but only a few can be accessed by roads or ship and even fewer by railways. For example, Nunavut is not connected to the rest of Canada by surface transportation, but air service is available to all 26 communities. A second characteristic is the north/south orientation of the North's transportation system. Unlike southern Canada, which has an east/west transportation axis, northern Canada has a 'feeder'

Figure 1.2   Aklavik, Near the Arctic Ocean

The small Native community of Aklavik, with a population of 597 in 2006, is located at the northern edge of the boreal forest in the Mackenzie Delta. In this winter scene, the Peel Channel is frozen, the ice road is operating, and, in the distance, lakes form a high percentage of the boreal forest landscape. Beyond is the Arctic coastline where tundra vegetation forms the natural vegetation. Access to the outside world is by air, by small boat to Inuvik in the summer, and by ice road in the winter. Foods, building materials, and other goods from southern Canada reach Aklavik by river barges that depart from Hay River. Hay River, on the southern shore of Great Slave Lake, has road and rail links to Edmonton, which serves as the 'gateway' city to the Northwest Territories.

Source: David Boyer/National Geographic/Getty Images.

system that allows resources to flow from the North to Canada's main transportation system. As well, the movement of supplies and food to northern communities and mining sites takes place on these north/south routes.

The Political North

Political geography is dominated by the division of the North into territorial and provincial governments and by the existence of First Nations reserves. The federal government plays a more central role in the three territories and within Native communities than in the provinces. For example, Ottawa provides most funding for territorial and First Nations governments. Provinces, on the other hand, receive much of their funding from royalties from resource development, thus making them less dependent on Ottawa. This independence allows provinces to treat their northern areas as provincial hinterlands somewhat along the lines of the core/periphery model discussed later in this chapter. Provincial governments have unlocked their northern resource hinterlands by promoting railway and highway construction within their provincial boundaries. The strategy was simple: by lowering the cost of transporting natural resources such as minerals and timber to world markets, hinterland development became possible. British Columbia provides one example of this strategy. The provincial government built and operated BC Rail (formerly the Pacific Great Eastern Railway) until it was sold to CN Rail in 2003. This provincial railway linked the vast northern interior of British Columbia with the port of North Vancouver. As a result, the forest and other resources of the northern interior of British Columbia were exploited and their products exported to southern markets by means of BC Rail.

The North's dual political structure also creates a mental divide for many Canadians, who believe that the three territories represent Canada's North. Margaret Johnston (1994: 1) correctly observed that 'there has been a tendency to view much of this [provincial] part of the north in Canada as less northern than Yukon and the Northwest Territories, and consequently it has received considerably less attention as a northern region.' Indeed, Coates and Morrison (1992) describe the northern areas of seven provinces as 'the forgotten north'. In this text, as noted earlier, the North includes the three territories and the cold lands found in the northern sectors of seven provinces. The political division between territories and provinces warrants identifying the three territories as the 'Territorial North' and the northern parts of the seven provinces as the 'Provincial North'.

Sparked by global warming, nations around the world are looking at the prospects of ocean shipping through the Northwest Passage and at the unclaimed portions of the Arctic Ocean. Canada is seeking to defend its interests by claiming control of (or at least the right to manage) the Northwest Passage and by gaining its share of the unclaimed seabed of the Arctic Ocean. The stakes are high because of the vast energy deposits in the Arctic Ocean and because of the potential economic/environmental significance of ocean shipping through the Northwest Passage. These complicated political matters are examined more fully in Chapter 8, and they signify a greater federal presence in, or awareness of, the North.

Last but not least, internal political realignment is taking place, with the prospect of a new third level of government in the provinces. The traditional political arrangement of territories and provinces has been shaken by the emergence of Aboriginal governance, which has flowed from comprehensive land claim agreements. Aboriginal governance takes two forms—ethnic and public governments—with Nunavut and Nunavik representing 'public' governments and First Nations adhering to ethnic governance.

Most public-sector employment is found in capital cities and regional centres. However, Aboriginal people often reside in small settlements. Two factors pose serious barriers to Aboriginal workers gaining access to employment. One is the geographic mismatch between jobs and small communities, i.e., few jobs are available where the majority of Aboriginal workers reside. Another is the mismatch between the job requirements and the skill/experience found in the Aboriginal workforce, i.e., employers seek workers with professional and trades qualifications and experience.

Common Characteristics

Even though the geographic extent of the Canadian North is vast, this region has many common characteristics (Table 1.2). A cold environment, sparse population, and extensive wilderness areas with limited but unique biophysical diversity are the rule. Another characteristic is the high percentage of Aboriginal peoples in the northern population, unlike in southern Canada, where the Aboriginal population forms a very small proportion of the total population. In some areas, they form a majority of the local population. In Nunavut and Nunavik, the Inuit make up around 85 per cent of the population. Like other hinterlands within the core/periphery structure of the world economy, the North has a resource economy. While forestry and mining activities account for most of the value of economic production in the North, these industries employ relatively few workers. In fact, the vast majority are employed by public agencies. In the next pages, three common characteristics—the North's winter, the concept of nordicity, and the core/periphery model—are explored more fully.

Table 1.2 Common Characteristics of the North

Physical Characteristics	Human Characteristics
Cold environment	Sparse population
Limited biophysical diversity	Population stabilization
Wilderness	High cost of living
Remoteness	Few highways
Permafrost	Aboriginal population
Vast geographic area	Settling of land claims
Fragile environment	Financial dependency
Slow biological growth	Resource economy
Importance of wildlife	Reliance on imported foods
Global warming	Country food
Continental climate	Economic hinterland

Northern Winters

Winter, its length and intensity, is one measure of a cold environment (Vignette 1.3). Canadians living in the North are confronted by one of the longest and coldest winters in the world. For Aboriginal peoples, the cold environment is accepted as a normal part of their way of life; for newcomers, the length and intensity of northern winters are beyond their experiences in southern Canada.

Vignette 1.3   Long Nights and the Arctic Circle

The **Arctic Circle** (see Figure 1.3) marks the location on the Earth's surface where, at the time of the winter solstice (21 December), night lasts for 24 hours. The Arctic Circle is an imaginary line that circles the globe at 66° 33'N, the northward limit of the Sun's rays at the time of the winter solstice. At this latitude, the Sun does not rise above the horizon for one day of the year. However, the polar night does not begin immediately because twilight occurs for a short period of time when the Sun is just below the horizon (Burn, 1995: 70). Interestingly, the position of the Arctic Circle is not constant because its location on the Earth's surface is determined by the angle of the Earth's axis of rotation relative to the plane of the Earth's orbit around the Sun. At the moment, the Arctic Circle is shifting towards the North Pole by about 15 metres per year.

The key physical element controlling Canada's cold environment is the low quantity of solar energy received in Canada's high latitudes, which roughly translates into the mean annual air temperature isotherms shown in Figure 1.3. The moderating influence of the warm Pacific Ocean breaks the latitudinal pattern of winter as measured by these isotherms. For example, the direction of the five-degree isotherm runs in a north/south direction along the coastal area of British Columbia while the same isotherm takes an east/west direction as it enters the interior of British Columbia where the influence of the Pacific Ocean diminishes.

The cold environment is also associated with extremely long winter nights (see Vignette 1.3). This phenomenon of darkness increases towards the North Pole. Except for a brief twilight at the Arctic Circle on the winter solstice, the sky is otherwise dark. Northerners living at the Arctic Circle are subjected to this 'darkness' phenomenon for at least one day a year while those living in higher latitudes are subject to an even longer period without seeing the Sun. Within this 'zone of winter darkness', the number of days in which the Sun does not rise above the horizon varies. At Alert (82° 29'N) there are nearly five months of continuous darkness, while at Inuvik (68° 21'N) there is just over one month of continuous darkness. The reverse conditions occur in the summer: Alert has nearly five months of continuous sunlight and Inuvik has just over one month of continuous sunlight—providing there is no cloudy weather. At the North Pole, the Sun appears on the horizon at the spring equinox (21 March) and does not set until the fall equinox (21 September). With each additional day, the Sun rises higher in the sky until the summer solstice (21 June), when it reaches its maximum height above the horizon.

There are economic, social, and psychological implications of this 'darkness' phenomenon in the North for southern Canadians, who are not accustomed to the lack
of light (and the lack of darkness) for long periods of time. For those living in high
latitudes, the continuous darkness of the Arctic winter imposes severe restrictions on
outdoor work and recreational activities. Even more important, the combination of
darkness and a reduction of normal activities may adversely affect the physical health,
emotional equilibrium, and motivation of southern Canadians living in high latitudes.
This phenomenon is commonly called 'bushed'. The reappearance of the Sun on the
horizon usually provokes a favourable response from 'transplanted' Canadians. In
Inuvik, residents celebrate the return of the Sun with a 'Sunrise Festival' featuring a
huge bonfire and a variety of community events to celebrate the Sun's return.

While temperatures have warmed in recent years, just how is this warming affecting the cold environment? It is difficult to substantiate 'impacts' on the environment

Figure 1.3 The Cold Environment

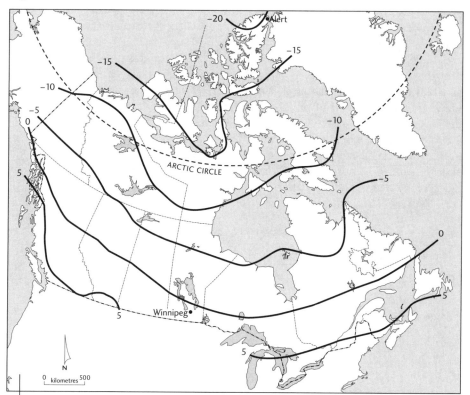

A measure of the North's cold environment is provided by mean annual air temperatures in degrees
Celsius. Each of the six isotherms indicates that the mean annual air temperature becomes progressively lower as latitudes increase. The energy deficit zone is defined as having an annual negative energy
balance. This zone lies north of the zero degree isotherm, which also closely approximates the southern
boundary of permafrost. Since the bulk of the temperature readings determining the location of the
isotherms shown here were recorded in the twentieth century, one question yet to be resolved is whether
global warming has affected or is affecting the cold environment, thus shifting the isotherms further north
and perhaps even eliminating the coldest isotherm (–20°C).

because of the complexity of measuring such change over time. Ice, however, responds more quickly to warmer temperatures, and it is relatively simple to measure change based on the extent of Arctic sea ice during the late summer. Glaciers, too, provide a baseline for measuring change. For sea ice, satellite photography produced by the National Snow and Ice Data Centre has revealed that the extent of sea ice on the Arctic Ocean as of 8 August 2007 was 5.8 million km^2 compared to the 1979–2000 August average of 7.7 million km^2 (Miller, 2007). The annual summer melting of Arctic sea ice varies. However, since 2000, the annual melt has not only exceeded the long-term average, it has also replaced much of the thicker, harder polar ice with thinner one-year ice. In the coming years, additional satellite photography of the Arctic Ocean will demonstrate if the area of annual summer melt is expanding, thus suggesting a long-term trend rather than a short-term variation.

Nordicity

While the North has a cold environment, this environment varies from place to place (Slocombe, 1995: 161). The human landscape also varies. **Nordicity**, a concept introduced by the Quebec geographer Louis-Edmond Hamelin (1979), combines human and physical factors to measure the degree of 'northernness' at specific places. It provides a quantitative measure of 'northernness' for any place based on 10 selected variables that aim to represent all facets of the North. These variables, called polar units, are a combination of physical and human elements such as summer heat (or the lack of) and accessibility (or the lack of). The North Pole has a nordicity value of 1,000 polar units, which is the maximum value possible. The southern limit of areas in Hamelin's classification system occurs at 200 polar units. At that point, the North ends (as defined by nordicity) and the South begins. Table 1.3 indicates the nordicity rating of selected Canadian centres.

Table 1.3 Nordicity Values for Selected Canadian Centres*

Centres in Southern Canada	Polar Units	Centres in Northern Canada	Polar Units
Halifax	43	Thompson	258
Montreal	45	Fort Nelson	282
Timmins	67	Whitehorse	283
Calgary	94	Schefferville	295
Winnipeg	111	Uranium City	396
St John's	115	Kuujjuarapik	414
Edmonton	125	Aklavik	511
Chibougamau	151	Iqaluit	584
The Pas	185	Old Crow	624
Grande Prairie	198	Sachs Harbour	764

*Centres with 200 or more polar units are defined as 'northern'.

Source: Hamelin (1979).

The physical elements measure 'coldness' while the human elements measure accessibility/development (see Appendix I for the complete list of variables, the assigned values, and the method of calculating nordicity). This approach permits the classification of the North into three regions (Middle North, Far North, and Extreme North) as shown in Figure 1.4. It also provides a dynamic quality that may change the nordicity rating of a place. For example, some of the variables determining the level of nordicity reflect human activities such as population size and transportation accessibility. If such variables change, then a place may acquire a lower or higher rating of nordicity (Vignette 1.4).

Hamelin is describing the North from a 'southern' perspective that reflects attitudes, beliefs, and values held by people residing in that part of Canada. An underlying assumption of such an ethnocentric viewpoint is that 'development' of the North reduces its nordicity and therefore makes it more like southern Canada. Another assumption is that the North is viewed as a hinterland of southern Canada. Southern

Figure 1.4   Canadian Nordicity

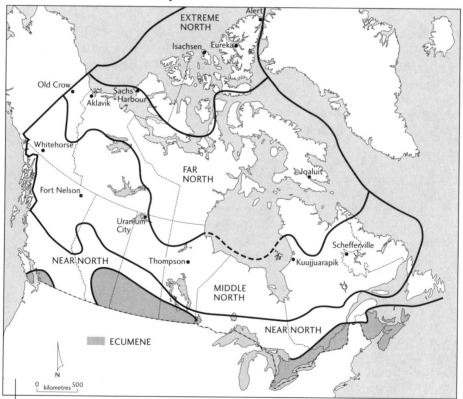

For geographers, 'nordicity' permits the creation of three northern regions: the Middle North, the Far North, and the Extreme North. The southern limit of nordicity is marked by the 200 polar unit line. Yet, do the concept of nordicity, the spatial place names in the North, and the notion of a population ecumene provide a southern, northern, or balanced interpretation of the North?

Source: Adapted from Hamelin (1979: 150).

Vignette 1.4   **The Dynamic Nature of Nordicity**

The concept of nordicity incorporates change caused by development as defined by population increase, greater accessibility, and more industrial development. As a result, the measure of nordicity for a particular place can change over time. Hamelin (1979: 35) stated that, in 1881, Saskatoon had a polar value of over 200. Over the next 75 years, Saskatoon grew in size and formed a transportation hub in western Canada. Consequently, by 1975, its nordicity rating fell to 116, placing the city in a 'southern' location. Similarly, Hamelin described the evolution of nordicity at Chibougamau, Quebec. In the 1880s, by this classification, its nordicity was around 400 on a scale of 1,000, but in 1979 he gave it a value of 151. At that time, he speculated that, given the role Chibougamau was expected to play in the James Bay Project, its nordicity could drop below 100. Now that global warming is a fact, physical elements may change, adding another dynamic feature to the concept of nordicity. For instance, if the summer ice pack retreats sufficiently, then ocean vessels could sail through the Northwest Passage.

Canadians see the North quite differently from those living in the North. For instance, each descriptive label—Middle, Far, and Extreme Norths—may seem strange and out of place to a Canadian born north of the sixtieth parallel. A northerner, for example, might have a different mental map of Canada, with the North described as the 'centre' and southern Canada as the 'distant land'. Furthermore, place names are a critical reflection of a region's cultural and historical experience. Ludger Müller-Wille (2001: 33) describes this importance:

> In Canada, as elsewhere, toponymy mirrors the country's cultural history and socio-economic and political structures, and it represents the various aspirations and goals of the different components that have shaped this particular spatial entity. The Canadian toponymic landscape, along with its glossary, is continuously evolving and thus also functions as a barometer for rapid changes that have occurred throughout the Canadian territory. Since the 1960s, these changes relate predominantly to the assertion of cultural and territorial rights by either Aboriginal or Québécois populations, using, among other elements, toponyms as strong cultural and political symbols to express distinctness.

Yet, Hamelin's scheme for determining nordicity, with its southern perspective, has a practical use. The federal and territorial governments have used this concept to determine isolation allowances for their employees, including administrators, nurses, and teachers. Back in the 1970s, the vast majority of these employees had come from southern Canada. For them, living in the North, but especially in a small community, represented a sacrifice warranting additional pay and staff housing. For the government, these allowances remain a necessary means of luring skilled and professional workers to relocate in isolated northern communities. A shortage of skilled and professional labour still exists and high salaries and housing allowances continue to attract those workers to the North.

Core/Periphery Model

Northern development takes place within a global economy. The **core/periphery model**—a product of Western thinking—best describes the economic relationship between industrial cores and periphery hinterlands within a capitalist economy (Wallerstein, 1979). Friedmann (1966: 76–98) turned this abstract economic theory into a geographic model with one core and three peripheral regions. In turn, Bone (2011: 17–20) has applied these four regions—core, upward transition, downward transition, and resource frontier—to Canada. In this text, the Canadian North is classified as a resource frontier.

Canada's North has another core/periphery nuance, namely that its Aboriginal peoples find themselves relatively powerless within another state. This political relationship sparked the term the 'Fourth World', introduced by the Canadian Aboriginal leader and author George Manuel (Manuel and Posluns, 1974) to describe the poverty, internal **colonialism**, and subsequent statelessness experienced by indigenous peoples within developed nation-states. For a long time, the relationship between the core political centre and the Aboriginal hinterland has been associated with the imposition of law and order, with the arrival of the Royal Canadian Mounted Police (RCMP detachments) in remote northern communities, and with social engineering, such as the notorious residential schools and the relocation of Aboriginal peoples into larger settlements to facilitate service provision. Only towards the close of the twentieth century did Aboriginal peoples, largely through land claim settlements, gain a modicum of independence, the best example being the creation in 1999 of the new territory of Nunavut. These changes in governance have helped to break some aspects of the political and economic confinement intrinsic to the concept of a 'Fourth World'.

The economic heart of the core/periphery model is resource extraction. Resource hinterlands are found in many parts of the world. Each is far from the manufacturing centres of the world but is connected to these larger centres and the market economy through trade. In Wallerstein's theory, one key assumption was that prices for manufactured goods increased over time more rapidly than agriculture and resource prices. In this way, the terms of trade became more and more favourable to the industrial core, thus allowing wealth to accumulate in this favoured region. But in recent times, prices for resource commodities have risen sharply, suggesting that a 'Super Cycle' for resource commodities has emerged, with higher commodity prices reversing the terms of trade, i.e., resource-rich countries and regions are benefiting from unusually high and seemingly sustainable commodity prices (Cross, 2007). The so-called Super Cycle first affected oil. Since 1972, world prices for oil have increased from US$2 a barrel to more than $90 a barrel by June 2011. Other commodity prices also have experienced this increase in demand and prices due to growing demand from China, India, and other rapidly industrializing countries. While the business cycle dictates that prices will fluctuate, only time will tell if the current range of energy and commodity prices will hold. In the meantime, resource hinterlands are enjoying an unprecedented 'boom'. In fact, Cross (ibid.) goes so far as to suggest that the fear of a return to lower commodity prices, as was experienced in the late twentieth century, is overstated because the expanding Chinese and Indian economies will require more and more energy and resources.

Two other important theoretical issues are embedded in the core/periphery theory. First, Wallerstein assumed a homogeneous labour force. Such an assumption means that the core/periphery model is silent on the critical issue of the place of indigenous peoples in global developments and, more specifically, of Aboriginal peoples in northern development. Second, Wallerstein saw no hope for diversification of the world's periphery because the core extracted all the wealth from trade with hinterlands. On the other hand, Friedmann depicted the periphery differently, and one hinterland area in his typology—the upward transition region—showed promise. While his work predated some of the specifics in this theory, Harold Innis (1930) presented a Canadian perspective on regional development through the lens of history, i.e., by examining the economic history of resource (staple) development in each region of Canada. In this way, Innis saw regional development occurring as a consequence of resource exploitation because it triggered a series of related economic activities that eventually led to regional economic diversification. In the case of the Canadian North, however, economic diversification is virtually impossible because its economy is subject to such a high level of economic leakage (the outflow of dollars to purchase outside goods and services not available in the North) and because its economy is based on non-renewable resources. Because of these two factors, Watkins (1977) feared that the North would fall into a 'staple trap', that is, when the non-renewable resources are exhausted, the North's economy would collapse.

For our purposes, the North is a resource hinterland, more specifically a resource frontier periphery, while the rest of Canada and the world are the industrial core. Companies in the industrial core dominate the economy in the resource hinterland and control the hinterland's pace of resource development. Northern development has been subject to 'good times and bad times' due to the variation in demand for resources. Put differently, much of the economic destiny of resource hinterlands is controlled by external forces and is extremely sensitive to fluctuations in world commodity prices. These fluctuations magnify the **economic cycle**, and may lead to boom/bust conditions in resource hinterlands. If Cross is correct, then the Super Cycle affecting resource commodities means that prices, while continuing to fluctuate, do so at a high range. But for the North, with its small business community, scattered population, and weak infrastructure, the fruits of development leak to other regions. For this reason alone, government intervention in the marketplace is warranted to ensure so-called 'northern benefits' from resource extraction remain in the North. Northern benefits, therefore, are crucial to northern development and lessen the North's dependency. Northern benefits and dependency are more fully discussed in Chapters 5 and 7.

Critical Issues

Within the two visions of the North lie several conflicting issues. Over the past 30 years, the balance of power among three key variables in the northern development equation has shifted (Table 1.4). To a large degree, this shift in power is related to the rejection of the old style of development so popular in the mid-twentieth century and its replacement with a development model that is more accommodating to the people who consider the North their home. This shift was driven by Supreme Court decisions, media coverage of events like the community hearings of the Berger

Table 1.4 Evolution of the Impact of Megaproject Proposals on Aboriginal and Environmental Issues

Year	Megaproject	Aboriginal Issues	Environment Issues
1949–54	**Iron Ore Project:** Iron Ore Company of Canada developed the ore deposits in northern Quebec and Labrador by constructing a resource town, railway, power plant, and extensive port facilities at Sept-Îles.	Ignored	Ignored
1971–85	**La Grande Rivière Hydroelectric Project:** A 1973 court ruling forced Hydro-Québec to negotiate with the Cree and Inuit of northern Quebec over Aboriginal title before completing the construction of the first phase of the James Bay Project. The result was the James Bay and Northern Quebec Agreement (1975).	James Bay and Northern Quebec Agreement (JBNQA)	Recognized in the JBNQA
1974–7	**Mackenzie Valley Pipeline Proposal:** The Berger Inquiry and the National Energy Board examined the social and environment aspects of this controversial proposal. Construction project did not proceed.	Discussed	Discussed
1981–8	**Oil Exploration in the Beaufort Sea:** As part of the National Energy Program, Ottawa announced the Petroleum Incentive Program, which provided $7 billion in federal funds for Canadian firms drilling for oil in frontier areas. Over 200 wells were drilled in the Beaufort Sea.	Ignored	Ignored
1982–6	**Norman Wells Oil Expansion and Pipeline Project:** The construction of this mega-project followed the 1981 federal environment report, Norman Wells Oilfield Development and Pipeline Project.	Federal social funding	Specific federal requirements
1991–2008	**NWT Diamond Mines:** Each mining proposal underwent a detailed environmental review and impact and benefits agreements were reached with local Aboriginal groups. More proposals for diamond mines are expected.	Impact and benefits agreements	Specific territorial and federal requirements
1993–2005	**Voisey's Bay Nickel Mine:** The Voisey's Bay mine proposal underwent a detailed environmental review and impact and benefits agreements were reached with local Aboriginal groups.	Impact and benefits agreements	Specific provincial requirements
2000–?	**Mackenzie Delta Gas and Pipeline Proposal:** Extensive environmental and social reviews by the National Energy Board and the Joint Review Panel were completed by the end of 2009. Aboriginal participation is strong, with the Aboriginal Pipeline Group set to own a third of the proposed pipeline. Yet, construction remains on hold because of unfavourable market condition for natural gas.	Federal social funding of $500 million over 10 years.	Specific territorial and federal requirements.

pipeline inquiry of the 1970s, and the rise of the environmental movement. In sharp contrast, much of the twentieth century was in the grip of modernization theorists, who supported the market economy and who viewed traditional practices as obstacles to change. This perspective of economic and social change provided a rationale for policy-makers, company executives, and Canadian society.[4] The relocation of Aboriginal peoples to settlements in the 1950s and 1960s exemplified a kind of social engineering with its philosophical roots in the modernization mentality—it seemed rational and cost-effective to relocate groups of people, but this did not account for the cultural impacts of relocation.

The first change saw resource companies forced to take into consideration Aboriginal cultural and environmental concerns.[5] The second change revolves around some Aboriginal groups taking an active role in the market economy, especially those groups that have negotiated comprehensive land claim agreements and that therefore have a financial base from which to begin. Furthermore, the link between comprehensive land claim agreements and self-government became more transparent with the Nunavut Land Claims Agreement because it included a commitment by Ottawa to create the territory of Nunavut.[6] Third, respect for the environment is now more widespread within Canadian society, various governments, and resource companies.[7] Along with the recognition of the importance of the environment to our future world, a growing number of regulations have made resource companies much more conscious of the need to minimize damage to the land and waters affected by their projects. Resource assessments and management are now central elements in northern development (Hanna, 2009; Noble, 2010; O'Faircheallaigh, 2007; Fitzpatrick, 2008), and these topics are discussed in Chapter 7. These three powerful shifts have had and will continue to have profound implications for the shape of northern development in the twenty-first century and may well lead to a greater diversification of the economy.

In spite of progress, however, the North still faces many issues and challenges related to the interaction of resource development, the environment, and the place of Aboriginal groups in resource development and Canadian society. In broad terms, all of these challenges relate to the effect of the resource economy on the northern environment and peoples. In particular, four issues are central:

1) Can the resource economy be a driving force for diversification of the northern economy that leads to sustainable development?
2) Have Aboriginal peoples successfully established a foot in both the market economy and their traditional economy?
3) Have resource industries sufficiently limited their impact on the environment?
4) How critical is Arctic sovereignty, but specifically the Northwest Passage and the international areas of the Arctic Ocean, for the North's economic future?

Since these issues are interrelated and are sometimes at cross-purposes, the search for solutions becomes extremely complex and inevitably must result in compromises. The linkage of these critical issues forms the crux of this book. Given the physical nature of the North, questions of economic diversification and Aboriginal self-determination may be resolved differently in the **Subarctic** than in the Arctic. For example, in comparison with the Subarctic, the Arctic at present has few resources that warrant commercial development. Similarly, given the federal jurisdiction and the differing political cultures in the Territorial North, these issues may be approached differently in the territories than in the provinces.

The next two chapters outline the physical and historical geography of the North. This background information equips the reader to better understand the physical limitations imposed on the resource economy, the historical process of northern development, and contemporary economic issues facing northerners. Northern development and its impact on Aboriginal peoples and the fragile environment form the main thrust of the remainder of this text. To be sure, past human activities have damaged the northern environment and the road to economic diversification is far from assured. We will see that the Aboriginal peoples of northern Canada face an uphill (but not impossible) struggle to find a place within Canada where they have a foot in both the market economy and their land-based economy.

Challenge Questions

1. While political scientists often define the North as consisting of the three territories of Yukon, Northwest Territories, and Nunavut, geographers do not. Why?
2. In your opinion, do most Canadians see the North as a resource frontier or as an Aboriginal homeland?
3. If you were to move from southern Canada to a small Inuit community in the Arctic, what adjustments might you face in order to develop a sense of place?
4. When Louis-Edmond Hamelin conceived of nordicity in the 1970s, he assumed that the North's physical geography was 'stable' while its human geography was not. Is this still true? If not, why not?
5. What is the connection between Inuvik's Sunrise Festival, the zone of winter darkness, and the sense of isolation?
6. Figure 1.3 illustrates the North's cold environment as measured by isotherms (mean annual air temperatures for many locations taken over a long period of time). If you created a similar map based on mean annual air temperatures for the same set of locations but for only the last 10 years, would you expect the same spatial results?
7. How is the Super Cycle for resource commodities challenging the basic premise of the core/periphery theory and what are the implications for the North's economy?
8. From Table 1.4, by examining two megaprojects, the Iron Ore Project and the Voisey's Bay nickel mine, what have been the principal changes over the past 50 years in regard to Aboriginal and environmental issues?
9. Do you agree or disagree with the last sentence in Vignette 1.2 that what is given up in land claim agreements is not easily equated to a sum of money, however large? Explain.
10. What two factors caused Watkins to forecast that the North would fall into the 'staple trap'?

Notes

1. The 'Producer Group' consists of Imperial Oil, ConocoPhillips, Shell, and ExxonMobil. Each company has a subsidiary as a member of the Producer Group: Imperial Oil Resources Ventures Limited; ConocoPhillips Canada (North) Limited; Shell Canada Limited; and ExxonMobil Canada Properties. In July 2011, Shell decided to sell its gas reserves, indicating that, in the company's judgement, the Mackenzie Gas Project is no longer viable.

2. Eldorado is a mythical place abounding in great wealth. This word is derived from the Spanish *el dorado*, meaning 'the gilded'. In the sixteenth century, Spanish explorers believed that a city of gold existed in the Americas. They named this fabled city El Dorado.

3. 'Sense of place' is a term used by geographers to denote the special and often emotional feelings that people have for the region in which they live. These feelings are derived from a variety of personal and group experiences; some are due to natural factors, such as climate, while others result from cultural factors, such as the economy, language, and religion. Whatever its origin, a sense of place is a powerful psychological bond between people and their region.

4. Development was seen as an upward trending linear progression of society and its economy within a capitalistic system. One of the chief contributors to this modernization view was Walt Rostow, who wrote *The Stages of Economic Growth: A Non-Communist Manifesto*. In the twentieth century, while most sociologists and political scientists subscribed to the modernization theory from the perspective of social and political changes, economists led by Rostow focused on the economic side of modernizatization. Modernization theory reflected Western values at the expense of other cultural values and practices.

5. The erosion of resource companies' power over the nature of development resulted from the loss of public support and the ensuing change in government policy. A new paradigm for northern development emerged from the Berger Inquiry of 1974–7 on a proposed Mackenzie Valley pipeline, and with this paradigm shift came a rethinking of the costs/benefits of megaprojects. Major resource projects were no longer automatically approved by Ottawa and provincial governments but had to run the gauntlet of public scrutiny. One result has been the emergence of corporate social responsibility to ensure that more benefits flow to the North; another has been a shift in power from Ottawa to the North, especially through the territorial governments and land claim agreements. Essentially, this process of social change results in a devolution of control, management, and ownership of the land and its resources from the centre (resource companies and Ottawa) to the hinterland (provincial and territorial governments and Aboriginal peoples).

6. While Nunavut represents one approach, this political option is not available for the vast majority of Aboriginal peoples. The search for a place within Canadian society continues and this search is led by Aboriginal leaders. Assimilation failed and integration seems stalled. Aboriginal issues reached national attention in 1969, when Ottawa proposed to 'enable the Indian people to be free—free to develop Indian cultures in an environment of legal, social and economic equality with other Canadians' (*Statement of the Government of Canada on Indian Policy*, 1969: 3). This 'White Paper', which called for the abolishment of the Department of Indian Affairs, the elimination of legal distinction between Indians and other Canadians through the amendment of the Indian Act, and the rejection of the legal validity of Aboriginal rights, caused a storm of protest from Indian leaders. Their reaction centred on the proposed loss of special rights embedded in the treaties and prompted Harold Cardinal to write *The Unjust Society: The Tragedy of Canada's Indians*. Yet, many Aboriginal peoples remain stranded in Third World (or Fourth World) living conditions. In 2006, Calvin Helin's *Dances with Dependency: Indigenous Success through Self-Reliance* challenged the status quo, arguing that the Indian Act is keeping First Nations in a state of dependency. Helin argues that Indian

people need to break away from dependency and become self-reliant. Whether or not his ideas are tenable, or acceptable among the Indian community, Canadian society, and the federal government, remains to be seen. However, as Aboriginal leaders become more dissatisfied with present relations with Ottawa, radical political change may be on the horizon. In July 2011, for example, as discussed in Chapter 7, the Assembly of First Nations called for a complete revamping of the institutions that form the basis of the relationship between the federal government and Aboriginal peoples (see AFN, 2011).

Unlike in other areas of Canada, Aboriginal peoples in the North often form a sizable minority or a majority in different areas. This demographic fact adds considerable political weight to the Aboriginal position in resolving these contentious issues. Northern solutions are tempered somewhat by the limitations of the region's geography, which encourages people to work together in a co-operative manner, and by the realization that change through negotiations is already underway. Nunavut represents a major political realignment within Canada, while a series of comprehensive land claim agreements have altered the political geography of the North. Both provide hope for economic and social gains by Aboriginal northerners within Canadian society (see Chapter 7 for more details). In turn, such events have affected Canadian society, making it more aware of minority issues and the social cost of resource development.

7. Hidden costs of development became more apparent in the post-World War II period. By the 1950s, the environmental movement emerged in the United States and it quickly spread to Canada. One of the first signs that Canadians were paying attention to environmental issues took place in 1958 when the Resources for Tomorrow Conference brought together administrators, developers, and scientists to discuss the management of resources. From that point on, non-governmental organizations (NGOs) mobilized public opinion against plans for resource development.

References and Selected Reading

Assembly of First Nations (AFN). 2011. *Pursuing First Nation Self-Determination: Realizing Our Rights and Responsibilities*. At: www.afn.ca/uploads/files/aga/pursuing_self-determination_aga_2011_eng%5B1%5D.pdf.

Bone, Robert M. 2011. *The Regional Geography of Canada*, 5th edn. Toronto: Oxford University Press.

Burn, Chris. 1995. 'Where Does the Polar Night Begin?', *Canadian Geographer* 39, 1: 68–74.

Canada. 1969. *Statement of the Government of Canada on Indian Policy*. Ottawa: Queen's Printer.

Cardinal, Harold. 1969. *The Unjust Society: The Tragedy of Canada's Indians*. Edmonton: Hurtig.

Coates, Ken, and William Morrison. 1992. *The Forgotten North: A History of Canada's Provincial Norths*. Toronto: James Lorimer.

Cross, Phillip. 2007. 'The New Underground Economy of Subsoil Resources: No Longer Hewers of Wood and Drawers of Water', *Canadian Economic Observer* (Oct.). Statistics Canada Catalogue no. 11–010. At: www.statcan.ca/english/freepub/11-010-XIB/01007/feature.htm.

Fitzpatrick, Patricia. 2008. 'A New Staples Industry? Complexity, Governance and Canada's Diamond Mines', *Policy and Society* 26, 1: 87–103.

Friedmann, John. 1966. *Regional Development Policy: A Case Study of Venezuela*. Cambridge, Mass.: MIT Press.

Hamelin, Louis-Edmond. 1979. *Canadian Nordicity: It's Your North, Too*, trans. William Barr. Montreal: Harvest House.

Hanna, Kevin S., ed. 2009. *Environmental Impact Assessment: Practice and Participation*, 2nd edn. Toronto: Oxford University Press.

Helin, Calvin. 2006. *Dances with Dependency: Indigenous Success through Self-Reliance*. Vancouver: Orca Spirit.

Innis, Harold A. 1930. *The Fur Trade in Canada*. Toronto: University of Toronto Press.

Johnston, Margaret E., ed. 1994. *Geographic Perspectives on the Provincial Norths*, vol. 3, Centre for Northern Studies, Lakehead University. Mississauga, Ont.: Copp Clark.

Manuel, George, and Michael Posluns. 1974. *The Fourth World: An Indian Reality*. Don Mills, Ont.: Collier Macmillan Canada.

McNiven, Chuck, and Henry Puderer. 2000. *Delineation of Canada's North: An Examination of the North–South Relationship in Canada*. Geography Working Paper Series No. 2000–3. Ottawa: Statistics Canada.

Miller, Bill. 2007. 'Arctic Sea Ice Retreats to Record Low by End of Summer with Ominous Consequences for Global Warming'. At: www.desmogblog.com/arctic-sea-ice-retreats-to-record-low-by-end-of-summer-with-ominous-consequences-for-global-warming.

Müller-Wille, Ludger. 2001. 'Shaping Modern Inuit Territorial Perception and Identity in the Quebec–Labrador Peninsula', in Colin H. Scott, ed., *Aboriginal Autonomy and Development in Northern Quebec and Labrador*. Vancouver: University of British Columbia Press, 33–40.

Noble, B.F. 2010. *Introduction to Environmental Impact Assessment: A Guide to Principles and Practice*, 2nd edn. Toronto: Oxford University Press.

Northwest Territories, Bureau of Statistics. 2007. Aklavik. At: www.stats.gov.nt.ca/Infrastructure/Comm%20Sheets/Aklavik.html.

O'Faircheallaigh, C. 2007. 'Environmental Agreements, EIA Follow-up and Aboriginal Participation in Environmental Management: The Canadian Experience', *Environmental Impact Assessment Review* 27, 4: 319–42.

Rostow, W.W. 1960. *The Stages of Economic Growth: A Non-Communist Manifesto*. Cambridge: Cambridge University Press

Slocombe, D. Scott. 1995. 'Understanding Regions: A Framework for Description and Analysis', *Canadian Journal of Regional Science* 18, 2: 161–78.

Statistics Canada. 2005. 'Land and Freshwater Area, by Province and Territory'. At: www40.statcan.ca/l01/cst01/phys01.htm.

Wallerstein, Immanuel. 1979. *The Capitalist World-Economy*. Cambridge: Cambridge University Press.

Watkins, Melville H. 1977. 'The Staple Theory Revised', *Journal of Canadian Studies* 29: 160–9.

2

The Physical Base

An understanding of the physical geography of the Canadian North provides the essential background necessary to appreciate the delicate balance among the North's physical base, human population, resource development, and fragile environment. An important feature of the North's physical base is the interrelationship between its various physical elements, but especially between climate, natural vegetation, soils, and permafrost. The five geomorphic regions, which provide much diversity within the North, are another notable characteristic of northern geography. Finally, the Arctic Ocean, its ice cover, and its role in global oceanic circulation and climate change are considered in this chapter.

Climate change has turned the usually stable feature of physical geography into something much more dynamic. For centuries, ice shelves on Ellesmere Island were at an equilibrium state and thus represented stability. This has changed rather suddenly. As reported in August 2010 (Smith, 2010), Professor John England observed a large piece of ice, perhaps the size of Bermuda, breaking off from the Ward Hunt Ice Shelf on Ellesmere Island in Nunavut. England concluded that the ice shelf itself is badly fractured and rapidly failing.

Of all the regions of Canada, the North has the narrowest range of natural resources, the most demanding physical conditions for settlement and resource development, and the most delicate environment. Unusually warm summers in recent years have affected the North, causing glaciers and sea ice to melt and permafrost to thaw. If this warming trend continues into the rest of the twenty-first century, the North's physical base will be greatly modified, with much of the permafrost and Arctic ice pack disappearing, the boreal forest extending far northward, and the Northwest Passage becoming ice-free for much of the year.

Overview

The North is a vast area consisting of more than three-quarters of the land mass of Canada. Within this cold **land mass**, the latitude ranges from 50° to 83°N. Not surprisingly, the North's physical character varies from place to place. To appreciate and at the same time to simplify these variations, our attention is focused on the broad spatial

Figure 2.1   Northern Edge of the Boreal Forest in the Mackenzie Delta

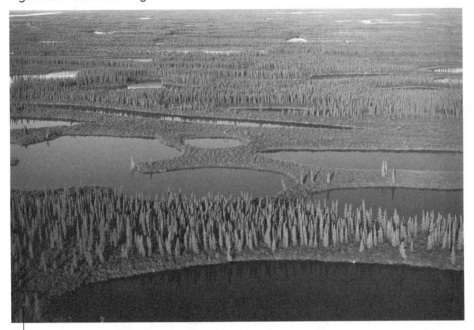

At the northern verge of the boreal forest, wetlands (lakes, muskeg, and peat bogs) are interspersed with black and white spruce trees. Under a warmer climate induced by global warming, longer and warmer summers could, by the end of the twenty-first century, change this forest/tundra transition zone into a closed boreal forest zone and extend the forest/tundra transition zone into some of the islands of the Arctic Archipelago.

Source: Staffan Widstrand/Nature Picture Library.

Vignette 2.1   Climate Change and Global Warming

Global warming refers to higher average annual temperatures that are leading to the warming of the planet and these higher temperatures are caused by the increase in greenhouse gases in the atmosphere. Known as the **greenhouse effect**, these gases, which trap solar energy and thus warm the Earth, are increasing due to the burning of fossil fuels. **Climate change** is a much broader term that encompasses all aspects of weather. Global warming, if sustained, could result in dramatic climatic changes. For example, higher temperatures could alter the global energy balance (see Vignette 2.2). By doing so, global warming could cause climatic change because, as the Earth's average temperature increases, winds and ocean currents move that heat around the globe, triggering long-term changes in weather conditions and, eventually, altering the geographic extent of natural vegetation zones, permafrost, and sea ice. Predictions of much warmer global temperatures for the late twenty-first century suggest a dramatic climatic change perhaps exceeding temperatures thought to have occurred in Climatic Optimum some 5,000 to 9,000 BP (Table 2.2).

patterns making up the North's physical base. These broad spatial patterns are represented by two natural regions or biomes, the Arctic and Subarctic (Figure 2.2), and these two biomes provide a context for discussing the complexities and diversity of the physical geography of the North. But first we must consider the basic natural characteristics of the North. Without a doubt, its polar climate dominates and shapes other northern elements. This climate is split into Arctic and Subarctic components that are associated with the Arctic and Subarctic biomes. Within each biome, natural vegetation, soils, wildlife, and humans have had centuries to adapt to the relatively stable climatic conditions. One challenge facing scientists and northerners is the impact that global warming will have on the biophysical nature of the North and the geographic extent of the Arctic and Subarctic biomes.

Climate determines the particular natural rhythm of the North, including the spring migration of the caribou herds to the tundra, the extreme variation in amount of daylight from summer to winter, and the dramatic release of water from ice-locked lakes and rivers every spring. Climate, too, determines the location of the treeline, beyond which summer growing conditions are too cold for tree growth. The treeline marks the natural boundary separating the Arctic and Subarctic biomes.

Figure 2.2 Arctic and Subarctic Biomes

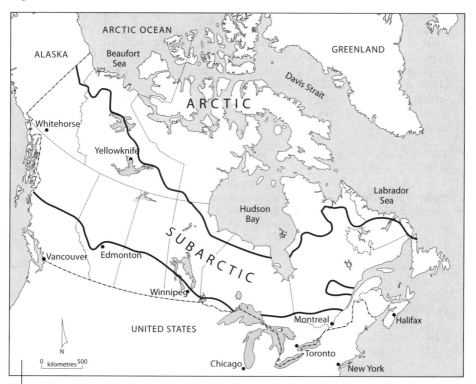

The natural division of the Canadian North into the Arctic and Subarctic biomes has biological, economic, and cultural implications. These implications focus on biodiversity and ecozones, the impact of resource companies on the environment, and the concept of traditional homelands for Indians and Inuit.

But the North's physical geography was also affected by past climates. Some 16,000 to 25,000 years ago, extremely low world temperatures associated with the Late Wisconsin ice advance dramatically altered the North's physical geography by freezing the ground to great depths, thus creating permafrost; by covering land with a continental ice sheet whose enormous weight depressed the Earth's crust; and, as the huge ice sheet slowly advanced, by scraping and scouring the landforms beneath the ice sheet. From this extremely cold climate, unique natural features emerged, such as patterned ground, permafrost, the polar ice cap, and pingos. With the post-ice age climate warming over the last 15,000 years, the Arctic and Subarctic biomes gradually expanded and eventually reached their present geographic arrangement.

The Arctic and Subarctic form the broad natural framework within which ecosystems and ecozones are found and, combined, constitute the northern environment. As a natural hierarchical structure, the two biomes provide the first division of the cold environment, followed by ecosystems and ecozones, which are simply smaller and more precise natural units. These natural units took hundreds, if not thousands, of years to develop their unique biological communities that interact with one another within a

---

Vignette 2.2   The Global Energy Balance

The Sun is the primary source of energy for the Earth. The flow of energy from the Sun to Earth and then into our atmosphere is an extremely complex process. Most solar energy reaches low latitude areas while high latitude areas like the Canadian North receive much less energy. The former are described as having an energy surplus while the latter have an energy deficit. The **global circulation system** transfers heat from low to high latitudes. This spatial differentiation in solar energy and the resulting global circulation system form the basis of world climates. Air and water move from tropical areas to polar areas. Besides transferring energy, the global circulation system also transfers pollutants from industrial complexes to the northern lands and water.

Three processes—radiation, energy storage, and energy movement—account for the variation in annual mean surface temperatures across the Earth's surface. Solar energy (short-wave radiation) heats the Earth's surface and, in turn, the Earth emits long-wave radiation into our atmosphere. The atmosphere is warmed mainly by long-wave radiation. Some long-wave radiation is lost because it passes through our atmosphere to outer space. Solar energy received at different latitudes on the Earth's surface varies because of the angle of the Sun's rays. The angle of these rays striking the Earth's surface ranges from right-angle (intense energy) at the equator to zero at the North and South poles for half of the year (no energy) and at a low angle for the rest of the year (low energy). As a result, more solar energy is received in equatorial areas (low latitude areas) than in polar areas (high latitude areas). This differential heating results in an energy surplus in equatorial areas and an energy deficit in polar areas. In low latitude areas, storage of energy occurs in the oceans and atmosphere. The global circulation system is comprised of atmospheric winds and ocean currents. These winds and currents transfer surplus energy to high latitude areas. However, because of the global pattern of atmospheric winds and ocean currents, this transfer is uneven, with more energy being received in the high latitudes of the west coasts of North America and Europe. At the same time, cold deepwater flows from the Arctic Ocean into the North Atlantic Ocean.

particular geographic area. Another important factor related to the northern environments is that, over this long time period, the number and variety of species inhabiting a particular geographic area or habitat have increased. Known as **biodiversity**, this has become one of the major environmental issues of our time, with the fear being that change—induced by human and natural causes—could reduce the region's biodiversity. Human causes are related to resource development and the growth of settlements, both of which have a negative impact on the North's fragile environment and therefore on its biodiversity (Vignette 2.3). Among the principal human impacts are clear-cut logging in the boreal forest, mining activities that introduced toxic wastes to the environment, and hydroelectric projects that result in an industrial landscape by creating reservoirs and altering the seasonal flow of rivers. Natural causes, which, indirectly, are also largely of human origin, are related primarily to global warming, which may see the Arctic environment greatly reduced in size and its species, such as the polar bear, threatened.

Vignette 2.3   A Fragile Environment

The northern environment is often described as fragile, meaning that the risk of anthropogenic damage is much higher in the North than in other regions of Canada. The primary reason for this regional variation is the much greater length of time required for nature to repair human damage to a northern environment compared to the time required in more temperate regions of the world. One illustration of the delicate nature of this cold environment is revealed by the relationship of temperature and precipitation to plant growth. Plant growth varies with temperature and precipitation. A combination of low temperatures and meagre precipitation results in very low levels of biological activity and, hence, a longer period of time is required for nature to heal itself. The same rule applies within the North, where the Subarctic can recover more quickly from physical damage because it has a longer and warmer summer plus more precipitation than the Arctic.

Northern Biomes

The complexity of the northern environment is revealed in both the formation and nature of its northern biomes. With the retreat of the last great ice sheets covering much of North America, plants and animals advanced towards the North, gradually establishing habitat in northern regions. In time, these former ice-covered lands became the Arctic and Subarctic biomes. For these biomes to form, the climate has to be relatively stable, i.e., variations in temperatures and precipitation have to fall within a relatively narrow range (Table 2.2). For example, the black spruce, one of the hardiest trees of the boreal forest, requires specific growing conditions to permit regeneration. Generally speaking, the average summer temperature must reach 10°C or higher for the black spruce to survive. Not surprisingly, then, the geographic position of the treeline has shifted in accordance with climate changes. For example, in the 400 or so years of the **Little Ice Age** (c. 1450–1850), the treeline retreated southward, thus expanding the area of the Arctic.

Another indicator of the complexity of the northern environment is found in ecosystems. Environment Canada (2007) has classified Canada's ecosystems, which include the Arctic and boreal ecosystems or biomes, into 20 terrestrial and marine eco-zones. Ten of the 15 terrestrial ecozones and three of the marine ecozones are found in the North. So many ecozones are in the North for the simple reason that the North occupies over three-quarters of the land mass of Canada. These ecozones are subunits within the much larger Arctic and Subarctic ecosystems or biomes. A biome is a broad, continental type of ecosystem characterized by distinctive climate and soil conditions and a distinctive kind of biological community adapted to those physical conditions. Natural vegetation in the form of tundra and boreal forest serves to distinguish the Arctic and Subarctic biomes, though the biological complexity and diversity of each biome goes well beyond natural vegetation (Vignette 2.4).

Vignette 2.4   Ecology, Ecosystems, and Ecozones

Ecology is the study of the interactions of living organisms with one another and their physical environment. One of the characteristics of biological life is its high degree of complexity and interrelationship. For example, a group of individuals of the same species living and interacting in the same geographic area is defined as a population. Many populations may exist in the same geographic area or habitat and, collectively, these populations form a biological community. Ecosystems or biomes represent large, continental units while ecozones represent smaller units within eco-systems. Both units express the spatial extent of this complexity and interrelation-ship. In Canada, there are 15 terrestrial and five marine ecozones. A map of Canada's ecozones is found at an Environment Canada website (www.ec.gc.ca/subsnouvelles-newsubs/0C1F54C7-6D14-9CE1-44D3-6F01FE67498B/f2-en.gif).

The Arctic

Lying north of the treeline, the Arctic is the coldest biome in Canada. Its cool summers and permafrost control the type of natural vegetation and soil development. Tundra is found over much of the Arctic, while extremely thin and immature **cryosolic soils** are associated with continuous permafrost. In this zone of permafrost, frozen ground remains close to the surface in the short summer, thus inhibiting soil development.

The Arctic is found primarily in Nunavut and the Northwest Territories, but the Arctic also exists in Quebec, Newfoundland and Labrador, Ontario, Yukon, and Manitoba. The Arctic Ocean is part of the Arctic, though the ownership of these wat-ers is still unsettled. Much of the Arctic Ocean is covered by a permanent ice cap. Since 1979 (the date of the NASA satellite photograph shown in Figure 2.5), the minimum geographic extent of the Arctic ice cap has decreased, replaced by open water (Figure 2.5 and Figures 8.2 and 8.3). The implications of more open water for Arctic sover-eignty are discussed in Chapter 8.

The Arctic is characterized by a very cold climate where the warmest month has a mean temperature of less than 10°C. Under such climatic conditions, normal tree growth and soil formation are not possible. Instead, tundra vegetation and thin soils known as cryosols (soils formed in areas of permafrost that have a shallow active layer in the summer) are found in the Arctic. Even when summer air temperatures thaw the top few centimetres of the ground, the presence of permafrost beneath this thin active layer acts not only as a cooling agent but also as a barrier to water. In such a cold landscape, soil-forming processes only work in the short summer when soil temperatures are often just above the freezing point.

Harsh Arctic climatic and soil conditions permit only a few varieties of plants, but the ones that do survive are true marvels of adaptability that can withstand long periods without sunlight. Arctic vegetation is divided into two subzones—Low Arctic and High Arctic. The Low Arctic occupies the mainland while the High Arctic is found in the northern reaches of the mainland and the Arctic Archipelago. The Low Arctic is associated with tundra vegetation. This subzone is characterized by nearly complete plant cover, including many shrubs, such as dwarf birch and willow, and sedges that appear in imperfectly drained lowlands. Tussock sedge and ground-hugging shrubs provide summer grazing for caribou, which are still a major source of food for Aboriginal peoples. Heath, herbs, and lichen are the typical plants. Reindeer moss, a grey-green, sponge-like lichen, is an important source of food for caribou, muskox, and other herbivores. While trees such as willows do grow, they reach a height of only a few centimetres. In the High Arctic zone, by contrast, little vegetation exists, though lichens are found on rock surfaces. Most of the land surface in the High Arctic consists of rock and unconsolidated material. Such barren Arctic lowlands are called polar deserts (Vignette 2.5).

The treeline represents the place where the last trees are able to grow and thus serves as the boundary between the Arctic and the Subarctic. Like many other natural features, the treeline in fact is a transitional zone between the closed boreal forest and exclusively tundra vegetation. While the treeline closely corresponds to the isotherm representing a 10°C monthly mean temperature for July, other natural factors, such as the depth of the active layer of permafrost, topography that protects trees from wind, south-slope radiation, and well-drained land, may result in patches of trees growing north of this isotherm or, conversely, tundra occurring south of it. All of these factors have produced a transition zone at the northern edge of the Subarctic that can be described as 'wooded tundra' and 'lichen woodland'. In this transition zone, the proportion of tundra to forest varies, but towards the southern edge of the Arctic the wooded tundra zone exists. Here, patches of bush-size spruce and larch are found in sheltered, low-lying areas while high, more exposed lands are treeless. Towards the south, the lichen woodland emerges, with stands of more mature trees mixed with open areas consisting of a thick ground cover of lichens (see Figure 2.4). This transition zone of the wooded tundra and the lichen woodland also represents a biological and cultural boundary. Polar bears and Arctic foxes prevail in the Arctic, while beaver and moose are restricted to the forest lands of the Subarctic.

The treeline (and therefore the Arctic) does not follow a latitudinal direction, but has a distinct northwest to southeast direction due to two factors. First, the continental effect causes the interior of the North to warm in the summer, thus allowing the treeline to reach the mouth of the Mackenzie River, well north of the **Arctic Circle**, which

Vignette 2.5 **Polar Desert**

Polar desert exists in the higher latitudes of the Arctic where extremely cold, arid conditions occur throughout the year. Low summer temperatures combined with permanently frozen ground greatly limit biological activity. For that reason, little vegetation is found in polar deserts. Lichens are by far the most important group of primitive plants found in polar deserts. Since the principal geomorphic process is a freeze/thaw cycle, a sterile, barren-looking landscape consisting of shattered bedrock, patterned ground, and unconsolidated materials prevails. An example of polar desert is provided in the photograph below of a barren landscape with no visible vegetation. Located on Melville Peninsula (around 69°N), frost action has produced a rugged surface of shattered rock fragments.

Figure 2.3 Polar Desert

Frost-heaved rock fragments along jointing sites, near Hall Beach, Melville Peninsula, Nunavut.

Source: Natural Resources Canada. 2010. 'Canadian Landscape s Photo Collection', Geological Survey of Canada, 2010, at: <gsc.nrcan.gc.ca/landscapes/details_e.php?photoID=807>.

Source: Reproduced with the permission of Natural Resources Canada 2011, courtesy of the Geological Survey of Canada (Photo 2002-552 by Dredge, Lynda).

is located at 66° 33'N and shown on Figure 1.3. Second, the cold waters of Hudson Bay and the Labrador Sea prevent tree growth along its coastline. The latitudes of three communities just south of the treeline are Aklavik (68° 13'N), Churchill (58° 48'N), and Cartwright (53° 36'N). Aklavik, located on a deltaic island at the mouth of the Mackenzie River, has a warmest-month average mean temperature of 14°C; Churchill, situated near the shore of Hudson Bay, has an average July temperature of 11.8°C; and Cartwright, lying along the Labrador coast, has a mean July figure of 12°.

Figure 2.4   Lichen Woodland Landscape

Close to the treeline, this transition zone between the Arctic and Subarctic consists of a sparse stand of black and white spruce with a lichen understory consisting of lichen, dwarf birch, dwarf willow, and labrador tea.

Source: Northwest Territories (2004). Photo: Dave Downey and Department of Environment and Natural Resources, Government of the Northwest Territories.

Figure 2.5   The Arctic Ice Cap, September 1979

The Arctic ice cap consists of pack ice and new ice. By September, the maximum ice melt has taken place with two critical areas containing open water. These areas are Lancaster Sound and the southern part of the Beaufort Sea. In September 1979, the section of the Northwest Passage between Lancaster Sound and the Beaufort Sea was blocked by thick ice. The data for the period 2002 to 2010 (Figure 8.2) indicated much more open water in September than was found in 1979, but the amount of open water, and therefore the retreat of the ice cap, varies from year to year.

Source: NASA (1979).

Figure 2.6   Arctic Ice Extent, September 2007

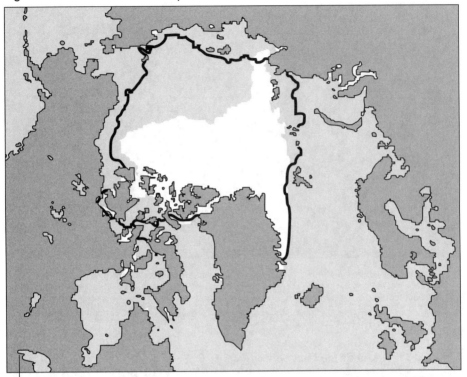

NASA satellite images reveal that, over the last 25 years, the Arctic ice cap has shrunk. Its extent on 25 September 2007 is shown in white while a dark line shows the median September monthly position based on data from 1979 to 2000. The change is remarkable. Compared with the NASA satellite image of September 1979 (Figure 2.5), the Northwest Passage from Lancaster Sound to the Beaufort Sea was open water in September 2007 but solid ice in 1979.

Source: NSIDC (2007b).

The Subarctic

The Subarctic is the largest natural region in North America. Its natural vegetation, the boreal forest or taiga, provides a clear indication of its geographic extent. The boreal forest extends in a continuous belt from the Rocky Mountains to Labrador. While its long winters are cold, summers are short but warm, allowing for a much richer vegetation cover than that found in the Arctic. Coniferous trees predominate. Variation in the size and density of forest cover south of the natural vegetation transition zone in the Subarctic is defined by two subzones: closed boreal forest and forest parkland. The closed boreal forest, a dense forest of mature fir, spruce, and pine, is found in southern Yukon, a small part of the Northwest Territories (principally the upper Mackenzie Valley), and the northern areas of the seven provinces. An example of this forest in the Northwest Territories is shown in Figure 2.7. Within this huge zone, the forest cover is broken by a variety of wetlands, including lakes, **muskeg**, and peat bogs. In this wet environment, black spruce and larch are the most common species. On well-drained

Figure 2.7   Boreal Forest in the Northwest Territories

The boreal forest extends across Canada in a green swath, reaching into high latitudes in the Northwest Territories. Just south of the Arctic Circle in the vicinity of Great Bear Lake and the Mackenzie River Valley, the boreal forest consists of mature stands of black and white spruce. In this photograph, the boreal forest extends to the shore of Oscar Lake.

Source: The Boreal Songbird Initiative, at: <www.borealbirds.org/>. © Ducks Unlimited, Canada.

land, species of spruce, fir, pine, and larch are common along with stands of poplar and birch. Towards its southern limits, broadleaf trees, particularly aspen and birch, are found. The forest parkland, a narrow transition area adjacent to the Canadian Prairies, is a combination of forest and mid-latitude grasslands. Small bushes, including blueberries, and grasses form the ground cover.

Podzolic and gleysolic soils are common in the Subarctic. In the Canadian North, thin, acidic **podzolic soils** are best formed under cool, wet growing conditions where the principal vegetative litter is derived from a coniferous forest. **Gleysolic soils** are associated with extremely poorly drained and often waterlogged land such as marshes and bogs. A low evaporation rate, immature drainage, and permafrost ensure an excess of ground moisture, resulting in severely leached soils and the widespread occurrence of ponds, lakes, and bogs. Yet, the longer summer temperatures in the Subarctic allow the better drained ground to thaw to a depth of several metres, promoting biological activity, plant growth, and chemical action. Unlike the cryosolic soils in areas of continuous permafrost, podzolic soils are associated with discontinuous and sporadic permafrost, both of which have relatively thick active layers (for more on the types of permafrost and their geographic extent, see the subsection on permafrost in this chapter).

The boreal forest is a rich zone of biodiversity with different types of wildlife: moose, deer, martens, rabbits, beaver, foxes, wolves, bears, eagles, and various birds.

Local medicinal plants for treating colds, diabetes, and heart and skin problems are known and used by First Nations and Métis. The biological diversity of the boreal forest includes thick layers of moss, soil, and peat that store huge amounts of organic carbon and thus play a significant role in regulating the Earth's climate. Boreal wetlands also filter millions of gallons of water each day that fill Canada's northern rivers, lakes, and streams.

Polar Climate

In the classification of climates throughout the world, the polar climate is the coldest climatic type. The polar climate is divided into four climatic subtypes: the Arctic, Subarctic, mountain, and ice cap climates. In this text, however, the mountainous areas of northern British Columbia, Yukon, and the Northwest Territories are treated as part of the Subarctic (Vignette 2.6). Similarly, the ice cap climatic type is merged with the Arctic climatic type. The ice cap climate is associated with glaciers found on Baffin, Devon, and Ellesmere islands. This climate has a mean temperature below freezing for all months. Those small areas covered by glaciers are considered part of the Arctic climate. Each climatic type is associated with an **air mass** that reflects the weather characteristics of that type. As these air masses move across the continent, they affect weather in other regions

Unlike other climatic types, the polar climate is characterized by extreme seasonal variations in the amount of solar energy. Summer days, for example, are long while winter days receive little to no sunlight. The Arctic Circle marks the latitude where the Sun remains above the horizon for one summer day (21 June, the summer solstice) each year and it remains below the horizon for one winter day (21 December, the winter solstice) each year. At latitudes well beyond the Arctic Circle, summer days and winter nights can last for months. At the North Pole, the Sun is above the horizon for six months and below the horizon for the rest of the year. A day of continuous darkness is referred to as a 'polar night' (Vignette 2.7).

Monthly receipts of solar energy vary widely throughout the year (Vignette 2.2). On a yearly average, the Earth's poles receive 40 per cent less radiation than the

---

Vignette 2.6 **The Subarctic Climate in the Cordillera**

Many factors, such as latitude, topography, the proximity of bodies of water, and the nature of the underlying surface, control climate. In Yukon, due to its mountainous nature, topography becomes very important. The territory benefits from Pacific airflows from the west, while high mountain ranges block Arctic air masses from the north. The mountain ranges also affect atmospheric circulation patterns, the amount, frequency, and type of precipitation, winds, atmospheric pressure, and the local radiation regime. Elevation in particular plays a major role in determining temperature. During the winter, a strong surface-based inversion develops in Yukon due to the net negative radiation balance. Thus, temperatures tend to increase with height, especially in the bottom 1,500 metres of the atmosphere.

Source: Adapted from Etkin (1989: 12).

Vignette 2.7 **The Polar Night**

The polar night is a period of continuous winter darkness. Twilight does not occur. Polar nights take place north of 72° 33′N, well beyond the Arctic Circle (66° 33′N). Why is this? We know that the Sun does not rise above the horizon at the Arctic Circle during the winter solstice. Yet there are not 24 hours of continuous darkness because diffused light from the sky is caused by the Sun's rays being reflected from a position below the horizon onto the atmosphere and then back down to the Earth. Twilight may last for an hour or more at the time when the Sun is below (but less than 6° below) the horizon, thus providing at that time sufficient light for outdoor activities. On 21 December, the latitude of 72° 33′N, not the Arctic Circle, marks the geographic point where 24 hours of continuous winter darkness occurs.

Sources: Burn (1995, 1996).

---

equator (Lawford, 1988: 144). Within the Canadian North, the major dividing line, the Arctic Circle, marks the point where solar radiation is reduced to zero for one day (21 December) and, at higher latitudes, for longer periods of time. Such low levels of radiation result in continuous cooling of the land and the buildup of masses of frigid Arctic air, which are associated with daily high temperatures of –40°C or lower and strong surface winds. In fact, the Arctic coast is one of Canada's windiest places, with annual average wind speeds exceeding 20 km/hr. Cold polar air is often associated with extreme wind-chill conditions that will freeze exposed flesh in a matter of seconds. Arctic winds drive frigid air masses southward, causing stormy and sometimes blizzard conditions in the Canadian Prairies, sub-zero temperatures in eastern Canada, and freezing temperatures in the southern United States. In the spring, solar radiation increases but much of its effect is lost due to snow-covered surface. In fact, up to 80 per cent of the spring solar radiation is reflected into space

---

Vignette 2.8 **The North's Energy Deficit**

The North is an area of energy deficit (Vignette 2.2). The length and intensity of winter in the North are indicative of its energy deficit. The principal factor is the low amount of solar energy received in high latitudes compared to that received in middle latitudes. Williams (1986: 6–7) defines a cold environment as having a negative annual heat balance due to a greater amount of long-wave radiation emitted to outer space than the amount of incoming short-wave radiation and energy transfer by global winds and ocean currents. Mean annual air temperature, defined as the total of daily mean temperatures for the year at one place divided by 365 days, provides one measure of the negative annual heat balance. Those areas with annual mean air temperatures of less than zero degrees Celsius have a negative heat balance. With the exception of a small area of the boreal forest that extends into the middle latitudes of Ontario and Quebec, the Canadian North has a negative energy balance (Figure 1.3).

by the snow-covered ground. Once the snow is gone, winter's grip is quickly broken. Temperatures recover rapidly and the ensuing warm weather quickly melts the ice from lakes, rivers, and the ocean. By early July, sea ice has disappeared from Hudson Bay, and a few weeks later the ice is gone from along the edge of the Arctic coast. The polar pack, no longer attached to the coastline, drifts around the Arctic Ocean. During the long summer days, massive amounts of solar radiation reach the northern lands, warming the ground. In response, plants quickly appear and flower. Daily summer temperatures in the Mackenzie Valley and southern Yukon can reach into the low thirties Celsius. By late August, however, summer is over. Within another month, ice has formed on lakes and rivers, marking a return to a frozen landscape.

The sharp seasonal shift of air temperatures is illustrated in Figures 2.8 and 2.9. The winter regime has low mean monthly temperatures. The coldest January temperatures are found in two places—the northern extremes of Ellesmere and Axel Heiberg islands and just south of Boothia Peninsula. On the other hand, the mean

Figure 2.8   Mean Daily Temperatures in Degrees Celsius, January

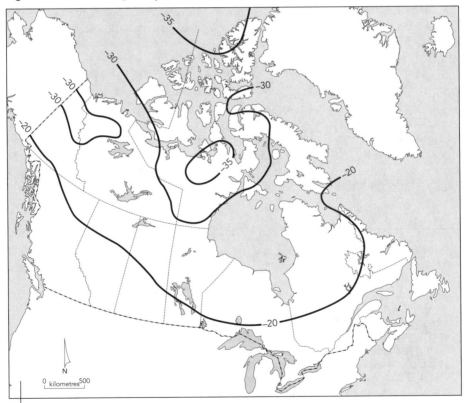

The Pacific and Atlantic oceans have a temperature-moderating influence on near-coastal areas so that continental areas are colder in winter and warmer in summer than are areas closer to the Atlantic and Pacific coasts. The –20°C January isotherm reaches nearly 50°N in Ontario while the same isotherm is north of the sixtieth parallel in Yukon and Nunavut.

Source: After Hare and Thomas (1979: 37).

Figure 2.9   Mean Daily Temperatures in Degrees Celsius, July

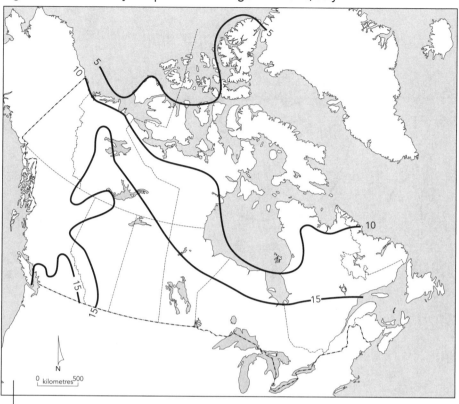

July isotherms indicate a northwest–southeast trend, with a warm corridor extending into the Mackenzie River Valley. This warm temperature corridor allows the treeline to reach the Mackenzie Delta (see Figure 2.1). The 15°C July isotherm reaches beyond 60°N in the interior of the Northwest Territories, indicating a continental warming effect.

Source: After Hare and Thomas (1979: 37).

daily temperature for July demonstrates both the warming effect of the northern land mass and the cooling effect of the Arctic Ocean, Hudson Bay, and the North Atlantic. A distinct northwest-to-southeast direction to the July isotherms exists. As well, a warm 'corridor' extends down the Mackenzie Valley to Norman Wells, where July temperatures are similar to those experienced in the Canadian Prairies.

Precipitation in the North is generally light, with the least amount falling in the Arctic due to the inability of Arctic air masses to absorb moisture from cold bodies of water. The isohyets shown in Figure 2.10 clearly indicate that the lowest amounts of annual precipitation are found in the Arctic Archipelago, where some islands record less than 100 mm annually. The lowest annual precipitation occurs in the ice-locked islands of the Arctic Archipelago (Vignette 2.9). Victoria Island, for example, has such scant rain and snowfall (often less than 140 mm annually) that the island is described as a 'polar desert'. Towards the northern edge of the Subarctic, annual precipitation increases. At Yellowknife, the annual precipitation is around 250 mm. The Subarctic,

Figure 2.10   Mean Annual Precipitation in Millimetres

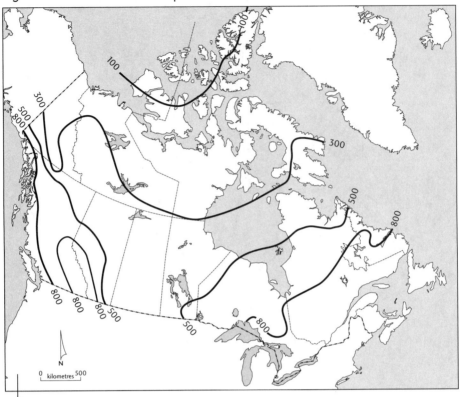

Air masses originating over the Atlantic and Pacific oceans bring most moisture to the North. Air masses forming over the cold Arctic Ocean usually contain little moisture. As a result, the highest levels of annual precipitation are found along the western and eastern edges of the North while the Arctic Archipelago is extremely dry, causing scientists to refer to this area as a polar desert (Vignette 2.5).

Source: After Hare and Thomas (1979: 41).

Vignette 2.9   Arctic Archipelago

The Canadian Arctic Archipelago (Figure 2.11) is a group of islands in the Arctic Ocean. Covering over 1.3 million km², they form the largest group of islands in the world. The largest islands are Baffin Island (507,451 km²), Victoria Island (217,290 km²), and Ellesmere Island (196,236 km²). Most islands have elevations below 200 metres and few topographic features. Elevations do rise above 2,000 metres in the eastern islands of the Arctic Archipelago. Mount Barbeau on Ellesmere Island, for example, reaches an elevation of 2,616 metres. At these high elevations, glaciers exist. The geological history of the Arctic Archipelago began some 3 billion years ago and Precambrian rock exists at the surface on Baffin, Devon, and Ellesmere islands. Most of the Arctic islands were formed much later and contain sedimentary rocks. Within these sedimentary rocks, vast oil and gas deposits exist.

on the other hand, generally receives more precipitation, usually over 300 mm annually. Precipitation does vary, however. The greatest amount of precipitation occurs in the Cordillera and along the Atlantic coast. In these two areas, the high terrain results in orographic precipitation (rain or snow caused when warm, moisture-laden air is forced to rise over hills or mountains and is cooled in the process). The south coast of Baffin Island receives around 400 mm annually, while the annual total for the southern coast of Labrador and northern Quebec exceeds 800 mm. Most precipitation falls as snow. In the spring, the runoff peaks when melting snow and ice flow into the streams and rivers. Some communities along the Mackenzie River are subject to spring flooding. Fort Simpson, situated on an island at the confluence of the Liard and Mackenzie rivers, has been inundated a number of times, and Aklavik, located in the delta of the Mackenzie River, is threatened by flood waters almost every spring. The occurrence of spring flooding at Aklavik is so regular that, in the late 1950s, the federal government decided to create the new town of Inuvik rather than expand the community of Aklavik.

The Arctic climate is defined as one in which the average mean temperature for the warmest month is less than 10°C. It has extremely long winters and a brief, cool

Figure 2.11   The Arctic Archipelago

Source: Adapted from Infoplease, at: <www.infoplease.com/atlas/region/nunavut.html>.

summer. The Arctic climate lies north of the treeline and includes all of the Arctic Archipelago, the coastal zone stretching from the Beaufort Sea to the coast of Labrador, and much of the interior of the northern territories, known as the 'Barren Lands', which stretch from Hudson Bay in the east to Great Bear and Great Slave lakes to the west. While the Arctic climate is normally associated with high latitudes, it does reach into the middle latitudes along the Labrador coast. Here, the chilling effect of the cold Labrador Current keeps summer temperatures along the east coast of Canada low, allowing the Arctic climate to extend along the Labrador coast to the northern tip of Newfoundland. Resolute and Iqaluit have Arctic climates and their mean monthly temperature regimes are shown in Table 2.1.

Unlike its Arctic counterpart, the Subarctic climate has a distinct but short, warm summer. Normally, this climate is found in continental or inland locations and is characterized by a wide range in seasonal and daily temperatures. This continental effect results in record daily cold temperatures being set in Yukon rather than in the Arctic Archipelago. The coldest temperature recorded in Canada for the period 1971–2000 was –62.8°C at Snag, Yukon. In contrast to the cold winter temperatures, a number of hot summer days can occur. In July 1989, for example, during a 'heat wave' in the Mackenzie Valley, Norman Wells recorded an all-time maximum daily high of 35°C. The climatic regions for all of Canada are shown in Figure 2.12.

Winter is the dominant season, and although an occasional summer day may be extremely hot, summers in the Subarctic are short, usually less than three months.

Table 2.1 Mean Monthly Temperatures for Chibougamau, Prince George, Iqaluit, and Resolute

Centre	Latitude	Jan.	Feb.	Mar.	Apr.	May	June
Chibougamau	49° 55'	–18.4	–10.5	–1.0	6.4	13.3	15.8
Prince George	53° 53'	–9.6	–5.4	–0.3	5.2	9.9	13.3
Iqaluit	63° 45'	–25.6	–25.9	–22.7	–14.3	–3.2	3.4
Resolute	74° 43'	–32.4	–33.1	–30.7	–22.8	–10.9	–0.1

Centre	July	Aug.	Sept.	Oct.	Nov.	Dec.	Year
Chibougamau	15.8	14.1	9.1	2.6	–4.7	–15.9	–0.8
Prince George	15.5	14.8	10.1	4.6	–2.9	–7.8	4.0
Iqaluit	7.6	6.9	2.4	–5.0	–13.0	–21.8	–9.3
Resolute	4.3	1.5	–4.7	–14.9	–23.6	–29.2	–16.4

The range of seasonal temperatures across Canada's North defies latitude. For instance, Chibougamau in the Subarctic of Quebec at latitude has a much lower mean January temperature (–18.4°C) than the BC Subarctic city of Prince George, which has a mean January temperature of –9.6°C. Yet, Prince George lies further north (54°N) than Chibougamau (50°N). The explanation for the warmer winter temperatures at Prince George is attributed to its proximity to the Pacific Ocean and its relatively warm winter air masses. Chibougamau, on the other hand, remains in the grip of Arctic air masses for most of January.

Source: Environment Canada (2002). Copyright © Her Majesty The Queen in Right of Canada, Environment Canada 2002. Reproduced with the permission of the Minister of Public Works and Government Services Canada.

Freezing temperatures can occur at any time. By late August, cool fall-like weather is common and there is an impending sense of winter. By November, the land, lakes, and rivers are frozen and winter, lasting until late April or May, has set in. During the winter, the warming influence of the Pacific keeps temperatures of the western Subarctic relatively high while the frozen Hudson Bay reinforces the cold continental effect on the winter temperatures of the eastern Subarctic. The average January temperatures for Prince George (–9.6°C) and Chibougamau (–18.4°C) demonstrate at the micro-level the impact of these differing climatic influences. As shown in Figure 2.12, Prince George is located at a more northerly location than Chibouogamau and yet Prince George has a milder climate (see Table 2.1).

Figure 2.12   Climatic Regions of Canada

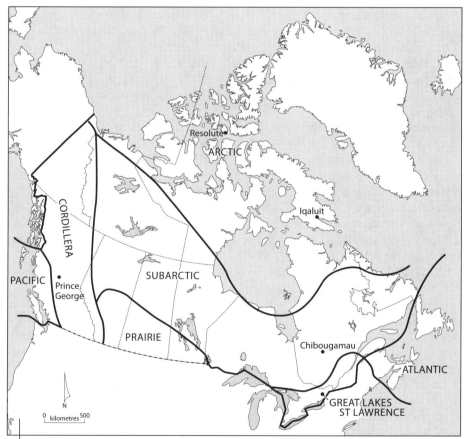

Three climatic regions—the Cordillera, the Subarctic, and the Arctic—are found in the Canadian North. The north/south orientation of the Cordillera climate is related to its high elevations The cooling effect of the waters of Hudson Bay and the Labrador Current have extended the Arctic climate much further south in the eastern half of Canada's North. On the other hand, the continental effect has allowed the Subarctic climate to reach into high latitudes in the Northwest Territories.

Source: After Hare and Thomas (1979: 17).

Past, Present, and Future Climates

Since its existence, the Earth has experience many climates. Each climate represents an average of relatively long and stable weather conditions over a significant geological time frame. Over long periods of time, nature responds by creating ecosystems with similar natural vegetation and soils. At the macro-scale, climate creates global zonal patterns such as the Arctic. From our perspective, the prospect of a much warmer climate compared to the present and near past has only emerged recently and is connected to the prospect of significant warming of the Earth within a relatively short time by the addition of increasing amounts of greenhouse gases to the atmosphere. Ironically, the Arctic is most likely to see the greatest changes in temperature because, as the snow and ice melt, more and more of the Arctic will have a surface that absorbs rather than reflects solar radiation.

What exactly is 'climate'? Climate is an average of complex natural phenomena that make forecasting climate change or reconstructing past climatic changes very difficult. For instance, while the source of the Earth's heat is constant emission of solar radiation from the Sun, many variables affect the warming of the Earth. One is the capacity of the ground to absorb (or reflect) solar energy; another is the reduction of solar radiation reaching the Earth's surface because of dust in the atmosphere caused by volcanic activity or large meteorites striking the Earth's surface. In addition, radiation from the Sun may vary slightly and that variation can account for changes in the Earth's average temperature. For example, sunspots (flares of solar energy) might increase solar energy to the Earth, but this theory is unproven. Yet, the **Maunder Minimum**, a period of relatively few sunspots, occurred at the same time as the Little Ice Age, suggesting that a low level of sunspot activity on the Sun's surface could result in slightly less solar radiation reaching the Earth.

Reconstruction of past climates is a challenge. The recording of surface temperatures and precipitation began around 1850 in Europe and North America and somewhat later in other parts of the world, and these measurements provide the basis for our understanding of climates. As a result, reconstruction of ancient climates involves proxy methods such as interpreting the geological record and relating temperature change to variations in oxygen isotopes found in ice cores. As for predicting future climates, global climate models (gcms) focus on estimating temperature increases based on the amount of greenhouse gases in the atmosphere. Since long-term variations in climate are measured in decades, centuries, and even millennia, ice in the form of glaciers and polar ice caps are our best indicators of climate change because they 'average' out the short-term weather variations and reflect longer-term temperature trends. (See Chapter 8 for a more detailed account of the impact of global warming on the Arctic ice cap.)

From the geological record, a general interpretation of climatic change over the last two millions years (the **Pleistocene Epoch** and the current **Holocene Epoch**) is possible, but even more detailed geological evidence over the last 25,000 or so years allows for more precise estimates regarding the chilling of the Earth, which caused ice sheets to form over practically all of Canada and northern portions of the United States (Vignette 2.10 and Table 2.2). After this cold era, which began to end about 15,000 years ago, the Earth warmed again, causing the ice sheets to melt and thus allowing biological life to re-enter the ice-free terrain. Geologists refer

Vignette 2.10 **The Pleistocene**

Going back in time, the last major glacial period began about 2,000,000 BP and is commonly known as the Pleistocene Epoch. During the Pleistocene, large ice sheets covered much of North America for long periods of time, but short interglacial periods occurred when the ice sheets retreated because of milder temperatures. At least four distinct ice advances and retreats have affected the land that today is Canada. The last ice advance, known as the Late Wisconsin Ice Age, began about 25,000 BP and was followed, around 10,000 years later, by a glacial retreat. Geologists call this present period the Holocene Epoch, which they believe will be followed by another glacial period.

to this current warm period as the Holocene and within that very short geological period, minor climatic variations have taken place; the warmest period, known as the Climatic Optimum, occurred between 9,000 and 5,000 BP. During the Climatic Optimum global temperatures were considerably higher than those currently experienced. Kaufman and others (2004) reconstructed Climatic Optimum temperatures for Canada's western Arctic. Their findings support the notion that temperatures in higher latitudes rose during this period, largely due to the albedo effect. In more recent times, two minor variations in global temperatures took place—the Medieval Optimum (c. ad 800–1300) and the Little Ice Age (c. 1450–1850). The Little Ice Age did impact the Thule, who had previously hunted the bowhead whale in the open waters of the Arctic Ocean. Within this colder environment, more extensive ice cover took away their opportunity to harvest these huge sea mammals and the Thule (and later their descendants, the Inuit) had to hunt seal and caribou (Fossett, 2001).

Table 2.2 Major and Minor Climatic Variations

Geological Period	Minor Variations in the Holocene Epoch	Time*	Temperature Change**
Late Wisconsin Ice Age		30,000–15,000 BP	−10°C
Holocene Epoch		12,000–present	
	Climatic Optimum	9,000–5,000 BP	+5°C
	Medieval Optimum	AD 800–1300	+1°C
	Little Ice Age	1450–1850	−1°C
	Present Warming Period	1850–present	+0.6°C
	Predicted Warming Period	2000–2100	+4°C

*Estimated data based on climate reconstruction research. For example, the start of the Little Ice Age is uncertain but is linked to extreme weather events in Europe. Such temperature 'measurements' provide only a rough approximation and hence the Little Ice Age may have begun as early as 1350. However, there is general agreement about its ending in the mid-nineteenth century.

**Estimated temperature changes relative to present climate are also derived from paleoclimatic reconstruction research based on surrogate measurements of regional temperatures such as ice cores.

Sources: IPCC (2007b); Houghton et al. (2001).

Climate Change

Since the mid-nineteenth century, the Earth began another warming cycle. Most climatologists believe that global temperatures are increasing because of anthropogenic actions, especially the burning of fossil fuels, which triggers the greenhouse effect (Vignette 2.1). Before the Industrial Revolution, natural forces alone affected climate change, including:

- known astronomical variations in the orbit and angle of the Earth vis-à-vis the Sun (the so-called Milankovitch theory) affect how much solar radiation reaches all parts of the Earth;
- changes in energy output from the Sun; and
- increases in volcanism that add huge amounts of fine volcanic particles into the stratosphere, thereby creating a dust veil that reduces the amount of sunlight reaching the Earth's surface and thus lowers temperatures.

World surface air temperatures have increased in the last century, but do these relatively minor increases fall within the range of 'normal' temperature fluctuations associated with our climate, or are they the forerunners of a climatic shift due to dramatic and permanent warming of the Earth? These warmer temperatures have caused glaciers to retreat or disappear and the summer melt of Arctic sea ice is increasing. The Intergovernmental Panel on Climate Change (IPCC) believes that human activities have been responsible for the rise in global temperatures in the twentieth century and that global temperatures will increase dramatically in the twenty-first century (Vignette 2.11). At the 2009 United Nations Climate Change Conference in Copenhagan, Stocker and Plattner predicted that the Arctic Ocean would be virtually ice-free by 2050. More specifically, the authors predicted that the geographic extent of Arctic ice would decrease from 9 million km^2 in 1950 to 1 million km^2 in 2050 (Stocker and Plattner, 2009: slide 9).

The fourth and most recent summary report of the IPCC (2007) stated that:

Warming of the climate system is unequivocal, as is now evident from observations of increases in global average air and ocean temperatures, widespread melting of snow and ice, and rising global average sea level. Changes in atmospheric concentrations of greenhouse gases (GHGs) and aerosols, land-cover and solar radiation alter the energy balance of the climate system. Global GHG emissions due to human activities have grown since pre-industrial times, with an increase of 70 per cent between 1970 and 2004. Eleven of the last twelve years (1995–2006) rank among the twelve warmest years in the instrumental record of global surface temperature (since 1850). The 100-year linear trend (1906–2005) of 0.74°C [0.56 to 0.92] is larger than the corresponding trend of 0.6°C [0.4 to 0.8] (1901–2000) given in the Third Assessment Report . . . . The temperature increase is widespread over the globe, and is greater at higher northern latitudes.

For the next two decades a warming of about 0.2°C per decade is projected for a range of SRES [The IPCC Special Report on Emission Scenarios (2000)]

Vignette 2.11   Intergovernmental Panel on Climate Change

The Intergovernmental Panel on Climate Change (IPCC) focuses its attention on potential global climate change. Formed in 1988 by the World Meteorological Organization and the United Nations Environment Programme, its role is twofold: (1) to assess the relevant scientific, technical, and socio-economic information dealing with human-induced climate change; and (2) to identify potential impacts and options for countering these impacts. The IPCC does not conduct research but bases its assessments on published scientific/technical literature, including global climate computer models that predict future climate conditions. Every six years, the IPCC issues its findings. The last report was published in November 2007.

Source: IPCC (2007a).

---

emission scenarios. Even if the concentrations of all GHGs and aerosols had been kept constant at year 2000 levels, a further warming of about 0.1°C per decade would be expected. Afterwards, temperature projections increasingly depend on specific emission scenarios. (IPCC, 2007b: 1, 4, 6)

Global Warming and Its Potential Impact on the North

Global warming could have a profound impact on the Arctic biome and, to a lesser degree, the Subarctic biome. If global temperatures increase substantially, why would the average annual temperatures in the high latitudes of the North increase more than temperatures in the lower latitudes found in southern Canada? The answer lies in the **albedo** effect (the proportion of solar radiation reflected from the Earth's surface). Under the global warming model, the annual extent and duration of snow cover will diminish. Under those two conditions, the North's surface would change from a highly reflective surface of snow and ice to a highly absorbing one of ground, i.e., a dark surface able to absorb more solar energy. In turn, the warmed Earth then re-radiates long-wave radiation to heat the lower portion of the atmosphere.

The potential impact on the Canadian North falls into three categories. First, the IPCC predicts that the polar ice pack will diminish in size and could possibly disappear. Under this scenario, natural conditions in the Arctic could return to those during the Medieval Climatic Optimum. During that time, bowhead whales frequented the Arctic Ocean because of the open water. In turn, the Thule based their harvesting economy on the bowhead whales. In modern times, an ice-free Northwest Passage would provide a much shorter shipping route between Asia and Europe. Second, the IPCC sees the boreal forest extending close to the shores of the Arctic Ocean (see also Bouchard, 2001). Such an advance would see the Arctic biome greatly reduced in geographic extent. The loss of tundra vegetation would affect the caribou calving grounds and their migration routes. Reduction of the duration and extent of sea ice would affect the polar bear population, which depends on seals as the main food source. Third, the IPCC believes that warmer temperatures will reduce the time that rivers and lakes are frozen and will result in the thawing of permafrost. One consequence for remote mining companies and communities that rely on winter ice roads for supplies would be to force them to turn to air transportation, which is much more expensive. Then, too, a warmer climate and the loss of tundra vegetation and river/sea ice would affect Inuit snowmobile travel. Already the Inuit recognize that travel by snowmobile on snow and ice has become less predictable and more dangerous. Such conditions could make access to country food more challenging and expensive and thus increase food shortages in Inuit communities, forcing them to purchase more store foods (Nickels et al., 2005). As well, warmer temperatures could cause the permanently frozen ground to thaw. Besides releasing water from the frozen ground, this would affect various buildings and other structures. Construction in permafrost areas has taken advantage of the frozen ground by placing piles within the frozen ground. These piles form the foundation for the various structures, whether they are government buildings, schools, or dwellings. If the permafrost melts, those structures would be at risk.

Geomorphic Regions

The North's physical geography is far from homogeneous. At a macro-scale, five geomorphic regions provide one insight to the North's heterogeneous nature.

Across the Canadian North, the surficial geology or landscape reflects the major geomorphic regions.[1] These regions have three major characteristics—they cover a large, contiguous area with similar relief features; they have experienced similar geomorphic processes that shaped the terrain; and they have had a similar geological history and therefore possess a common geological structure. These geomorphic regions extend over hundreds of km², making them easily identified from a high-flying aircraft or from satellite photographs.

The five geomorphic regions found in the North are the Canadian Shield, the Interior Plains, the Cordillera, the Hudson Bay Lowland, and the Arctic Lands (Figure 2.13). The Laurentide Ice Sheet of the Late Wisconsin period modified the surface of these geomorphic regions either directly, by glacial erosion or deposition, or indirectly, by the formation of glacial lakes and the invasion of depressed land by seawater. **Erratics**—huge boulders moved great distances and then deposited on the ground— are one example of the power of glacial erosion; **nunataks** are mountain peaks that rose above the alpine glaciers. The Canadian Shield, stretching from Labrador to the Northwest Territories, is the largest geomorphic region in the Canadian North. It is the geological core of the northern landscape. Its Precambrian rocks are more than 2.5 billion years old and are found under most of the more recently formed strata, such as the Interior Plains and the Hudson Bay Lowland.

Over most of the northern areas of Manitoba, Ontario, Quebec, Labrador, the northern half of Saskatchewan, the northeast corner of Alberta, and about half of the Northwest Territories, the Canadian Shield is exposed at the Earth's surface. Shaped like a saucer, its highest elevations are found around its outer limits while the central area lies beneath the waters of Hudson Bay. Much of the exposed Canadian Shield consists of a rough, rolling upland. Along the east coast from Labrador to Baffin Island, the Shield has been strongly uplifted. Glaciers moving downslope to the sea deeply scoured valleys and created fjords, giving the coast spectacular scenery.

Figure 2.13   Geomorphic Regions of Canada

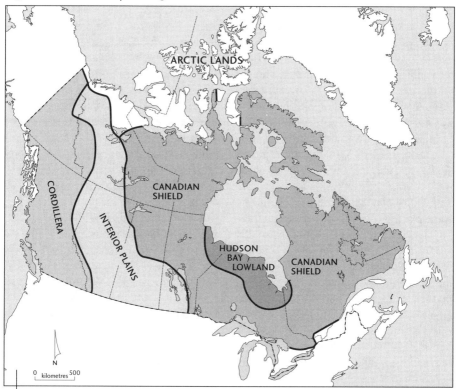

Five of Canada's seven geomorphic regions are found in the Canadian North. The other two are the Great Lakes–St Lawrence Lowland, located in southern Ontario and Quebec, and the Appalachian Upland in Atlantic Canada.

Source: After Bird (1972) and Slaymaker (1988).

The Interior Plains, a flat to gently rolling landscape, lie between the Cordillera and the Canadian Shield. The sedimentary rocks of the Interior Plains were formed after the end of the Precambrian era (some half-billion years ago) and include Cretaceous-age rocks formed some 100 million years ago. Within these Cretaceous-age rocks, vast oil and gas deposits exist. At the end of the last glacial advance, a mantle of glacially deposited debris or glacial till covered these sedimentary rocks. In other places, glacial lakes drained, forming flat lowlands consisting of **glaciolacustrine** material, while outwash plains resulted from meltwater streams that deposited **glaciofluvial** materials. At the same time, fast-flowing rivers fed by the melting ice sheet carved deep valleys known as glacial spillways. These and other glacial features on the landscape are indications of the last ice age. The Mackenzie River and its tributaries created a river valley system that extends to the Arctic Ocean. Two huge deltaic landforms exist. One is at the western end of Lake Athabasca, where the Peace and Athabasca rivers deposit their silt. The other is the Mackenzie Delta, located at the mouth of the Mackenzie River.

The Cordillera, a complex mountainous region, occupies much of British Columbia, Yukon, and a small portion of southern Alberta and the Northwest Territories west of the Mackenzie River. The Canadian Cordillera begins at the forty-ninth parallel. The Cordillera is about 800 km wide and extends to the Alaska border. Two geologic forces, known as faults and folds, formed its basic landform features while erosional agents, primarily ice and water, reshaped these landforms. The Cordillera contains a variety of mountainous terrain as well as plateaus, valleys, and plains. It also includes what are today remnant glaciers and ice fields. This geomorphic region has majestic mountains, including the world-famous Rocky Mountains. These mountains were formed by severe folding and faulting of sedimentary rocks. The highest mountain in Canada, Mount Logan, has an elevation of nearly 6,000 metres and is part of the St Elias Mountains in the southwest corner of Yukon.

During the Late Wisconsin Ice Age, the Cordillera was glaciated. Alpine glaciers created arêtes, cirques, and U-shaped valleys. Over most of the Cordillera, the land became ice-free some 10,000 years ago and in this post-glacial period, river terraces, alluvial fans, flood plains, and deltas were formed. While today many glaciers are retreating, some glaciers are still active, though their rate of movement is relatively slow.

The Hudson Bay Lowland is a low, flat coastal plain that has recently emerged from the Tyrrell Sea, the name of the prehistoric and considerably larger Hudson Bay of some 8,000 years ago. The surface of this lowland consists of recently deposited marine sediments combined with reworked glacial till. These deposits accumulated in the post-glacial period when the lowland was beneath the Atlantic Ocean (part of a much larger Hudson Bay). With the retreat of the ice sheet from this area some 8,000 years ago, the process of isostatic rebound came into play. More and more of the seabed became land as the surface of the Earth gradually rose after having been depressed by the massive weight of the ice sheet. With the receding waters of Tyrrell Sea, this process created a distinct geomorphic region known as the Hudson Bay Lowland.

The Precambrian bedrock underlying the Hudson Bay Lowland is masked entirely by glacial and marine sediments. The inland boundary of this region is marked by elevations of around 180 metres, which indicate that, with the removal of the weight of the ice sheet, the Earth's crust began to regain its former shape. This phenomenon is called **isostatic uplift** or isostatic rebound. Assuming that the rate of isostatic uplift remains around 70 to 130 cm/100 years, Professor Barr speculated that the Hudson Bay Lowland will continue to increase in size and Hudson Bay will recede further. Within 12,000 years, Hudson Bay could become a fraction of its present size (Barr, 1972). However, this hypothesis, presented in 1972, does not account for the possibility of rising sea levels caused by global warming.

The Arctic Lands are a complex geomorphic area centred on the Arctic Archipelago. Here, geological events and geomorphic processes have created lowlands, hilly terrain, and mountains. The cold, dry climate results in many periglacial features, such as tundra polygons, that provide a distinctive appearance to the land. Scattered bedrock and patterned ground are widespread in the lowlands, while alpine glaciers occur in the more mountainous landscapes. Rolling to hilly terrain underlain by permafrost is affected by slumping, caused by **solifluction** or **gelifluction**: 'soil flow' down a sloping frozen surface (Trenhaile, 1998: 84–5). A variety of landforms are found in the Queen Elizabeth Islands (those islands lying poleward of the Northwest Passage, including Prince Patrick Island in the west and Devon Island in the east). Most of Canada's glaciers are found in the mountainous terrain of the eastern section of the Queen Elizabeth

Islands, especially Ellesmere Island. These glaciers are retreating, signalling the start of a major shift in their geographic extent. If the warming trend continues, glaciers could disappear in time. A similar retreat is taking place with sea ice and permafrost, though changes in the latter are more difficult to measure. As well, ice in the ground is less exposed to sunlight and is insulated from surface heat by vegetation cover.

Glaciation

Glaciation is one of the principal erosional agents that have fine-tuned a variety of landscapes in Canada's North. Glaciation involves the formation, advance, and retreat of glaciers. The most recent phase of continental glaciation began some 25,000 years ago. Glacial formation took place as the climate cooled, causing snow to accumulate and eventually forming glaciers. Glacial retreat commenced when the climate began to warm, causing the front of the continental glaciers to melt. As the glaciers melt and recede, material contained in the ice is deposited on the ground. In the Canadian North, unglaciated terrain is limited to a small area of Yukon where there was not enough precipitation to nourish the expansion of glaciers.

During the last glacial advance of the Pleistocene Epoch, two huge ice sheets, the Laurentide and the Cordillera, covered most of Canada. These ice sheets reached a maximum thickness of 4,000 and 2,000 metres, respectively. In comparison, today the largest Canadian glaciers are found on Ellesmere Island and have a thickness of nearly 1,000 metres. Some time around 15,000 BP, the climate warmed, causing these huge ice sheets to melt, i.e., retreat. By 14,000 BP, an ice-free corridor appeared along the eastern edge of the Cordillera ice sheet, connecting the unglaciated areas in Yukon with the rest of ice-free North America. By about 10,000 BP, most of these two ice sheets had melted.

During this process of advance and retreat, two geomorphic processes took place. First, the advancing ice sheet caused glacial erosion; later, the retreating ice sheet deposited debris on the land. Glacial erosion took various forms, such as scraping off the unconsolidated material and plucking out huge chunks of bedrock. Where the bedrock was highly resistant, the rock was scraped and scoured. In the mountains, alpine glaciers moved quickly downslope, which had the effect of 'sharpening' mountain features, as shown in Figure 2.14. Alpine glaciers create mountain peaks formed by cirques (basins carved by the glaciers) and arêtes (razor-like ridges). Other evidence of the erosional power of alpine glaciers is found in U-shaped valleys in the Cordillera and the coastal fjords of Labrador and Baffin Island.

As the Earth warmed, both alpine and continental glaciers deposited vast amounts of debris. Glacial features of deposition consist of water-sorted deposits and unsorted deposits. As the ice sheets melted, some of the material contained in the ice was discharged into running meltwater and glacial lakes. The debris held in the ice sheets was deposited on the land; in some cases, glacial meltwaters sorted the debris or till into eskers and outwash plains. **Eskers**—long, narrow ridges of sorted sands and gravel—were deposited from melt streams within or beneath the decaying ice sheet (Figure 2.15). Some eskers are over 100 kilometres in length. The most common glacial deposit is **glacial till**, which consists of unsorted material deposited by a melting ice sheet or glacier. Glacial till extends over vast areas while **drumlins**, formed by

Figure 2.14 Alpine Glaciers

The St Elias Mountains, located in southern Yukon, contain many alpine glaciers with an erosional force that sharpens the features of mountains. The three shown in this photograph are descending in three parallel valleys from a much larger glacier located in the bowl-shaped basin (called a cirque) at the top of the mountain, Between the valleys are sharp-edged ridges or arêtes.

Source: Reproduced with the permission of Natural Resources Canada 2011, courtesy of the Geological Survey of Canada (Photo 2002-678 by Bélanger, Robert).

massive subglacial flooding, appear as clusters of low, elongated, whale-backed hills, shaped by the flow of the ice. Most drumlins are believed to have been formed a short distance behind the ice margin just prior to deglaciation and therefore record the final direction of ice movements. As the massive ice sheets melted, enormous quantities of water were released. These meltwaters either overtaxed the existing southern-flowing drainage system or formed glacial lakes. Often, these glacial lakes were created when the northward-flowing rivers were still blocked from reaching the sea by the remaining ice sheet. The largest glacial lake, Lake Agassiz, occupied much of Manitoba.

Glaciation has created different landforms in each of the five geomorphic regions. In the Cordillera, mountain features have been sharpened into arêtes and peaks; in the Interior Plains, deposits of glacial debris are everywhere; the Canadian Shield has been subjected to ice plucking, scraping, and scouring; the Hudson Bay Lowland was depressed by the weight of the ice sheet; and, in the higher elevations and latitudes of the Arctic Lands, disintegrating glaciers, including the remnants of the Ellesmere Ice Shelf, stand out.

Periglacial Features

Periglacial landforms, i.e., areas on the perimeter of glaciated or permanently frozen regions, are widespread in the Arctic Lands. The most distinctive periglacial landforms

Figure 2.15   Esker in Arctic Quebec

This esker is located near the Inuit settlement of Salluit, on Hudson Strait at the top of northern Quebec. In general, eskers range from small, sinuous ridges a few tens of metres long, to linear features up to 15 km long and even longer. They consist of post-glacial gravel and other sediments that were deposited by subglacial or englacial streams.

Source: Natural Resources of Canada, 'Esker', *Interpretation Guide of Natural Geographic Features*, 2008 Centre collégial de développement de matériel didactique, photo no 6432, at: <www.cits.rncan.gc.ca/site/eng/resoress/guide/esker/pg07.html>.

are ice-cored hills known as **pingos**. These hills are associated with permafrost and a cold, dry Arctic climate where frost action (the freezing-and-thawing cycle) is the dominant geomorphic process. They occur in lowlands where continuous permafrost exists, and are formed when an ice lens in permanently frozen ground is nourished by extraneous water. Over time, the ice lens grows and pushes itself upward, forming a mound and eventually a hill. However, the most common periglacial feature is **patterned ground**—symmetrical forms, usually polygons, caused by intense frost action over a long period of time (Figure 2.16). Patterned ground includes frost-sorted circles of polygon patterns of stones and pebbles. The general process of frost heave causes coarse stones to move to the surface and outward.

During the **Late Wisconsin glacial period**, these conditions were associated with the southern edge of the Laurentide and Cordillera ice sheets. As these ice sheets retreated, more periglacial features were formed. Hugh French (1996: 5) has described this process:

> During the cold periods of the Pleistocene, large areas of the now-temperate middle latitudes experienced intense frost action and reduced temperatures because of their proximity to the ice sheets. Permafrost may have formed, only to have been degraded during a later climatic amelioration.

Figure 2.16   Tundra Polygons

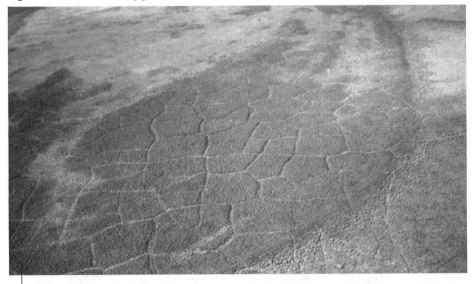

Tundra polygons rearrange the surface of the ground into polygon-like shapes. They are formed because of the freeze–thaw cycle, which is the principal erosional agent in areas of continuous permafrost. The freeze–thaw cycle causes the ground to contract and expand, thus rearranging the surface material into polygon-like forms. Relic tundra polygons still exist in areas in which continuous permafrost conditions once existed.

Source: Reproduced with the permission of Natural Resources Canada 2011, courtesy of the Geological Survey of Canada (Photo KGS-791 by Harrison, J.M.).

French estimates that periglacial features are found over one-fifth of the Earth's surface. Today, relic periglacial features exist great distances from Canada's cold, dry Arctic climate, but the formation of new patterned ground and other periglacial features continues to take place in the Arctic Lands.. Tundra polygons are a common periglacial feature found in the Arctic Lands.

Permafrost

Permafrost, or perennially frozen ground, is found in almost all of the Canadian North. It is defined as ground remaining at or below the freezing point for at least two years. Permafrost reaches its maximum depth in high latitudes where its frozen state extends to several hundred metres or more into the ground; at more southerly sites its depth may be less than 10 metres. The greatest recorded thicknesses in Canada, on Baffin and Ellesmere islands, are over 1,000 metres.

Permafrost has existed for thousands of years in the Canadian North. The upper layer of permafrost (called the active layer) thaws each summer. The thickness of the active layer varies from a few centimetres in the Arctic to several metres in the Subarctic. The southern extent of permafrost is associated with the mean annual air temperature isotherm of 0°C (Williams, 1986: 3).

Since permafrost varies in depth and geographic extent, it is divided into four types—continuous, discontinuous, sporadic, and alpine (Figure 2.17). An area is classified as having continuous permafrost when over 80 per cent of the ground is permanently frozen; for discontinuous permafrost, 30–80 per cent must be frozen; and for sporadic permafrost, less than 30 per cent of the ground in an area is permanently frozen. The sporadic permafrost zone represents a transition area between permanently frozen and unfrozen ground. Alpine permafrost is not defined by the percentage of permanently frozen ground but by its presence in a mountainous setting. Such permafrost is found in British Columbia and Alberta. In terms of geographic distribution, continuous permafrost is found where the mean annual temperature is around –7°C while discontinuous permafrost lies between –5°C and –7°C (see Figure 1.3). Sporadic permafrost is often found between 0°C and –5°C.

Permafrost affects the land surface by slowing the growth of vegetation, impeding surface drainage, and creating periglacial landforms such as pingos and thermokarst features. While permafrost and periglacial landforms are often found in the same geographic area, permafrost refers to permanently frozen ground while periglacial features are created by the action of freezing and thawing. The melting of permafrost has only become a problem since efforts have been made to develop the North by building

Figure 2.17   Permafrost Zones in the Canadian North

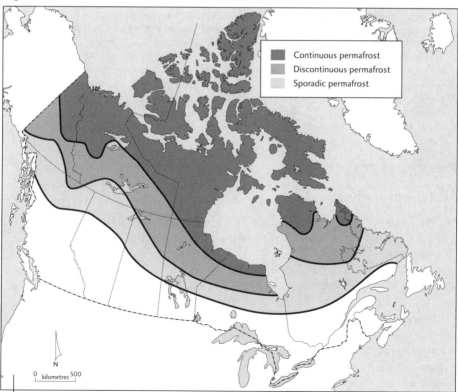

The three permafrost zones extend across the Canadian North. In addition, occurrences of permafrost are found at higher elevations in the Cordillera.

Source: After *The Atlas of Canada*, at: <atlas.gc.ca>.

roads, pipelines, and towns, and, more recently, by warmer summer temperatures (Vignette 2.12). The design of these human-made features must take into account the presence of permafrost. Exposure of ice-rich ground during construction can result in retrogressive thaw slumps (Burn and Lewkowicz, 1990), and such slumps can prove costly to the project and to the environment. Permafrost has made northern construction 'a matter of geotechnical science and engineering' (Williams, 1986: 27).

Vignette 2.12 Melting of Permafrost

Disturbance or removal of vegetation cover by construction projects can result in the melting of permafrost and the resultant subsidence or sinking of the ground. How does this melting come about? The natural vegetation cover provides insulation from the Sun's radiation. When removed, the darker ice-rich ground then becomes warmed by the Sun and the higher soil temperatures melt the ice contained in the ground. The result is a shift from stable terrain to unstable terrain. One effect is subsidence, with the ground settling to a lower level. Another is gelifluction in hilly areas where slumping takes place on the steep slopes. Finally, subsidence results in thermokarst topography where the melting of ice embedded in the ground forms an irregular or hummocky landscape. For many decades, construction firms have taken special measures to ensure the thermal regime of the ground is not altered by the removal of vegetation. One measure is to cover steep slopes with wood chips and thus insulate the ground from the Sun's rays. Another is to undertake construction projects in the winter, which minimizes the impact on the frozen ground.

Global warming presents a new and wider concern as the southern limit of permafrost retreats and continuous permafrost changes into discontinuous permafrost and discontinuous permafrost into sporadic permafrost. However, with natural vegetation cover intact, this retreat is slowed. On the other hand, when ice in permafrost is exposed, such as along the shoreline at Tuktoyaktuk, its disintegration is rapid and is accelerated by wave action.

Northern Hydrology

Northern Canada's rivers and lakes have two phases—active (spring/summer) and inactive (winter). This hydrological cycle is most active in the spring due to the release of water stored as snow and ice. As well, most precipitation falls during the spring. With such a rush of water, flooding often occurs. For the rest of the year, the cycle is inactive, that is, the flow of water is well within the capacity of streams and rivers. Though annual precipitation in the North is low, water resources in the North make up the bulk of Canada's water reserves. These resources are found in four drainage basins—the Arctic, Hudson Bay, Atlantic, and Pacific basins (Figure 2.18). The river systems found in each drainage basin empty into the three oceans surrounding Canada. The Arctic and Hudson Bay drainage basins form nearly 75 per cent of the area of Canada and, despite relatively low precipitation over much of this area, these two basins account for almost 50 per cent of the stream flow (Table 2.3).

In terms of runoff volume, the Mackenzie River is the most important river in the North. Since most of its headwaters are located in British Columbia, Alberta, and

Figure 2.18   Drainage Basins of Canada

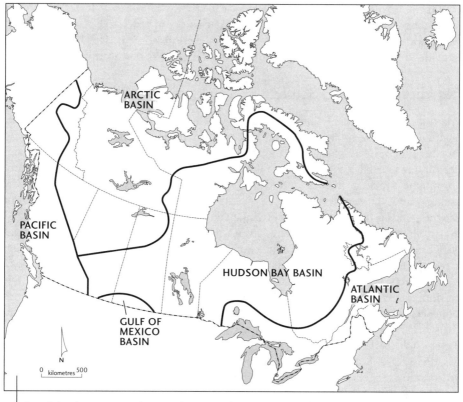

ARCTIC
BASIN

PACIFIC
BASIN

HUDSON BAY BASIN

ATLANTIC
BASIN

GULF OF
MEXICO
BASIN

N

0 kilometres 500

Canada has four major river basins and one minor basin, the waters of which flow to the Gulf of Mexico. A more detail account of river basins is shown in *The Atlas of Canada*.

Source: After *The Atlas of Canada*, at: <atlas.gc.ca>.

Saskatchewan, spring melting first occurs south of 60°N while the lower reaches of the Mackenzie River are still frozen. Ice jams frequently occur, causing widespread flooding. The Mackenzie River discharges vast quantities of freshwater into the Beaufort

Table 2.3   Drainage Basins of Canada and Their Streamflows

Drainage Basin	Area (million km²)	Streamflow (m³)
Hudson Bay	3.8	30,594
Arctic	3.6	20,491
Atlantic	1.6	21,890
Pacific	1.0	24,951
Total	10.0	97,926*

*The drainage basin of the Gulf of Mexico extends into a small portion of southern Alberta and Saskatchewan, and accounts for an area of 21,600 km² or 0.003 per cent of Canada's territory. Accordingly, the Canadian section of the Gulf of Mexico drainage basin contributes a similarly small amount of streamflow: 209 m³.

Sea. Other northern rivers, such as La Grande and Churchill, also empty freshwater into the sea. One impact of this freshwater is to stratify the ocean waters, that is, river water tends to overlie the colder, saltier, and denser ocean waters. This stratification of ocean waters results in surface layers having a lower salt content. This freshwater also provides a suitable estuarial habitat for marine life, such as bowhead whales.

Northern Seas and Sea Ice

The seas of the Canadian North extend from the Beaufort Sea to the Labrador Sea and include the waters of Hudson Bay. These cold water bodies have less impact on the northern hydrological cycle than might be expected. The reasons are twofold: (1) cold seas and sea ice have a lower evaporation rate than warm seas, thus limiting moisture forming in Arctic air masses; and (2) the circulation of the Arctic Ocean is relatively isolated from the rest of the world's currents because the **Beaufort Gyre circulation system** tends to limit flows to the Atlantic and Pacific oceans and because of the cold water. However, with the warming of the waters of the Arctic Ocean and the loss of its ice cover, a major question facing climatologists is whether this circulation system is weakening, thus leading to greater interaction with the Atlantic and Pacific oceans.

The Arctic Ocean has a permanent ice cover known as the Arctic ice pack.[2] By definition, the Arctic ice pack is ice that has not melted for at least two years and is not attached to land and therefore moves in a clockwise motion within the Arctic Ocean. This circumpolar movement within the Arctic Ocean is called the Arctic Oscillation. Polar pack ice is much thicker and harder than new ice, which forms and melts within one year. **Land-fast ice** is a common form of new ice. Land-fast ice is attached to coasts and extends offshore over shallow parts of the continental shelf. Unlike the polar pack ice, land-fast ice remains fixed in location while pack ice drifts, driven by currents and winds.

Arctic ice and waters in relatively small amounts flow into the Atlantic Ocean around Baffin Bay, where it mixes with the warmer Atlantic waters, forming Subarctic water. Baffin Bay is connected to the Arctic Ocean through Nares Strait and Jones and Lancaster sounds, and to the Labrador Sea through Davis Strait. Most icebergs are formed from the Greenland Ice Sheet. The Labrador Current carries them to the waters off Newfoundland, where they represent a hazard to both ocean shipping and offshore drilling rigs. During winter, when there is extensive ice cover, a **polynya** or open-water area (called 'North Water') exists in the northern part of Baffin Bay. The explanation for the natural factors that create open water in a frigid environment still eludes physical scientists (Vignette 2.13).

Sea ice is one of the unique features of northern oceans. Another is icebergs. In late winter, sea ice extends from the Arctic Ocean to the North Atlantic waters off-shore from Labrador and the island of Newfoundland. Icebergs are floating masses of freshwater ice that broke away from glaciers. Most icebergs in the North Atlantic have calved (broken away) from glaciers as the glaciers enter the sea. Each year, 10,000 to 15,000 icebergs enter the North Atlantic, with most originating from glaciers along the west coast of Greenland. A few come from glaciers in the eastern Canadian Arctic islands. Icebergs float in the Labrador Current beyond the edge of the land-fast sea ice. This zone, known as 'Iceberg Alley', extends from Baffin Bay to the waters off the coast

Vignette 2.13 **Polynyas**

The Arctic Ocean has a thick cover of ice, but a dozen or more areas of open water do occur each winter. Many recur each winter. The largest recurring open water lies between Baffin Island and Greenland. Such open water is known as a polynya, which is a derivation of the Russian word for open water surrounded by ice. The explanation remains unresolved but most hydrographers believe that the basic mechanism involves various combinations of tides, currents, ocean-bottom upwellings, and winds that keep the surface waters moving and thus prevent them from freezing. Polynyas may be as small as 50 metres across or as large as the famous North Water polynya, which often extends over 130,000 km². These biological 'hot spots' serve as Arctic oases where marine animals, polar bears, and birds congregate. Not surprisingly, archaeologists have uncovered Thule settlement sites near the North Water. One 'hot spot' is Lancaster Sound. The North Water polynya extends into Lancaster Sound, which is considered the 'Serengeti of the Arctic' because most of the world's narwhals and a large number of whales, walruses, and seals frequent these waters. As well, polar bears roam the perimeter of the open water, hunting for ringed and bearded seals, though these bears also will feed on the carcasses of beluga whales, grey whales, walruses, narwhals, and bowhead whales

of Labrador and Newfoundland. Once the sea ice has melted, icebergs can be found close to land. Before disintegrating, icebergs may float as far south as 40°N, which is approximately the same latitude as New York City. Iceberg viewing is an important late spring tourist activity in Newfoundland.

Sea ice varies in thickness and duration. The most durable and thickest ice is found in the **Arctic ice pack**. In the summer, sea ice disappears first in the Great Lakes and offshore of Atlantic Canada and last in the Arctic Ocean. Accordingly, the length of time for open water in the Arctic Ocean is very short—perhaps only weeks—while in mid-latitude waters open water exists for most of the year.

Sea ice has a dynamic element that affects climate. A NASA ice expert, Josefino Comiso (NASA, 2010), expresses this relationship:

> The oceans are crucial to Earth's climate system, since they store huge amounts of heat. Changes in sea ice cover can lead to circulation changes not just in the Arctic Ocean, but also in the Atlantic and Pacific oceans. If you change ocean circulation, you change the world's climate.

More specifically, the increasing summer reduction in ice cover in the Arctic Ocean has altered its currents and thus the amount of ice/water/energy flowing into the North Atlantic Ocean through Baffin Bay.

The trend over the last three decades is clear. The Arctic Ocean is warming. Satellite imagery and on-site measurements reveal that the ice cover has diminished in both extent and average thickness. The significance of the Arctic ice cover having a greater proportion of new ice is that this thinner ice melts more quickly in the summer, thus creating a longer period of open water. In September 2007, for instance, satellite

imagery indicated that the extent of open water in the Arctic Ocean was more extensive than in the previous decades, with the extent of sea ice diminished to 4.3 million km² (National Snow and Ice Data Center, 2007b). The most recent data, from
29 September 2010, revealed that Arctic sea ice reached its third lowest minimum
extent at 4.6 million (the 2008 minimum was 4.7 million km², and the 2009 figure
of 5.4 million km² reversed the trend for one season (NASA, 2010). For more on this
subject, see the section 'Changing Geography of the Arctic Ocean' in Chapter 8.

Grand Themes

From this discussion of the North's physical geography, we can see that the complex
but changing nature of this cold environment provides many challenges and calls
for continued research to monitor its changing nature, especially that of the Arctic
Ocean and its potential impact on the global climate, as well as on the people and
communities for whom the North is their homeland. Reflecting back to the impact
of climate warming on the Ward Hunt Ice Shelf mentioned at the beginning of this
chapter, this impact symbolized the enormous change taking place in all aspects of
Arctic geography as a result of a warmer climate. At this juncture, the reader is alerted
to four of the grand themes in geography, and these themes will reverberate in the
following chapters:

1) Is our physical world changing and, if so, how will it impact the North's biomes?
2) Can human activities, but especially those related to resource development, reduce
   their stamp on the physical environment?
3) Climate change as a global political and environmental issue is a new phenomenon,
   and its impact on the North, especially as recorded by retreating sea ice, glaciers,
   and permafrost, is remarkable. But why is climatic warming taking place more
   robustly in the North than in other areas of the world?
4) Are the Inuit benefiting from Arctic warming or is their way of life threatened?

The answers to these questions will determine to a large extent not only what happens
in and to Canada's North but also, quite possibly, the future course of humankind.

Challenge Questions

1. What is the basis of the relationship between climate and natural vegetation?
2. What natural feature provides a visible mark on the landscape suitable for determining zonal boundaries?
3. Why is the North considered a fragile environment?
4. Temperature increases are predicted to be much greater in the Arctic than in
   southern Canada. What role does the albedo effect play in these temperature
   increases?
5. Based on the information in Table 2.2, which past climatic variation in the
   Holocene best describes our current climate?

6. Why are most periglacial features found in the region known as Arctic Lands?
7. Why did the glaciers create different landforms in the Cordillera than in the Interior Plains?
8. How would the loss of Arctic Ocean ice stand out as a pivotal change that would affect the global heat balance?
9. Why has the amount of greenhouse gases in the atmosphere increased over the last century?
10. Why is ice, in the form of glaciers, ice shelves, and Arctic pack ice, more likely to melt than ice contained in permafrost?

Notes

1. Sauderson, Smith, and Woo (2000) summarize recent developments in Canadian geomorphology. Much of their discussion deals with research advances in geomorphology occurring in northern Canada.
2. The Arctic ice pack is sometimes referred to as the polar ice pack. However, since there are ice packs in both the Arctic and Antarctic oceans, the term 'Arctic ice pack' is preferred.

References and Selected Reading

Asplin, M.G., J.V. Lukovich, and D.G. Barber. 2009. 'Atmospheric Forcing of the Beaufort Sea Ice Gyre: Surface Pressure Climatology and Sea Ice Motion', *Journal of Geophysical Research* 114, C00A06, doi:10.1029/2008JC005127.

Barber, D.G., R. Galley, M.G. Asplin, R. De Abreu, K.-A. Warner, M. Pućko, M. Gupta, S. Prinsenberg, and S. Julien. 2009. 'Perennial Pack Ice in the Southern Beaufort Sea Was Not As It Appeared in the Summer of 2009', *Geophysical Research Letters* 36, L24501, doi:10.1029/2009GL041434.

Barr, William. 1972. 'Hudson Bay: The Shape of Things to Come', *Musk-Ox Journal* 11: 64.

Boon, S., D.O. Burgess, R.M. Koerner, and M.J. Sharp. 2010. 'Forty-Seven Years of Research on the Devon Island Ice Cap, Arctic Canada', *Arctic* 63, 1: 13–29.

Bouchard, Mireille. 2001. 'Un défi environnemental complexe du XXIe siècle au Canada: l'identification et la compréhension de la réponse des environnements face aux changements climatiques globaux', *Canadian Geographer* 45, 1: 54–70.

Burn, Chris R. 1995. 'Where Does the Polar Night Begin?', *Canadian Geographer* 39, 1: 68–74.

———. 1996. *The Polar Night*. Scientific Report No. 4. Inuvik: Aurora Research Institute, Aurora College.

——— and A.G. Lewkowicz. 1990. 'Retrogressive Thaw Slumps', *Canadian Geographer* 34, 3: 273–6.

Cohen, J. Stewart. 1997. *Mackenzie Basin Impact Study Final Report*. Downsview, Ont.: Environment Canada.

Draper, Dianne, and Maureen Reed. 2004. *Our Environment: A Canadian Perspective*, 3nd edn. Toronto: Nelson.

Environment Canada. 2002. 'Canada Climate Normals 1971–2000'. At: <www.msc-smc.ec.gc.ca/climate/climate_normals/index_e.cfm>.

———. 2007. 'Ecosystems'. At: <www.mb.ec.gc.ca/nature/ecosystems/index.en.html>.

Etkin, Dave. 1989. 'Elevation as a Climate Control in Yukon', *Inversion: The Newsletter of Arctic Climate* 2: 12–14.

Fossett, Renée. 2001. *In Order to Live Untroubled: Inuit of the Central Arctic, 1550–1940*. Winnipeg: University of Manitoba Press.

French, Hugh M. 1996. *The Periglacial Environment*, 2nd edn. London: Longman.

———— and Olav Slaymaker, eds. 1993. *Canada's Cold Environments*. Montreal and Kingston: McGill-Queen's University Press.

Hare, F. Kenneth, and Morley K. Thomas. 1979. *Climate Canada*, 2nd edn. Toronto: Wiley.

Houghton, J.T., Y. Ding, D.J. Griggs, M. Noguer, P.J. van der Linden, X. Dai, K. Maskell, and C.A. Johnson. 2001. *Climate Change 2001: The Scientific Basis*. Contribution of Working Group I to the Third Assessment Report of the Intergovernmental Panel on Climate Change. Cambridge: Cambridge University Press. At: <www.ipcc.ch/pub/reports.htm>.

Intergovernmental Panel on Climate Change (ipcc). 2007a. 'About the IPCC'. At: <www.ipcc.ch/about/about.htm>.

————. 2007b. 'Summary for Policy Makers', *Climate Change 2007: Synthesis Report*. At: <www.ipcc.ch/pdf/assessment-report/ar4/syr/ar4_syr_spm.pdf>.

Kaufman, D.S., et al. 2004. 'Holocene Thermal Maximum in the Western Arctic', *Quaternary Science Reviews* 23: 529–60.

Lawford, R.G. 1988. 'Climatic Variability and the Hydrological Cycle in the Canadian North: Knowns and Unknowns', *Proceedings of the Third Meeting on Northern Climate*, Canadian Climate Program. Ottawa: Minister of Supply and Services, 143–62.

MacDonald, Brian, ed. 2007. *Defence Requirements for Canada's Arctic*. Vimy Paper 2007. Ottawa: Conference of Defence Associations Institute.

NASA. 2005. 'Arctic Sea Ice Imagery'. At: <www.nasa.gov/centers/goddard/images/content/134724main_seaice_min_2005_lg.jpg>.

————. 2006. 'Arctic Sea Ice Thickness Model v1 (1940–2060)', *Visible Earth*. At: <visibleEarth.nasa.gov/view_rec.php?id=10080>.

————. 2009. 'Feature: Arctic Sea Ice Extent is Third Lowest on Record', 6 Nov. At: <www.nasa.gov/topics/earth/features/seaicemin09_prt.htm>.

————. 2010. 'Arctic Sea Ice', Earth Observatory, 19 Oct. At: <earthobservatory.nasa.gov/Features/WorldOfChange/sea_ice.php>.

National Snow and Ice Data Center. 2007a. 'Arctic Sea Ice at Record Low', 11 Sept. At: <edition.cnn.com/2007/TECH/science/09/11/arctic.ice.cover/>.

————. 2007b. 'Overview of Current Sea Ice: September 2007', *Arctic Sea Ice News* (Fall). At: <nsidc.org/news/press/2007_seaiceminimum/20070810_index.html>.

Natural Resources Defence League. 2007. 'The Boreal Forest: Earth's Green Crown'. At: <www.nrdc.org/land/forests/boreal/page5.asp>.

Nickels, S., C. Furgal, M. Buell, and H. Moquin. 2005. *Unikkaaqatigiit—Putting the Human Face on Climate Change: Perspectives from Inuit in Canada*. Ottawa: Joint Publication of Inuit Tapiriit Kanatami, Nasivvik Centre for Inuit Health and Changing Environments, and Ajunnginiq Centre.

Northwest Territories. 2004. 'Vegetation on the Taiga Plains'. At: <forestmanagement.enr.gov.nt.ca/forest_resources/eco_land_classification/taiga07_vegetation.html>.

Renfrow, Stephanie. 2006. 'Arctic Sea Ice on the Wane: Now What?', NASA: Earth System Science Data and Service. At: <nasadaacs.eos.nasa.gov/articles/2006/2006_seaice.html>.

Sauderson, Houston C., Dan J. Smith, and Ming-Ko Woo. 2000. 'Progress in Canadian Geomorphology and Hydrology', *Canadian Geographer* 44, 1: 56–66.

Slaymaker, Olav. 1988. 'Physiographic Regions', in James H. Marsh, ed., *The Canadian Encyclopedia*, vol. 3. Edmonton: Hurtig, 1671.

Smith, Dan, ed. 2010. 'UAlberta's John England on Fracture of Ward Hunt Ice Shelf', *CAG GeogNews*, 27 Aug. At: <www.geog.uvic.ca/dept/cag/geognews/geognews.html>.

State of the Planet: Arctic. 2005. At: <www.thewe.cc/weplanet/poles/mastpole.html>.

Stocker, Thomas, and Gian-Kasper Plattner. 2009. 'The Physical Science Basis of Climate Change: Latest Findings to be Assessed by WG I in AR5', United Nations Climate Change Conference, COP 15 IPCC Side Event, IPCC Findings and Activities and Their Relevance for the UNFCCC Process. Copenhagen, 8 Dec.

Trenhaile, Alan S. 2009. *Geomorphology: A Canadian Perspective*, 4th edn. Toronto: Oxford University Press.

Williams, Peter J. 1986. *Pipelines and Permafrost: Science in a Cold Climate*. Ottawa: Carleton University Press.

———— and Michael W. Smith. 1991. *The Frozen Earth: Fundamentals of Geocryology*. Cambridge: Cambridge University Press.

Young, Steven B. 1994. *To the Arctic: An Introduction to the Far Northern World*, 2nd edn. New York: Wiley.

3

The Historical Background

History is essential for an understanding of present events taking place in Canada's North. The human story of Canada's North began some 15,000 to 30,000 years ago when the first Old World hunters entered Alaska and then Yukon, having crossed the land bridge known as Beringia created by lower sea levels during the last ice age. The post-contact history is examined in more detail here because it provides insights into the changing relationship between Canada's first peoples and the newcomers—first, the European explorers, traders, and settlers, and later, Canadians who have come from practically everywhere in the world. In recent years, Aboriginal issues have come to the fore and a challenge to Canada's Arctic sovereignty has surfaced. To simplify matters, the North's history is set into three historic phases:

1. the pre-contact and early contact period;
2. the fur trade era;
3. the resource development period.

Overview

Old World hunters passed through the ice-free corridor into the heartland of North America as the Laurentide Ice Sheet melted, and these Paleo-Indians began to occupy the lands all the way to the tip of South America as early as 11,000 years ago (Dickason with McNab, 2009: 8). The Arctic, however, remained uninhabited until 5,000 years ago when the first wave of migrants from the Bering Strait area, the Paleo-Eskimos, began to spread across the Arctic coastline of Canada. The Paleo-Eskimos had the necessary technology to live in the Arctic. At the time of contact, the **Indian** and **Inuit** peoples of the Canadian North numbered close to 60,000. First known contact, though a fleeting one, was with the Vikings on the northern tip of Newfoundland at the beginning of the eleventh century. More continuous contact began in the sixteenth century when explorers landed on the east coast of Canada and Basque fishers established summer fish-processing camps on the island of Newfoundland.

The second phase, characterized by much greater interaction with European fur traders, brought many changes to Aboriginal cultures and economies. The fur trade brought great riches to the Hudson's Bay Company (HBC) based in London and to the North West Company in Montreal. The fur trade reached its zenith in the eighteenth century when **Native peoples** across the Subarctic were engaged in trapping beaver and trading beaver pelts for European goods. At first, Indians brought their furs to the HBC posts along Hudson and James bays, but by the end of the eighteenth century both companies had established fur posts in the rich fur country of the Mackenzie Basin. With the amalgamation of the two companies in 1821, the inland Montreal fur trade route via the Great Lakes was abandoned in favour of the more economical Hudson Bay route. By the late nineteenth century, the Hudson's Bay Company made use of American and later Canadian rail service to transport its trade goods and fur bales. The fur trade exposed Native peoples not only to Western goods but also to another culture. In the early days of the fur trade, the relationship between the trappers and the fur traders was more of a partnership, but gradually the relationship saw the Indian trappers grow more and more dependent on the traders and, eventually, on the federal government. At the beginning of the twentieth century, the Arctic white fox pelt was highly prized in the European fur market, and the HBC opened fur-trading posts in the Arctic, thereby bringing the Inuit into the fur economy.

The last historical phase began when Canada gained political control over its northern territories in 1870 (the HBC lands, known as Rupert's Land) and 1880 (the Arctic Archipelago). While the fur economy continued to hold sway, the late nineteenth century saw the emergence of resource development. In 1896, the discovery of gold in the Klondike transformed Yukon from a fur-trapping economy to a resource economy. While other natural resources were exploited in the following years, no resource development approached the magnitude of the Klondike gold rush until World War II. Seeking secure supplies of oil and the reliable transport of men and materials, the US Army financed several large-scale development projects, including the building of the Alaska Highway from Dawson Creek, BC, to Fairbanks, Alaska, the expansion of oil production at Norman Wells, NWT, and the construction of the Canol Pipeline from Norman Wells to Whitehorse, Yukon, and then to Fairbanks.

With the conclusion of World War II, the United States faced a shortage of energy and minerals. US industry turned to Canada for energy and raw materials. By the 1970s, other industrial countries followed. Gradually, the North was transformed into a resource hinterland for the world economy. At the same time, great social changes were taking place among the Aboriginal community. One change was their relocation to small settlements, which exposed Aboriginal peoples, but especially their children, more intensely than ever before to Canadian culture and the Canadian economy through the institutions established by the state in these settlements. Other changes were a weakening of their land-based economy, a growing participation in the wage economy, and a greater dependency on government.

The First Phase: Pre-Contact

Old World hunters who crossed the Beringia land bridge into Alaska were in all likelihood the first people to set foot in North America (Vignette 3.1). But just when these paleolithic people reached Alaska and then the heartland of North America remains in doubt. Archaeological evidence confirms that Paleo-Indians occupied what is today

---

Vignette 3.1 Beringia

During part of the last ice age, a land bridge some 1,500 km in length joined Asia to North America. This land is called Beringia. Now submerged some 100 metres below sea level, the land bridge allowed Old World hunters to move into Alaska. Around 18,000 years ago, the ice sheets reached their maximum extent. At that time, huge quantities of water were locked into the ice sheets, causing the sea level to drop by as much as 150 metres. Later, as the climate began to warm, two major physical events took place. The ocean waters rose, covering the land bridge once again, and an ice-free corridor appeared between the Laurentide and Cordillera ice sheets. This corridor made it possible for the Old World hunters to migrate southward along the eastern slope of the Rocky Mountains.

---

the southwest United States some 11,000 years ago, suggesting that their ancestors migrated along the ice-free corridor between the Cordillera and Laurentide ice sheets some 12,000 years ago or earlier (Vignette 3.2).

Yet, there is a nagging feeling that Old World hunters may have reached the interior of North America before the ice-free corridor appeared (Bonnichsen and Turnmire, 1999). Archaeological evidence suggests humans arrived in Yukon before the last ice advance—some 25,000 to 30,000 years ago. Archaeological findings such as those at Bluefish Caves near Old Crow in Yukon support this hypothesis, although the dating of the Bluefish Caves artifacts remains controversial (Cinq-Mars and Morlan, 1999).

Vignette 3.2 The Clovis Theory: The Original Paradigm

Since the 1930s, the Clovis theory has served as the central paradigm for archaeologists studying the peopling of the Americas. According to the Clovis theory, some 12,000 years ago, Old World hunters from Alaska and Yukon found an ice-free corridor through the vast ice sheet to the unglaciated areas of the New World. Armed with spears that had a distinctive fluted projectile point, these Paleo-Indians quickly spread across North and South America. Archaeological evidence clearly demonstrates that Paleo-Indians were hunting the woolly mammoth and other big-game animals in the Great Plains and the Southwest of what is today the United States as early as 11,000 years ago. At the same time, primitive hunters had reached the southern tip of South America. The empirical support for the Clovis theory is based on archaeological finds of flaked-stone projectile points. These distinctive spear points were first unearthed in 1932 at a New Mexico mammoth kill site near Clovis, New Mexico. Since then, similar finds dated to the same early period have been made in other parts of North America. However, more recent archaeological findings, especially the Monte Verde site in Chile, which apparently dates to 12,500 years ago, suggest that Old World hunters reached South America well before the Clovis people. Does this mean that the original paradigm is no longer valid?

Sources: Adapted from Thomas (1999); Bonnichsen and Turnmire (1999).

One theory espoused by Wright (1995) suggests that two major migrations from Siberia to the temperate areas of North America took place—one before the last ice advance (25,000 years ago or more), the other when the ice sheets began to retreat (12,000 to 15,000 years ago).[1] Another theory gaining more and more support suggests that Old World hunters spread along the Pacific Coast by island-hopping before the ice-free corridor opened. This theory assumes that, with the lower ocean level, more offshore islands beyond the reach of the Cordillera Ice Sheet were available to these ancient peoples. In addition, volcanic islands that no longer rise above sea level, in both the South and North Pacific, might have made such possible island-hopping more feasible many thousands of years ago than would be the case today.

Occupying Northern Canada

Once south of the ice sheet, the Old World hunters gradually spread across the Americas (Vignette 3.2). Their main source of food remained the woolly mammoth and other big-game animals. Around 10,000 years ago, the woolly mammoth became extinct and its hunters, now described as Paleo-Indians, had to adjust to new circumstances. At the same time, the Laurentide and Cordillera ice sheets had retreated, allowing the people to occupy these new hunting grounds. The archaeological record for this period is very sketchy, leaving no clear picture of the pattern of occupancy by various Paleo-Indian groups until contact. Arid regions, such as the US Southwest, parts of the Pacific coast of South America, and the Far North, have provided more archaeological evidence of human occupancy than is the case for more humid regions with greater precipitation.

Roughly a thousand years later, around 9,000 BP, a second wave of Old World hunters, the Paleo-Eskimos, crossed the Bering Strait from Siberia to Alaska. By this time, ice sheets had retreated from the Arctic coast, allowing the sea-based hunting economy of the Paleo-Eskimos to flourish. These early migrants are named Denbigh after an important archaeological site in Alaska. Their culture, the Arctic Small Tool Tradition, was based on the use of flint to shape bone and ivory into harpoons and other tools. As sea hunters, they occupied Arctic coastal lowlands and eventually reached the coast of Labrador and Newfoundland. Over time, the culture of these people either evolved or was replaced by a new wave of Paleo-Eskimo migrants—the Dorset—who in turn were replaced by the Thule. While the archaeological story is incomplete, two major cultural changes have been identified. The first one saw the Denbigh culture evolve into or replaced by the Dorset culture. The Dorset lived in semi-permanent houses built of snow and turf, heating them with soapstone oil lamps. Some archaeologists believe that the Thule, with their superior technology and weaponry, overwhelmed the Dorset and maybe even the Viking colony on Greenland (Thompson, 2010).

Around AD 900, the Thule culture appeared in coastal Alaska. Here, the Thule hunters had developed the technology and skills necessary to hunt the large bowhead whales. They migrated eastward, perhaps driven by population pressures or encouraged by a warming of the Arctic climate that resulted in an increase in open water in the Arctic Ocean. In any case, the Thule spread across the Canadian Arctic using their innovative transportation system of skin boats (kayak and umiak) and dogsleds to harvest seal, walrus, caribou, and the bowhead whale. By AD 1000, the Thule reached Greenland, where they probably encountered the Vikings.

After AD 1300, the climate began to cool, making ice cover more extensive. The climate continued to cool and the period between 1450 and 1850 is referred to as the Little Ice Age. With the more extensive ice cover, sea areas suitable for the bowhead whale became more limited. Unable to secure food from their favourite source, the bowhead whale, the Thule had to shift their hunting strategy and, therefore, became more mobile and formed smaller groups to hunt the seal and caribou. About this time, Europeans, such as the English explorer Martin Frobisher, came into contact with these people. Their descendants are the present-day Inuit. Ironically, the Little Ice Age coincided with the era of European exploration in the Arctic, and the colder temperatures and more extensive ice cover doubtless made such exploration more difficult, and likely was instrumental in the ill-fated Franklin expedition of 1845–7 (Vignette 3.5).

Terra Nullius

The New World was not, as Europeans believed, *terra nullius*—empty lands. The Eurocentric perspective regarded lands without permanently occupied land, such as signs of agriculture and settlements, as empty, though not necessarily uninhabited. Yet Aboriginal peoples inhabited these lands. Dickason (2009: 10) portrays this pre-contact world:

> When Europeans arrived, the whole of the New World was populated not only in all its different landscape and with varying degrees of density, but also with a rich cultural kaleidoscope of something like 2,000 or more different societies.

Exactly how many lived in Canada, let alone northern Canada, in these early days is unknown.[2] Unlike the more populated lands to the south, the North must have had a small population spread over a vast territory. One estimate suggests that some 50,000 Indians and Inuit occupied the Subarctic and Arctic areas of Canada at the time of contact (Crowe, 1974: 20), while Mooney (1928) and Kroeber (1939) estimated a slightly higher figure of 56,000, with 16,000 Inuit in the Arctic and 40,000 Indians in the Subarctic. Such numbers show a population density of 130 km² per person, indicating that the hunting economy could only support a relatively small population due to the limited availability of game and its seasonal nature. In fact, the Arctic and Subarctic, as might be expected, had the lowest population density of all the Aboriginal cultural regions in North America, suggesting lower carrying capacity, which is a measure of the capacity of the land/sea to support people.

In 1857, the first comprehensive estimate of the Indian and Inuit population of British North America was made by Sir George Simpson for the Special Committee of the British House of Commons on the affairs of the Hudson's Bay Company (*Census of Canada*, 1876, vol. 4: ix–xiv). Simpson's figure of nearly 139,000 for British North America was based on reports made by local HBC officials on the number of Indians trading at their posts. Approximately 61,000 Indians were reported as trading at posts in the Subarctic. Added to these numbers was Simpson's guess that 4,000 Eskimos occupied the Arctic. In 1881, the census indicated about 108,000 Indians, Métis, and Inuit residing in the provinces and territories of Canada (Table 3.1).

Table 3.1 Indian Population, 1881

	Indian Population	Total Population	Indian Percentage of Total Population
Prince Edward Island	281	108,891	0.3
Nova Scotia	2,125	440,572	0.5
New Brunswick	1,401	321,233	0.4
Quebec*	7,515	1,359,027	0.6
Ontario*	15,325	1,923,228	0.8
Manitoba*	6,767	65,954	10.3
British Columbia	25,661	49,459	51.9
North-West Territories	49,472	56,446	87.6
Canada	108,547	4,324,810	2.5

*In 1881, the geographic size of Quebec, Ontario, and Manitoba was smaller than today. As well, the North-West Territories included lands in present-day Alberta, Saskatchewan, much of Manitoba, Ontario, and Quebec as well as the three territories. After 1881, Manitoba, Ontario, and Quebec expanded their geographic size by obtaining land from the North-West Territories. In 1905, the provinces of Alberta and Saskatchewan were formed and the North-West Territories became the Northwest Territories.

Source: Statistics Canada, 1880–81 Census of Canada.

Where did Aboriginal peoples live? The geography of Aboriginal people at the time of contact is based on the diaries and accounts of early explorers, fur traders, and missionaries. The broad pattern involves a threefold geographic division, with the Algonquian Indians living in the eastern Subarctic while the Athapaskan Indians occupied the western Subarctic. To the north lived the Inuit. This same broad geographic settlement pattern exists today.

Determining the geographic location of individual **tribal groups** is much more problematic, but these peoples likely had occupied the territory shown in Figure 3.1 for some time, perhaps centuries. The geographic stability of the pre-contact hunting societies can only be guessed at because archaeological evidence is so limited. Only in the Arctic is there archaeological evidence and oral history documenting the forced replacement of one group by another—the Dorset by the Thule, who were the ancestors of the Inuit.

These small but highly mobile Indian and Inuit hunting groups had become finely attuned to harvesting the particular food resources in their geographic territory. Within that territory, hunters followed a seasonal rhythm of movement as they sought wild game. Through this seasonal movement, they exerted a form of political control over the territory. The main source of food was big game. For many, caribou was the preferred food; whale blubber and seal meat were common food sources for the Inuit. These hunting people moved according to the seasonal migrations of wild animals. Such a migratory hunting strategy is a form of sustainable resource use because the necessity of moving from one area to another limits the pressure on local resources. Limited weapon technology and cultural taboos also reduced the possibilities of over-hunting or overfishing.

Figure 3.1   Geographic Areas Occupied by Aboriginal Peoples at Contact

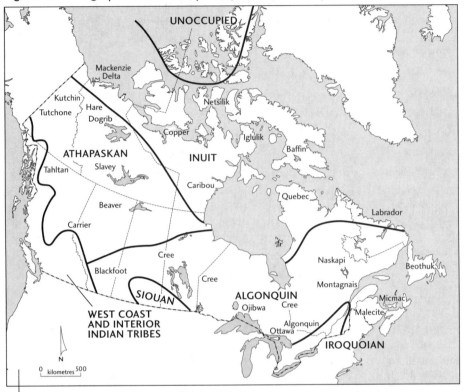

Indians and Inuit had spread across northern Canada well before contact with Europeans. The geographic arrangement of these peoples at the time of contact has been reconstructed by anthropologists and ethnographers. Yet, it is important to note that, prior to contact with Europeans, Aboriginal peoples were not static in geographic terms. For instance, the Dorset occupied the Arctic prior to the Thule, but by around AD 1000 the Thule had occupied their lands and then came into contact/conflict with the Viking colonies on the eastern coast of Greenland.

Source: After Harris (1987: Plate 11).

The pre-contact hunting economy had many attractions but it also contained an element of risk. For example, an unforeseen reduction in game, such as a shift in the annual migration route of caribou, could expose hunters and their dependants to severe shortages of food and perhaps starvation. As Ray (1984: 1) observed, 'opinions are divided whether hunting, fishing, and gathering societies generally faced a problem of chronic starvation—the more traditional viewpoint, or whether they were the original affluent societies.' Key to resolving this puzzle is reliability of food sources and the size of the population. Some geographic areas had a more readily available source of food than other areas. The west coast of British Columbia, for example, had abundant fish stocks, which permitted relatively dense and sedentary pre-contact Indian populations (Harris, 1997: 30). Caribou hunters, who depended on the seasonal migration patterns of caribou, had to deal with a more problematic food source. Over the long run, archaeological evidence suggests that the population in each cultural region attained a size that was in balance with its food sources, but short-term food shortages occurred.

The Arrival of Europeans

The first Europeans to land on North American soil were the Vikings. During the tenth century, the Vikings regularly travelled from Greenland to Ellesmere and Baffin islands to trade with either the Dorset or Thule peoples. They also attempted to establish settlements along the coast of Labrador and Newfoundland. The first and only Norse site found in Canada is located at L'Anse aux Meadows on the northern tip of Newfoundland's Great Northern Peninsula.

Some 500 years later, the Viking discoveries were all but forgotten, leaving Europe's geographic knowledge of the New World scanty at best. European explorers sought to reach the rich spice lands of Cathay by sailing across the Atlantic Ocean but reached North America. Reports of the new land's great riches, based on information obtained from local Indians, may have been misinterpreted or embellished in hopes of securing financial support for later voyages. In 1536, Jacques Cartier returned to France with a report of a golden 'Kingdom of Saguenay' (Trudel, 1988: 368). This report was based on information supplied by Indians and supported by the Iroquois chief Donnacona, whom Cartier had captured and taken back to France. Cartier thought that he had discovered diamonds and gold from this mythical kingdom. These riches, upon careful examination in France, turned out to be quartz and iron pyrites—'fool's gold'.

In a similar vein, Martin Frobisher, under orders from Queen Elizabeth I, set sail from England to find the Northwest Passage to the Orient. Instead, he discovered Baffin Island and the promise of great mineral wealth—gold. Like the Spaniards, the English sought gold and silver from this New World. Would Frobisher's discovery give England a source of gold to rival Spain's New World gold mines? Over 1,200 tonnes of rock from Baffin Island were transported to England, but this rock was found to contain not an ounce of gold! Nevertheless, Frobisher's two sea voyages in 1576 and 1577 mark the beginning of British exploration of Arctic Canada. As with most early voyages, contact with the local peoples was established and trade often followed. However, contact between these two very different peoples did not always end on a friendly note. For instance, after landing on the shore of Baffin Island in 1576, five of Frobisher's crew were captured by local Inuit during a skirmish and were never seen again (Cooke and Holland, 1978: 22–3). The following year, Frobisher lured an Inuk in his kayak to come near his ship. Captured and taken to England, the Inuk's skills at handling his kayak while hunting the royal swans were observed by the Queen (Crowe, 1974: 65). After a few years, he died of natural causes in England.

During the remainder of the sixteenth century and in the early seventeenth century, many explorers, including Davis, Hudson, and Button, continued to search for a route to the Orient. In the course of this search they explored the waters of Baffin Bay and Hudson Bay. These explorers failed to find a route to the Orient but the 'casual' trade with Indians for furs marked the beginnings of a highly profitable enterprise and led to the formation of an English fur-trading company—the Hudson's Bay Company—in 1670.

The Second Phase: The Fur Trade

The story of the fur trade in Canada began in the fifteenth century with the Basque fishers who traded with Indians along the shores of Newfoundland. From these humble beginnings, the fur trade became the centrepiece of New France's economy and then the most profitable enterprise of the Hudson's Bay Company. As early as the sixteenth century, French fur traders, including the coureurs de bois located along the St Lawrence River, recognized that the best furs came from Quebec's Subarctic, where the long, cold winter resulted in long-haired fur pelts. Geography provided easy access to these northern Indians by means of southward-flowing rivers, such as the St Maurice and Saguenay.[3] Late in the next century, the British sought to reach this rich fur region by a sea route through Hudson Strait and then by establishing a series of trading posts at the mouths of rivers flowing into Hudson and James bays. For over a century, the French and English competed for the highly prized furs of the northern Indians, and by 1760 the fur trade was certainly the dominant economic activity in New France and Rupert's Land (Figure 3.3). Even when French Canada fell to the British in 1763, two centres, London and Montreal, continued to compete for northern furs (Figure 3.5). Montreal fur traders of the North West Company, now a blending of French-Canadian and Scottish traders, established posts on the Saskatchewan River in order to intercept Indians on their way to trade at the Bay posts. After decades of fierce and sometimes bloody competition, the two companies merged in 1821 (Figure 3.6). Over the next 100 years or so, the Hudson's Bay Company dominated the Canadian fur economy.

The Hudson's Bay Company

Like all colonial powers, Britain and France looked to their colonies for riches. Wealth often took the form of gold and silver, but in the case of Canada's North, wealth took the form of beaver pelts. The fur trade drew Canada's North and its Indian inhabitants into the European market, where the demand for furs was dependent on the European fashion industry (Vignette 3.3). The political implications of the fur economy are far-reaching and take the form of a core/periphery relationship within a mercantile economic system. In this historical period, the core nations were Britain and France while the North served as the fur hinterland.

Vignette 3.3   Beaver Hats: The Necessary Attire of Gentlemen

For several centuries, the fur trade was based on the beaver skin, highly valued in Europe for hats. This demand, driven by fashion, was both the strength and weakness of the fur trade. The beaver hat was a sign of upper-class status. As shown in Figure 3.2, the beaver hat had a number of styles, reflecting occupation/status in life. Back in Canada's trading posts, prices of beaver and other furs were subject to sharp fluctuations—often caused by variation in the demand for these pelts—and these price changes could greatly alter the number of furs required for European goods. The beaver was sought not for its entire pelt but for fur-wool, the layer of soft, curly hair growing next to the skin, which had to be separated from the pelt and from the layer of longer and stiffer guard hairs. This fur-wool was then felted for cloth or hats. The beaver hat went out of fashion in the 1850s and was replaced by 'top hats' made of silk and other materials.

Source: Adapted from Wolfe (1982: 159–60).

Figure 3.2   Beaver Hats

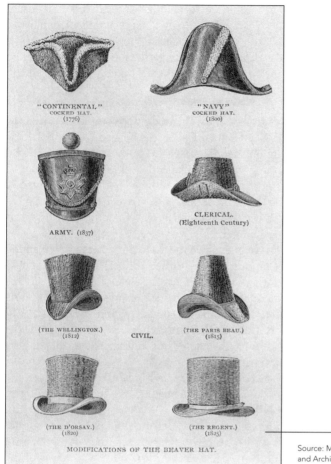

Source: Martin (1892: 125). Courtesy Library and Archives Canada.

For over 300 years, the HBC dominated the Canadian fur trade. This British company was the most powerful economic, social, and political force in the Canadian North. Within the British mercantile system the HBC was an instrument of its foreign policy and a means of acquiring wealth, just as was the English East India Company that was formed 70 years earlier. The role of these companies was twofold: to function as commercial enterprises and to exert British sovereignty.

Ironically, the British did not conceive of a northern sea route to the rich fur lands of Canada's North through Hudson Bay. Such a route would outflank the French route to the south and, in the context of British/French rivalry, strike an economic blow against their enemy. The idea of a northern route came from two French traders, Medard des Groseilliers and Pierre Radisson. After failing to gain support from French officials, Groseilliers and Radisson approached the British and convinced Prince Rupert that a great deal of money could be made by trading with Indians for furs along the Hudson Bay coast. These northern Indians currently traded their furs with the French, who were based on the St Lawrence. Groseilliers and Radisson knew that ships anchored near important northern rivers would soon attract local Indians and trade would ensue. Prince Rupert persuaded his cousin, Charles II of England, and some merchants and nobles to back the venture. In 1670, the King granted wide powers to the 'Governor and Company of Adventurers' of the Hudson's Bay Company, and Rupert became the company's first governor. These powers included exclusive trading rights to all the lands whose rivers drain into the Hudson Bay. These lands were named Rupert's Land.

At first, the company sent ships to trade with the Indians who gathered at the mouths of rivers draining into Hudson Bay. Once the trade was completed, the ships returned to England. The Hudson's Bay Company soon established permanent trading posts, encouraging Indians living in the interior of the country to trade there. By locating at the mouths of rivers draining western Canada, the HBC extended its control over lands far into the interior, thereby increasing the number of Indians involved in trapping and, of course, maximizing the number of furs received. The HBC was a great commercial success from the start. It was also a political success, for its Bay forts provided an important foothold for the British.

The harsh climate around Hudson Bay prevented the fur traders from supplementing their food supply with agricultural products. They turned to local Indians to supply them with game and with other useful products of Aboriginal technology, such as snowshoes and canoes. These Indians who maintained permanent camps near the HBC forts became known as the Homeguard. Indian middlemen became traders who exchanged European goods for furs with inland Indians. This arrangement provided great profits to the middlemen and allowed them to control the inland Indians. Often, the middlemen Indians would not even permit other Indians to travel to the coastal trading posts.

The trading posts along the coasts of Hudson Bay and James Bay all followed the same pattern, being established at the estuaries of the major rivers draining the interior of the forested lands of what are now Manitoba, Saskatchewan, Quebec, and Ontario. The most important of these trading posts was York Factory. Situated at the mouth of the Hayes River, it offered the easiest river access to the rich fur lands of the interior. This post, because of its access to the river network leading into the interior of the Subarctic, soon began drawing furs from Indians as far distant as the Mackenzie River Basin (Zaslow, 1984: 6). By the early eighteenth century, then, the HBC's fur hinterland extended into the 'unclaimed' territories indicated in Figure 3.3. A similar case may be made for the Russian fur trade zone because Indians in the Yukon Basin may have made their way to the Pacific coast trading posts.

The Time of Competition

The hold of the Hudson's Bay Company on the fur trade in the western interior remained firm until French traders established inland posts on the Saskatchewan River. By 1750, French competition had caused a sharp decline in the number of pelts reaching York Factory. After the fall of New France in 1763, Montreal became the strategic headquarters for a number of fur traders, first known as the 'Pedlars' and, in 1784, as the North West Company. In spite of a long supply route from Montreal to the western interior of Canada, these Montreal fur traders were determined to intercept Indians on their way to trade furs at the HBC posts on the shores of Hudson Bay

Figure 3.3   European Spheres of Influence, 1760

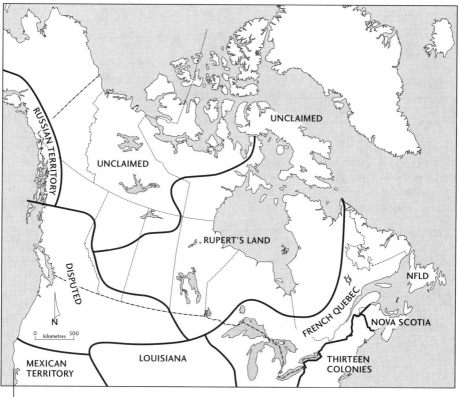

The struggle to gain control over lands in North America is part of the much larger story of European colonization of the world. By 1760, most of North America had been claimed by Britain, France, Russia, and Spain. The last unclaimed land lay in the northwest part of Canada and Alaska. In the coming century, British and Russian fur traders would collide, with the Russians penetrating inland from the Pacific and the British moving inland using the Mackenzie River system.

and thus to obtain the highly prized furs from Canada's Northwest. The strategy of the Montreal traders was simple—reduce the distance Indian fur trappers had to travel to reach a trading post. The North West Company built a number of fur posts and connected them by its two main fur brigades. The Montreal brigade would bring supplies from Montreal to Grand Portage (later Fort William), which served as the supply depot, where trade goods were shipped for later distribution to its inland posts; the northern brigade originated at Fort Chipewyan, bringing fur bales to Grand Portage that would be taken to Montreal by the Montreal brigade (Vignette 3.4).

The strategy of permanent inland posts, however, had one serious problem—provisioning. The North West Company solved this problem by supplying its inland posts with pemmican, a food made by Plains Indians, and by having its **Métis** employees and Indian trappers supply the post with country food. Pemmican was light and nutritious and, most importantly, would not spoil. It consisted of dried meat (usually buffalo) pounded into a coarse powder and then mixed with melted fat and possibly dried Saskatoon berries. In this way, the fur trade of the late eighteenth century

Vignette 3.4 Geographic Advantages of Fort Chipewyan

In 1788, the North West Company established a fort on the northwestern shore of Lake Athabasca in present-day northeast Alberta. It was named Fort Chipewyan after the Indians who hunted in the area. Fort Chipewyan was an ideal place for the furthermost post on a fur-trading route. Not only was the fort located in prime beaver country where the pelts were of the highest quality, but its location allowed the northern brigade to leave Fort Chipewyan at breakup, reach Grand Portage, and return to Fort Chipewyan with a new supply of trade goods before freeze-up. From the east, the Montreal brigade brought a new supply of trade goods to Grand Portage and returned to Montreal with fur bales. Another geographic advantage of Fort Chipewyan's location was its ready access to the Athabasca River, which breaks up in the spring a month before the lake. This gave the western fur brigade an extra month for their canoe trips to rendezvous with their eastern counterparts.

Source: Donaldson (1989: 10).

became dependent on food supplies from another natural region, the grasslands of the Canadian Prairies. This arrangement indirectly involved the buffalo-hunting Plains Indians and Métis in the fur trade.

Figure 3.4 Governor George Simpson on a Tour of Inspection in Rupert's Land, Nineteenth Century

Following the merger of the two great fur-trading companies in 1821, Governor George Simpson of the Hudson's Bay Company went on a tour of inspection of trading posts. His objective was to make the new company more efficient, which meant reducing the number of posts and workers. Governor Simpson had his own, narrow-beam, eight-metre 'express canoe' for travelling. He travelled with a crew of Iroquois voyageurs and a personal Scottish piper.

Source: HBCA, PAM P-390 (N9370), HBC's 1926 calendar illustration, Hudson's Bay Company Archives; Provincial Archives of Manitoba.

The success of the Nor'Westers' inland trading posts forced the Hudson's Bay Company to meet the competition by moving inland. In 1774, the English company established its first inland post at Cumberland House on the Saskatchewan River. This action caused the Montreal traders to move further west, and in 1776 Louis Primeau established a post on the Churchill River near the present-day settlement of Île-à-la-Crosse (Cooke and Holland, 1978: 95). In 1787, the remaining individual traders either joined forces with the now powerful North West Company or left the country. Competition between the North West Company and the HBC intensified, leading to the strategy of 'matching' trading posts, i.e., if one company established a post in a new area, the other company built one nearby.

For the Aboriginal peoples, there were several consequences of this competition. First of all, the new posts allowed inland Indians direct access to traders. Second, it weakened the power base of the Cree, who lost their role as middlemen. The role of

Figure 3.5   Fur Trade Posts, 1760

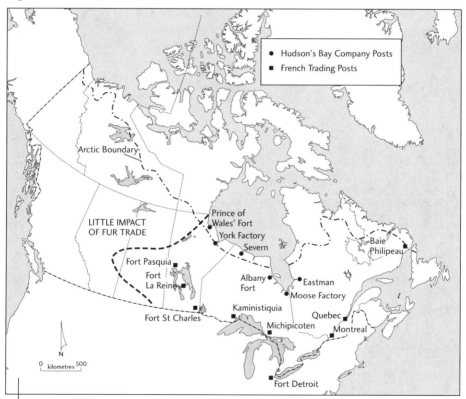

British and French control over the land in northern Canada was determined by the fur trade. Fur-trading posts operated by subjects of these two countries marked the limits of their colonial empires. In what is now Quebec, the conflict between France and England was over beaver pelts—the main form of wealth in the North. The British, through the Hudson's Bay Company, developed a northern fur trade route through Hudson Strait and the company established trading posts at the mouths of rivers flowing into James Bay and Hudson Bay, thus allowing inland Cree Indians easy access to their trading posts. This strategy effectively redirected the trade in beaver pelts from posts in New France to Hudson's Bay trading posts, thus damaging the main economic activity in New France and thereby weakening New France's economy.

Figure 3.6   Fur Trade Posts and Main River Routes, 1820

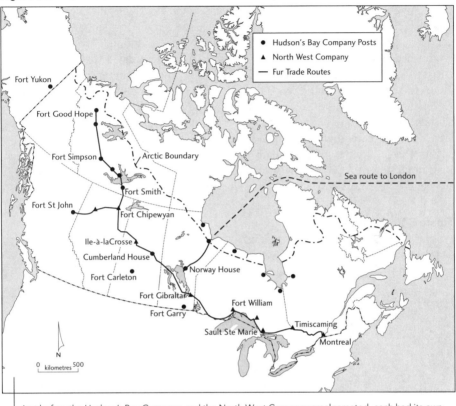

Just before the Hudson's Bay Company and the North West Company amalgamated, each had its own river routes to supply their trading posts and to ship fur pelts. Neither company had an interest in the Arctic because Europeans valued beaver and other pelts found in Canada's Subarctic. Fashions were changing, however, and by the mid-nineteenth century the beaver hat was no longer popular in Europe (see Vignette 3.3).

middlemen passed to the Chipewyans, who traded with the northern Indians, namely, the Yellowknife and Dogrib Indians. Finally, the competition between trading companies drew local Indians more fully into the fur trade and so made them more dependent on trade goods.

Amalgamation: A Time of Monopoly

The next phase in the fur trade saw a return to a monopolistic situation. With profits down and frequent bitter struggles between rival fur traders, the British Colonial Office wanted the North West Company and the Hudson's Bay Company to settle their feuding. In 1821, a parliamentary Act of the British government attempted to placate both parties by devising a coalition. The British government authorized that the name, charter, and privileges of the HBC would be assigned to the new firm but that its fur traders would include those of both the North West Company and the Hudson's Bay Company.

Figure 3.7   York Factory, Hudson Bay, 1925

From the late seventeenth century to the late nineteenth century, the depot at York Factory was the central base of operations for the Hudson's Bay Company. Indians brought fur pelts great distances by canoe to trade for British goods. The company shipped its fur pelts to London, England, through Hudson Strait, and York Factory received trade goods and other supplies in the same manner. By the late nineteenth century, the HBC began to ship fur pelts by river boat up the Red River, by rail to New York, and then by ship to London.

Source: Library and Archives Canada/PA–041571.

Under the direction of George Simpson, the 'new' Hudson's Bay Company began a rationalization of its trading posts. Posts were abandoned, staff reduced, and prices increased. The North West Company's southern route was discontinued, leaving the Bay route via Hudson Strait to London. Most furs were shipped by York boats and canoes to York Factory and then loaded on sailing ships for England. Soon profits rose, reaching unimagined levels. The company controlled not only Rupert's Land but also the North-Western Territory that extended further west to British Columbia and north to the Arctic Ocean.

While his main objective was to increase HBC profits, Simpson realized that such profits could only be achieved if the Indian way of life also 'prospered', i.e., if Indians devoted most of their time to trapping. He supported the concept of supplying Indians with food when country food—fish and game—was in short supply. Short supply was a product of the fur trade, largely because the demand for more game to meet the needs of fur traders resulted in overexploitation around fur posts and because trappers, faced with depleted game stocks in the more accessible areas, had less time for hunting in their traditional hunting grounds. Another factor was an advance in hunting technology from bows and arrows to muskets and then rifles, thus tipping the balance in favour of the hunter.

A new challenge to the Bay route appeared in the mid-nineteenth century. A combination of steamboat and rail service to New York first came into play in 1859 when the steamboat, the *Anson Northrup*, left the upper reaches of the Red River in the United States and arrived at Fort Garry (Winnipeg). Within a decade, the fur trade shifted from the Bay route to the US steamboat/ox trail route to St Paul, Minnesota. Fur brigades now followed a southern route to Fort Garry. From there, the fur bales were loaded on steamboats bound for St Paul and then the bales were transferred to rail cars bound for New York. The last stage in the journey to London, England, was by transatlantic steamer. Later, when the Canadian Pacific Railway was completed, the furs were shipped along this Canadian railway to Montreal and then on to London.

Impact on Indians

The fur trade, by introducing European goods, institutions, and values to the Aboriginal people, caused many changes to the traditional Aboriginal way of life and rearranged the geographic distribution of many Indian tribes. Traditional Aboriginal material culture was greatly diminished by substituting European goods for those produced from the local environment. Indian women found an iron needle much easier than a bone needle for sewing skins. Similarly, iron kettles and copper cooking pots replaced the traditional wooden and bone ones, making domestic life easier, and wool Hudson's Bay blankets[4] became a popular trade item among the Native peoples themselves. European firearms, iron knives, and axes were in great demand, replacing traditional weapons such as bows and arrows. Such an advance in weaponry often tipped the balance of power between those Indian tribes with muskets and those with bows and arrows. The fur trade had implications, therefore, for tribal relations, control of territory, and access to fur-trading posts.

In a similar fashion, Europeans challenged Aboriginal customs and spiritual beliefs. Most spiritual changes occurred when missionaries introduced Christianity to the North. These missionaries accelerated the process of cultural change. By the middle of the nineteenth century, Anglican and Roman Catholic missions had been established throughout the North. Usually, Anglican and Roman Catholic priests located their missions adjacent to the fur posts. Their objective was simple—to convert the Indian peoples to their version of Christianity. It was not long before the shamans/medicine men no longer held a central position in Aboriginal society. In the nineteenth century the introduction of boarding-school 'European-style' education by the two churches was an important step towards assimilation and the erosion of traditional Aboriginal beliefs and languages. Until the post-World War II years, the federal government encouraged the Anglican, Roman Catholic, and other Christian missions to educate Aboriginal children. The languages of instruction were English and French and the children were not allowed to speak their own language. Regardless of the 'good intentions' of the missionaries, their efforts further weakened the confidence of Aboriginal peoples in their own cultures, causing them to be more dependent on but at the same time not part of Western culture.

The fur trade drew the Indians and, much later, the Inuit into a new pattern of social and economic relations. Yet the degree of involvement of Aboriginal people varied over time and space. While the exchange of furs for trade goods provided the Indians and Inuit with access to European technology, the main impact was on those Aboriginal people involved in direct trading with Europeans. Others, more distant from the fur traders, were only marginally involved. Relations between trappers and traders were not constant over time. Whenever trapping and/or hunting was unable to satisfy the needs of Aboriginal people, they turned to their fur trader for help. In this way, a rather balanced, mutually advantageous relation in which each partner was more or less equal and certainly in need of the other changed into an unbalanced dependency, with Native people trapped into bartering for Western tools and weapons now essential for hunting, fishing, and trapping. Goods were also needed to run the household. This dependency culminated in an Aboriginal welfare society, which Ray (1984) has argued was not a sudden event associated with the decline of the fur trade but rather was rooted in the early fur trade and the role of the Hudson's Bay Company:

> The Hudson's Bay Company was partly responsible for limiting the ability of Indians to adjust to the new economic circumstances at the beginning of [the twentieth] century. Debt-ridden, repeatedly blocked from alternate economic opportunities, and accustomed to various forms of relief for over two centuries, Indians became so evidently demoralized in the twentieth century, but the groundwork for this was laid in the more distant past. (Ray, 1984: 17)

The geographic expression of dependency on the fur trader and its consequences for Indian economies is illustrated in an 1820 map constructed by Heidenreich and Galois (Harris, 1987: Plate 69). They classify the Indian economies into seven types, depending on their involvement in the fur trade. The disruption to local Indian economies was magnified as the level of involvement in the fur trade increased. The more disrupted Indian economies were so entangled in the fur economy that the pressure on wildlife, but especially the beaver, often resulted in an ecological collapse. The pressure to overhunt and trap came from the demands of the fur trader for country food and from the need for more trade goods.

The fur trade brought new technology to Indian communities but it also drew them into a different economic system and cultural world. Concepts of private and public property were foreign concepts. In 1870, the transfer of Hudson's Bay Company lands to Canada sent shock waves through the Métis communities in the Red River Valley who feared the loss of their lands and way of life. While the Hudson's Bay Company supported the fur economy and discouraged agricultural settlements, Ottawa was determined to settle the West. To prepare the land for agricultural settlement, Ottawa began to survey the land in order to determine ownership. Land surveys sparked two Métis rebellions, and led to treaties and the assignment of Prairie Indians to reserves where they became marginalized on their own land. Fortunately for the Woodland Indians, the forested lands of the Subarctic were not suitable for agriculture and therefore the Crown lands remained available for hunting and trapping. While landownership had changed with the transfer of Rupert's Land and the North-Western Territory to Canada,[5] the daily lives of the Native peoples remained firmly tied to the fur trade.

Trapping Comes to the Arctic

Until the beginning of the twentieth century, fur trading was confined to the Subarctic. When European fur buyers decided that the Arctic white fox pelt had commercial value, the Hudson's Bay Company quickly spread its operations into the Arctic and soon had a series of Arctic trading posts, including Wolstenholme (1909) on the Ungava coast, Chesterfield (1912) on the Hudson Bay coast, Aklavik (1912) near the mouth of the Mackenzie River, and Padlei (1926) in the Barren Lands. As with the Indian tribes, the fur economy changed the traditional Inuit ways. In the Repulse Bay area, Inuit camps became sites of winter trapping, while sealing at the ice edge began to replace breathing-hole sealing as the main winter activity (Damas, 1968: 159).

Exploration and the Fur Trade

The fur trade had always been tied to exploration, as the traders continued to search for new trapping areas. Fur traders were eager to explore new lands and to make contact with distant Indians who might supply them with furs. As always, the push into new lands was driven by the need to reach the Indian hunters before a rival fur trader.

Fur traders occasionally undertook exploration for other reasons. For example, Samuel Hearne's remarkable overland journey from Fort Prince of Wales at the mouth of the Churchill River to the Arctic Ocean in 1771–2 was motivated by the desire to substantiate reports of a 'rich' copper deposit along the Arctic coast. In 1768, northern Indians had brought several pieces of copper to the Hudson's Bay trading post of Fort Churchill (Morton, 1973: 291). The governor of the post, Moses Norton, sent Samuel Hearne in search of the deposit. In his first two attempts, Hearne failed. Crossing the Barrens was no easy feat, even with Indian guides and company servants. Only with the help of Matonabbee and his Chipewyans was Hearne able to reach the mouth of the Coppermine River in 1771 and examine the copper deposit (Hearne, 1959; Rich, 1967: 298). The success of this amazing journey was due to Chipewyan knowledge of the land and animals. Hearne not only reached the shores of the Arctic Ocean but he also travelled west to Slave River. Near Great Slave Lake, Matonabbee traded European goods for furs trapped by Dogribs, which he would later exchange for more trade goods at Fort Churchill (Morton, 1973: 298).

Hearne made a number of important observations about the Chipewyans. For the first time, Europeans had some inkling of the enormous extent of travel by Indians living in the Subarctic. Matonabbee and his group ranged a vast hunting territory, extending from Hudson Bay to the Arctic Ocean to Great Slave Lake. These forested and tundra lands, interconnected by rivers and lakes, were not the exclusive hunting grounds of the Chipewyans, nor were they marked by fixed boundaries. For example, during the summer, some of these Indians moved into the Barrens, which also served as the hunting territory of the Inuit. But in these sparsely populated lands, the chance of contact was slim and direct contact could easily be avoided. Yet, at this particular time in history, the Chipewyans were the dominant group on the Barrens, partly because of their leader and partly because they were armed with muskets and other European weapons. As middlemen in the fur trade, they were able to terrorize the Inuit and exploit neighbouring Indian tribes such as the Dogribs and still profit from trade at Fort Churchill.

Vignette 3.5 The Search for Franklin

In 1845, a British naval expedition headed by Sir John Franklin set out to find the Northwest Passage aboard two ships, the *Erebus* and the *Terror*. At that time, the climate was still in the grip of the Little Ice Age, making ice conditions more challenging for ships than would be the case today. As well, much of the route was unknown. Franklin's challenge, then, was to find a sea link from Barrow Strait to the Arctic coast near the mouth of the Mackenzie River and from there to the Pacific Ocean. By 1848, Franklin's expedition was assumed lost and efforts to find him, his crew, and ships sparked the greatest search in Arctic history. In 1853, a Hudson's Bay Company search party headed by John Rae first learned of the fate of the Franklin expedition from local Inuit. Six years later, an expedition headed by Leopold McClintock and William Hobson, and sponsored by Lady Franklin, was successful in finding the remains of part of the Franklin party near King William Island (see Figure 2.11), as well as the only written account of the Franklin expedition.

Source: Cooke and Holland (1978: 212–16).

Figure 3.8 HMS *Investigator*

The HMS *Investigator* was one of many British naval vessels searching for the two missing ships of the Franklin expedition. In 1853, the ship was abandoned in thick ice in M'Clure Strait. This strait poses the most difficult ice conditions along the Northwest Passage. In 1969, the SS *Manhattan*, an American oil tanker refitted to test the viability of transporting oil from Alaska through the Northwest Passage to the east coast of North America, was unable to penetrate the ice in this strait.

Source: Lieut. S. Gurney Cresswell, *A Series of Eight Sketches in Colour, of the Voyage of H.M.S. Investigator During the Discovery of the North-West Passage* (London: Day and Son, and Ackermann and Co. 1854). Courtesy Library and Archives Canada, Acc. No. R9266-757.

Whaling and the Inuit

English, Scottish, and American ships were drawn to the Arctic waters in search of another form of wealth—whales. In the eastern Arctic, Baffin Bay and Hudson Bay were popular whale-hunting areas; in the western Arctic, whaling ships followed the Alaskan coast to the mouth of the Mackenzie River. These whalers and British naval expeditions of the first half of the nineteenth century helped find the elusive Northwest Passage. The tragic loss of the Franklin expedition (Vignette 3.5) indicates the perilous nature of these voyages, but the ensuing search for Franklin and his men resulted in one of the greatest naval rescue efforts in the nineteenth century. Over the course of nearly 12 years much of Canada's Arctic coastline was explored and mapped by upward of 30 separate expeditions mounted to search the Arctic for Franklin and his men (see, e.g., Berton, 1988).

Whaling began in Davis Strait in the seventeenth century. Later, whaling ships ventured further north and even into Lancaster Sound. Dutch, German, English, and Scottish whaling ships plied these Arctic waters in summer voyages. Between 1820 and 1840, whaling reached its greatest intensity with over a hundred ships involved. By 1860, American ships appeared in Hudson Bay, and several decades later the Americans were whaling in the Beaufort Sea. Wintering over by whaling ships was most common in the more inaccessible areas, such as Hudson Bay and the Beaufort Sea. By the time a sailing ship reached Hudson Bay or the Beaufort Sea, sea ice had begun to form, making whaling impossible. These ships, spending the winter months frozen in a sheltered harbour, would get an early start in the spring, leaving time to return to their home ports before the beginning of the second winter.

The impact of the whalers on the Inuit was greatest in places where wintering over was common. While the population of whalers was not great (perhaps 500 men annually wintered in the Beaufort Sea and 200 in Hudson Bay), this contact drew the Inuit into a new economic system. Engaged in the whaling industry as pilots, crewmen, seamstresses, and hunters, Inuit hunters, now armed with rifles, greatly increased their take of caribou in order to supply the whalers with game. In exchange for these services, the Inuit obtained trade goods. Contact with the whalers had its downside, however, counterbalancing economic gains. Excessive drinking sometimes led to violence, but by far the most negative impact of contact with Europeans and Americans was exposure to new diseases. Epidemics of smallpox and other contagious diseases swept through contact Inuit communities, reducing their populations. Table 3.2 shows the estimated population decline of one group of Inuit over the course of a century. Even so, the sudden disappearance of the whalers, upon whom many Inuit had come to depend, was a blow. The whaling industry saw its products lose favour with southern consumers as new products came into the marketplace. First, petroleum replaced whale oil in the late nineteenth century as a fuel for lighting. Then, a few decades later, the market for baleen collapsed when steel products were substituted. By World War I, commercial whaling in the Arctic had ceased.

The Third Phase: Resource Development

The Canadian North came into being as a political entity in 1870 when Rupert's Land and the North-Western Territory were transferred from Britain to Ottawa. At one stroke

Table 3.2 Mackenzie Eskimo Population, 1826–1930

Year	Population Estimate	Source
1826	2,000	Franklin (1828: 68–228)
1850	2,500	Usher (1971a: 169–71)
1865	2,000	Petitot (1876a: x)
1905	250	RCMP (1906: 129)
1910	130	RCMP (1911: 151)
1930	10	Jenness (1964: 14)

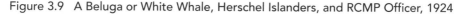

Source: Smith (1984: 349). National Museum of Natural History, cardwell@si.edu. Reprinted by permission of the publisher.

of the pen, the fur monopoly of the Honourable Company was broken, its southern, more temperate lands were opened for settlement, and its inhabitants were now under Ottawa's tutelage. A decade later, the British government transferred to Canada the rest of its Arctic possessions, the Arctic Archipelago (even though all the islands had not yet been discovered). By 1880, the geographic extent of Canada had been realized, though Newfoundland was to join Confederation much later.

Figure 3.9   A Beluga or White Whale, Herschel Islanders, and RCMP Officer, 1924

In 1924, Herschel Island had an HBC trading post and an RCMP detachment. In that same year, Herschel Island witnessed the first Arctic court case and subsequent hanging of two Inuit from the Central Arctic, who were convicted of killing several people, including an RCMP officer. In the 1920s, the Inuvialuit along the Beaufort Sea coast and at Herschel Island commonly harvested beluga whales. The island also served as Yukon's most northerly RCMP detachment, which was originally opened to monitor American whalers and thus exert Canadian sovereignty. Whaling had a huge impact on the Inuvialuit. The economic and social interaction between the Inuvialuit and the whalers, most of whom wintered over, was considerable. In 1895–6, for example, 12 whalers with 1,000 to 1,200 crew wintered at the mouth of the Mackenzie River and Herschel Island.

Source: Yukon Archives, Ernest Pasley fonds #9238.

Ottawa did little to integrate the North and its Aboriginal peoples into Canada's economy. The main reason was that the federal government had its hands full with nation-building in southern Canada. The national priority was the settlement of western Canada. The other reason is that Ottawa had no plan for the North except to wait for the private sector to undertake resource development and to allow the Indians, Métis, and Inuit to remain in the fur economy (Vignette 3.6). Then, too, Ottawa chose to keep its administrative costs to a minimum, meaning that education and health services for Aboriginal peoples and investments in transportation received little attention.[6] This laissez-faire policy of the federal government continued until the 1950s (Rea, 1968).

Still, Canada began to wonder what riches it had inherited. In 1888, a Senate Committee on the Resources of the Great Mackenzie Basin investigated this question. Later, the Senate published a 'highly enthusiastic report on the potential for agriculture, fisheries, forestry, mining, and petroleum, setting the precedent for the optimistic

---

Vignette 3.6   Aboriginal People: Who Are They?

'Aboriginal people' refers to Indians, Métis, and Inuit. The term 'Aboriginal' obscures their heterogeneous nature. Geographically, Aboriginal people are spread across Canada in relatively small groups; culturally, they have evolved in somewhat different historical contexts; and linguistically, they form more than 50 distinct language groups. While most now speak English (and/or French in Quebec), three Aboriginal languages—Cree, Ojibway, and Inuktitut—are still widely spoken. A few Métis still speak Michif, a mixture of Cree and French. Since language is an essential component of culture, the loss of native language does not bode well for their cultures. In addition, the growing number living in major cities and towns are often cut off from their cultural roots.

The Constitution Act, 1982, section 35(2), recognizes three Aboriginal groups— Indians, Inuit, and Métis. But this simple definition masks their diversity and the administrative classification of Canadians of Indian ancestry as status Indians, non-status Indians, and Métis. **Status Indians,** simply, are those who are members of treaty First Nations who are registered under the Indian Act. **Non-status Indians** are those who, prior to 1960, were enfranchised (allowed to vote), or, more generally, those not covered by treaty, such as the Innu of Labrador, who never signed a treaty with Britain, the colonial Newfoundland government, or Canada. Finally, the Métis are those of mixed ancestry (chiefly Indian–French and Indian–Scottish) who originated around the Red River Colony in what is today southern Manitoba, largely as a result of fur-trade 'relations', and who have developed their own unique culture. As noted, with the 1982 Constitution Act, the Métis achieved recognition as a unique Aboriginal people.

These legal differences are not based on biological criteria but can be explained through an understanding of the history of the Indian Act (1876). At that time, the British North America Act (s. 91[24]) assigned the federal government responsibility for 'Indians, and Lands reserved for the Indians'. With the Indian Act, Indians were placed in a different legal position from other Canadians. The Inuit have never been subject to the Indian Act. In 1939, however, a court decision ruled that they were a federal responsibility.

and promotional tone that has continued to this day [1970s] to pervade government pronouncements on northern resources' (Rowley, 1978: 79). While the Senate report sparked considerable interest, the national priority was to settle the Prairies. Little commercial activity took place in the Subarctic and northern life continued to revolve around the fur trade and the Hudson's Bay Company.

The Klondike Gold Rush

In 1896, the discovery of gold on the Klondike River in Yukon confirmed the optimistic position of the Senate Committee on the Resources of the Great Mackenzie Basin. With thousands flocking to the newly discovered gold fields, the population of Yukon quickly reached 30,000. Yukon was suddenly transformed from a fur-trapping economy to a resource economy.[7] Traces of gold had been discovered in Yukon as early as 1866, but the famous Klondike find of 1896 triggered a gold rush. Two Tagish brothers, Skookum Jim and Dawson Charlie, and George Carmack, a prospector married to their sister Kate, made the world's most famous gold strike on Rabbit Creek (later renamed Bonanza Creek), a small tributary of the Klondike River. As news spread to the outside world, thousands rushed to Yukon. According to Crowe (1974: 121), the coming of so many white men shattered the world of the Tagish, Tutchone, and other Indian tribes in Yukon. These Indians lost control of their traditionally occupied lands and became involved in the gold economy. Their participation varied from prospecting to packing supplies from the coast over the Chilkoot Pass (ibid., 122). Some found a place as wage earners, working as deckhands on the riverboats or even as carpenters in Dawson. Hunters sold game to the miners.

The dark side of the gold rush saw the outsiders occupying Indian lands, killing wildlife for food, and exposing local people to 'new' diseases. Perhaps even more significant to the Indians was the sudden imposition of another economic lifestyle, forcing major adjustments in their way of life. Some integrated as best they could while others attempted to continue their old ways.

With the sudden arrival of so many American prospectors, the issue of Canadian sovereignty arose. Prior to the gold rush, Indians and a handful of white traders and prospectors had inhabited Yukon. With so few white people, Ottawa had sensed no need to send its officials there, but now the federal government quickly dispatched detachments of the North West Mounted Police to Yukon to 'show the flag' and to enforce Canadian laws and regulations. The Mounties issued licences, collected taxes, and kept law and order. In this way, Canadian sovereignty was demonstrated. This was important because most miners were American citizens and fears of annexation were widespread in Ottawa.

The police also began to enforce Canadian conservation laws. For the most part, Indians and Inuit could avoid such interference with their ways by keeping away from settlements. However, concerns about wildlife mounted in Ottawa and during 1916–17 Parliament passed legislation—the Northwest Game Act and the Migratory Birds Convention Act—that greatly affected Aboriginal hunting activities. The police were obliged to apply these game laws to Aboriginal hunters, thereby adding another irritant to Aboriginal relations with federal officials (Figure 3.10).

Vignette 3.7   Changing Aboriginal Society in Yukon

As Ken Coates (1988: 73) has written, 'the Yukon Indians have faced the industrial frontier for over a century. Their situation provides a useful longitudinal study of Native reaction to the forces of cyclical mining development, seasonal industrial activity, and white encroachment on traditional hunting territories.' Change began with fur trading but the discovery of gold quickened its pace. A trickle of prospectors had been arriving in Yukon for three decades before the incredible Klondike gold rush at the end of the nineteenth century, when, '[b]y 1898, . . . the gold-seekers were desperate to cross the [Chilkoot P]ass, [and] the Indian packers could make $100 a day and more, carrying loads of up to 200 pounds across the mountains' (Crowe, 1974: 122). In the twentieth century, Yukon Native peoples experienced the encroachment of World War II highway and pipeline developments, and later saw lead/zinc and other mines open and close, pipeline proposals, oil exploration off the Arctic coast, and an ongoing influx of newcomers to their lands. Indians took advantage of wage employment, but for the most part they remained on the land as hunters and trappers. This mixed economy took hold with traditional hunting and trapping activities at its core but supplemented by seasonal wage employment.

Figure 3.10   Mounties at Their Herschel Island Depot, *c.* 1925

Source: Yukon Archives, Derek Parkes fonds, YA 95/53, #8.

Geopolitics Enters the North

During World War II, the North was transformed from a backwater on the international stage to a vital strategic region for the Allied Forces. The North became a military bridge between two theatres of war. Its geopolitical role in world affairs involved providing a safe, inland supply route to the European and Pacific theatres of war. Vast sums were spent to create a military infrastructure. While there was a lull in military activities after 1945, the North soon regained a geopolitical role as a buffer zone between the two superpowers—the United States and the Soviet Union. Now, it served an 'early warning radar' role for the United States, which feared a surprise Soviet air attack.

The military expenditures in the Canadian North during World War II were not to defend the northern territory but to develop supply lines to the major theatres of war, and these projects marked the beginning of a modern transportation system. The US government funded most of the projects because it was the most effective way of supplying its troops in Alaska and England. The four main projects were:

- *Alaska Highway and the Northwest Staging Route.* With the threat of a Japanese invasion of Alaska, a secure supply route to Alaska was considered essential by the American military—hence the construction of the Alaska Highway and the upgrading of the Northwest Staging Route, a series of airfields with paved runways that provided an air route for ferrying aircraft from the United States to its military bases in Alaska.
- *Norman Wells Oil Expansion and Canol Pipeline Project.* The United States wanted a secure supply of oil for its troops in Alaska. Since the ocean shipping route from California to Alaska was considered too vulnerable to Japanese submarines, the Americans elected to increase oil production at Norman Wells in Canada's Northwest Territories and ship that oil to Fairbanks, Alaska, by means of a pipeline. The Canol Pipeline connected Norman Wells with Whitehorse, Yukon, where the oil was refined. It was then shipped by pipeline to Fairbanks.
- *Project Crimson.* An air supply route to Britain was essential. At that time, the range of aircraft was insufficient to cross the Atlantic Ocean from New York or Halifax to London. The solution called for the construction of a series of landing fields in Canada's eastern Arctic that would allow Canadian and American military planes to fly to Baffin Island and then on to Greenland, Iceland, and finally to US military bases in the United Kingdom. Using a polar air route, short-range tactical aircraft could easily make the long journey by refuelling at landing strips in the Canadian North, Greenland, and Iceland. Airports and fuel depots were built along two eastern routes leading to Frobisher Bay (now Iqaluit). The Quebec route included the refuelling depot at Fort Chimo (now Kuujjuaq), while the Manitoba route involved The Pas and Churchill.
- *Goose Bay.* The military complex built at Goose Bay, Labrador, served as a major American air base during World War II. In 1941, Ottawa leased the western end of Hamilton Inlet from Newfoundland (at that time still a colony of Britain) for 99 years. This site was not troubled by fog and cloud and therefore had more than

twice as many flying days as the air base at Gander, Newfoundland. Both air bases were strategic links in the North American chain of defence and in supplying warplanes to the United Kingdom.

While this improved transportation system was designed to meet military needs, it also made northern resources more accessible to world markets. Gradually, more and more of the Subarctic became integrated into the market economy, supplying raw materials and energy to the populated areas of Canada and the United States. Like the impact of the permanent fur-trading posts of the Hudson's Bay Company in the late seventeenth century, towns based on primary industries and government services had an impact on the human geography of the North by extending a hierarchical network of trade centres across the North and thereby more closely integrating northern peoples and their activities into the Canadian economy and society.

A second effect of the military was the involvement of Aboriginal people in the wage economy and their subsequent movement into settlements. The Aboriginal workers found employment with the military an appropriate way to obtain Western products, such as steel knives, rifles, and building materials, and southern foods, medical services, and entertainment. As Aboriginal families became more reliant on the construction camps, they became known as 'settlement Eskimos' to differentiate them from those who continued to live on the land. These relatively small numbers of Eskimos lived in shacks made from surplus materials scrounged from the construction sites. As Crowe (1974: 180) described this scene:

> Canada and the U.S.A. joined in building weather stations, signal stations, and air-defence posts in the north from 1946 to 1956. Some, like the stations of the Queen Elizabeth Islands, were built far to the north of native settlements. Others, such as those at Cambridge Bay and Hopedale, attracted hunters and their families who hoped to find work, salvage 'goodies' and taste excitement.

During the Cold War, the United States saw the Canadian North as a buffer zone between it and its superpower rival, the Union of Soviet Socialist Republics. As the military capabilities of the USSR increased, Canada and the United States constructed three early-warning radar systems across Canada in the 1950s. The Pinetree Line (completed in 1954) extended across southern Canada, the Mid-Canada Line (completed in 1957) was situated along the 55th parallel, and the Distant Early Warning Line (completed in 1957) was along the 70th parallel. The Pinetree Line was jointly financed; the Mid-Canada Line was undertaken by Canada; and the Distant Early Warning Line was paid for by the United States. With the end of the Cold War, the strategic importance of northern Canada greatly diminished and any further expensive military investment for northern Canada was set aside.

Post-World War II Resource Boom

After World War II, American demand for resources outstripped their domestic supply. American industry sought resources beyond the United States, and Canada's North offered many opportunities. Megaprojects were the order of the day and the lack of transportation was the challenge. During the late 1940s, the grandest expression of a megaproject took place in the Subarctic of Quebec and Labrador—the development of iron ore mines and the building of a railway from Schefferville to Sept-Îles by the Iron Ore Company of Canada. The Iron Ore Company was composed of Canadian and US companies, including major US steel companies such as National Republic, Armco, Youngstown, and Wheeling-Pittsburgh. The purpose of this privately funded project was to supply much-needed iron ore, especially to United States steel plants. The principal shipping routes were (1) up the St Lawrence Seaway to Cleveland, Ohio, by lake carriers; (2) along the Atlantic seaboard by ocean carriers; and (3) across the Atlantic Ocean to Rotterdam (Gern, 1990: Appendix A). The mines and the resource town, in turn, needed power, which caused the development of the hydroelectric complex on the nearby Churchill River in Labrador. The first power station was built in 1954 to meet the needs of the Iron Ore Company at its Schefferville, Quebec, mine. Labrador's second hydro-generating station was built in the early 1960s to supply power to iron ore mines in Labrador City and Wabush. Watson (1964: 467–8) describes the transformation of a northern region from 'undeveloped to developed' and the link of such development to the American steel industry:

A railway was built from the little fishing village of Sept-Îles—now a thriving port—up over the high, formidable, scarp-like edge of the faulted Shield, across extremely rugged terrain, deeply eaten into by the rejuvenated sharply entrenched rivers, a distance of 360 miles to the Knob Lake iron field. Here the town of Schefferville soon grew up, a major outpost in the wilderness. The making of the port, the building of the railway, and the installation of nearby hydroelectric dams and works, were all major operations, but with American backing they were soon completed, and in 1954 the production of ore started, and over 2 million tons of ore were shipped out to the iron-hungry cities of the St Lawrence and Great Lakes lowland.

Watson fails to mention the social impacts of this project on the local Aboriginal people—the Naskapi and Montagnais (Innu)—or the environmental consequences of open-pit mining in an area underlain by permafrost. Unlike the James Bay Hydroelectric Project, no one raised the question of Aboriginal title to these yet unceded lands or the matter of environmental damage.[8] As Aboriginal leaders began to comprehend the effect on their people of the massive developments taking place in the North, their voices, along with those of environmentalists, were raised, challenging two major development projects proposed in the 1970s—the James Bay Project and the Mackenzie Valley Gas Pipeline Project.

Megaprojects signalled a new phase in northern resource exploitation. Ottawa supported such economic initiatives with new policies and programs designed to promote resource development. In 1957, Prime Minister John Diefenbaker presented a northern development concept, the 'Northern Vision', calling for the opening of the northland by building transportation routes and communication lines, thereby linking northern resources to southern markets. In 1958, a series of programs was put into place, including 'Roads to Resources'. The Roads to Resources program provided funds for new roads leading to potentially valuable natural resources in the northern reaches of the provinces. The federal share of these monies was determined by a formula of sharing costs with the provincial governments. The territorial counterpart to this program was the Development Road Program, with the federal government paying for all the construction costs.

During the next decade, northern development continued to hold Canadians' attention. In 1969, Richard Rohmer, a lawyer, author, Air Force general, and political adviser, presented a more complex version of public involvement in northern development. His Mid-Canada Development Corridor proposal called for the building of a northern railway across the mid-north, a transportation corridor meant to stimulate settlement and development similar in effect to the building of the CPR across the Canadian West in the 1880s. In many ways, this grand scheme was an elaboration of Diefenbaker's Northern Vision. According to Rohmer, investment and planning would come from the federal government, ensuring Canadian control and ownership. Rohmer envisioned this massive federal undertaking as a means to strengthen both Canada's economy and its national purpose. He saw the Mid-Canada Corridor concept as a counterweight to the American ownership of Canadian natural resources.

Canadians, particularly non-Aboriginal northerners, responded warmly to Rohmer's concept of a developed North. Federal officials, on the other hand, were cool to his idea of building a railway across the Subarctic, partly because of the potential drain to the treasury but mostly because Ottawa considered it 'unsound'. In the West, a handful of provincial officials reacted suspiciously, fearing that this transportation system would serve the interests of central Canada rather than those of the western provinces. The mixed response to this project indicated that while Canadians wished to see the North developed, there were differences of opinion as to how the goal was to be accomplished.

By the late 1960s, environmental and Aboriginal spokespersons challenged development schemes. What were the anticipated impacts on the fragile polar environment? How would such industrial projects affect Aboriginal land use? These issues stirred thoughts about social justice, laying the groundwork for economic, political, and social change in the Canadian North.

Pressure for Change: Food Security for Aboriginal Peoples

Living on the land had many attractions. However, the challenge to secure enough wild game throughout the year was considerable. For that reason, food security was a serious weakness in the hunting economy. When opportunities arose to obtain goods and foodstuffs in the cash economy by earning a wage from unloading supply ships or gaining temporary work at construction sites, Aboriginal people seized them. Such efforts were not new. In the past, exchanges took place with explorers, fur traders, and whalers. Work was simply a different form of obtaining goods and foodstuffs. Such activities complemented hunting and increased food security.

In the early 1950s, cases of starvation among Arctic hunting families in the remote interior of the Barren Lands shocked Ottawa and the Canadian public (Vignette 3.8). The fear was that starvation would be repeated because the caribou herds were declining. More than that, Ottawa recognized that the days of living on the land were over.

Vignette 3.8   Inuit Caribou Hunters Starve

Before the late 1950s, the Caribou Inuit (Ahiamiut) lived inland and hunted caribou. When their chief source of food, the barren ground caribou, became scarce, they fell into difficult times. Their remote location and nomadic lifestyle made communications with the outside world irregular, and therefore little help was available. In the 1950s, their main contact was through the Department of Transport weather station at Ennadai Lake. When word began to reach the outside world that the Ahiamiut were starving and that many had died, the federal government attempted to rescue the survivors by resettling them in coastal villages such as Eskimo Point. Williamson (1974: 90) reported that starvation had reduced the population of the Caribou Inuit from about 120 in 1950 to about 60 in 1959.

Moving Aboriginal people to settlements was believed to be the solution to food security. Beyond that, relocation was seen as a first step in integrating Aboriginal people into Canadian society. The new urban dwellers would live in public housing and their children would attend school. In time, Ottawa assumed that integration into the wage economy would occur. Unfortunately, this strategy had several flaws. First, settlements had no economic base and therefore few employment opportunities existed. Second, few newcomers to settlements were fluent in English and even fewer had job experience and skills. Third, returning to the traditional hunting economy was not an option and hunting around the communities soon resulted in the exhaustion of wildlife.

Life in settlements had a negative impact on Aboriginal culture, which was designed for life on the land. Adults found themselves trapped between two worlds while their children struggled to adjust to this new sedentary existence. Adjustment to village life was particularly difficult because most spoke little English. Living on the land presented hunters with few restrictions, but villages had many rules and regulations related to living in close proximity to your neighbours. Sled dogs, so essential for the hunting/trapping economy, became a nuisance in villages. Then, too, Native children, attending schools in the settlements, lost the opportunity to learn the ways of the hunting and trapping economy. Hunting/trapping grounds were far from settlements. Now Aboriginal men, not extended families, formed the hunting unit and the men spent much less time on the land than before.[9]

The outcome was a new form of dependence, this time on social welfare. At Port Harrison (now Inukjuak) on Quebec's Ungava Peninsula, local officials decided to relocate these Quebec Inuit to the faraway places of Grise Fiord and Resolute Bay in the Canadian Archipelago where game was more abundant. This move turned into a social disaster (Marcus, 1991). In addition, the cultural leap from the Aboriginal lifestyle practised on the land to settlement living was simply too great. The result was the so-called 'lost generation'. Those adults who were the leaders of the land-based economy found themselves misfits within the community. Ottawa also saw relocation as a response to the fear that the land-based economy could no longer support the Aboriginal peoples.

Aboriginal Peoples' Place in Development

The spread of the resource economy into the North, especially the provincial norths, resulted in conflict over land. But this conflict is over more than land—it is a struggle between conflicting goals, preferences, and values. The observations of Hicks and White (2000: 48) regarding the Inuit apply equally well to other northern Aboriginal peoples:

> It would be hard to underestimate the extent and the speed of social and cultural change experienced by the Inuit in recent decades. Most Inuit over the age of 40 were born on the land in snow houses or tents to nomadic families whose lives depended almost entirely on hunting, fishing and trapping, and who had almost no exposure to mainstream North American society. They now watch cable television in their living rooms while their children play video games or surf the Internet. Life in permanent communities built upon the wage economy, the welfare state and modern technology changed Inuit society fundamentally. Profound changes in economic activity were linked to other changes: traditional patterns of authority (for example, the respect accorded elders) were challenged by new forces, single-parent families (rare in traditional Inuit society) became common, and a range of traditional values and practices were weakened. The impaired capacity of Inuit to hunt was critical since hunting was not only the economic mainstay but also the cultural focus of traditional Inuit society.

The James Bay Project and the James Bay and Northern Quebec Agreement provide a microcosm of this struggle. On the one hand, the provincial government through its Crown corporation aspired to develop the water resources of northern Quebec. On the other hand, the Cree and Inuit sought to protect their traditional way of life. These two conflicting perspectives came to a compromise in the form of the James Bay and Northern Quebec Agreement. In the 1970s, similar battles took place in other parts of the North. The proposed Mackenzie Valley Gas Pipeline pitted the large gas companies against the Dene. Underlying both struggles were the issues of landownership and self-government. For the Dene, self-government took the form of a homeland, Denendeh, which was perceived as an ethnic political region encompassing much of the Northwest Territories south of the treeline. At the same time, the Inuit pursued a similar dream called Nunavut, which encompassed most of the Arctic region in the Northwest Territories.[10]

Two visions—homeland and hinterland—continue to govern Canadian thought and writings into the twenty-first century. Underlying these visions is the reality of different peoples holding differing values, languages, and hopes for the future. Indigenous peoples, for instance, subscribe to a holistic cosmos and therefore do not accept the Western notion that a direct relationship exists between the economy and the land. Often these differences reveal themselves as tensions between Aboriginal and non-Aboriginal peoples. Such tensions lurk beneath the surface of everyday life but break through when the economic or political stakes are particularly high. These tensions, whether small or large, underlie the ongoing struggle for economic and political power in the North. Often, this struggle becomes extremely stressful when a large resource project is planned for an area inhabited primarily by Aboriginal people. Such conflicts reached a critical phase in the 1970s when Aboriginal organizations opposed two major construction projects: the proposal to build a natural gas pipeline across northern Yukon and along the Mackenzie River (the Mackenzie Valley Pipeline Project) and the construction of hydroelectric dams in northern Quebec (the James Bay Project). Some 30 years later, Aboriginal organizations are entering the resource economy. By 2002, Quebec Cree and Quebec Inuit had signed resource development agreements with Quebec to allow further hydroelectric development in their traditional homelands; and, with the exception of the Deh Cho First Nation, Aboriginal groups in the Mackenzie Basin had proposed an Aboriginal natural gas pipeline. These agreements mark a significant shift in the strategy of Aboriginal leaders—from protecting the environment to ensuring partnership in resource development. These watershed events are discussed in the next two sections as the legal and media turning points.

The Legal Turning Point

The Royal Proclamation of 1763 provided the basis of Aboriginal land claims.[11] Treaties provided a small land base (reserves) but traditional lands—land used in the past for hunting—were classified as Crown lands. While Aboriginal peoples could continue to hunt and trap on these lands (usufructuary right), they had no other rights, such as an ownership right, to Crown lands. Since the 1970s, the legal case for traditional landownership has gained strength in areas where treaties were not made. As long as the North remained beyond the orbit of the market economy, Aboriginal peoples continued to live on Crown land without interference from outsiders. However, when the land became commercially valued, problems arose. Federal and provincial governments believed that Aboriginal peoples could hunt and trap on Crown land at the 'pleasure' of the Crown and therefore had no further claim to the land. Northern development ignored the original inhabitants. For example, in 1949, the Iron Ore Company of Canada undertook a huge resource development project in northern Quebec and the adjacent area of Labrador without a thought about Aboriginal title.

This position was challenged in the courts. In 1973, the Supreme Court of Canada ruled in the **Calder** case that Aboriginal title did exist in law and that where it was not extinguished Aboriginal title to Crown land must still exist (Vignette 3.9). Just what Aboriginal title meant was to be defined by negotiations. At the same time, the Cree and Inuit of northern Quebec were contesting in the courts of Quebec the right of Hydro-Québec to pursue the James Bay Project on their traditional hunting lands. In 1975, an out-of-court settlement resulted in the first modern treaty.[12] By this time, the federal government acknowledged the need for a negotiating process for determining those claims to land where no treaty existed. This process is known as comprehensive land claim negotiations. Much of the North fell into this category. Aboriginal peoples who reached modern land claim agreements—with the federal government in the territories and with the federal and provincial governments in the provinces—took control of their lands and established their own corporations. By 2007, the Cree, Inuit, and Naskapi of Quebec, the Inuvialuit, Gwich'in, Sahtu, and Tlicho of the Northwest Territories, the Inuit of Nunavut, the Yukon First Nations of Yukon, the Nisga'a of British Columbia, and the Labrador Inuit had concluded modern treaties, while the Labrador Innu had reached an agreement-in-principle (see Chapter 8 for a more detailed account).

Vignette 3.9 The *Calder* Decision and Aboriginal Title

The 1973 *Calder* decision opened up the debate on Aboriginal title by giving it legal credibility. The *Calder* case was a long legal battle. In 1967, the Nisga'a Tribal Council, the president of which was Frank Calder, launched legal proceedings against the government of British Columbia to obtain recognition of their Aboriginal claim to lands in the Nash Valley. The Nisga'a argument was based on two points: (1) they had not surrendered their lands to the province; (2) they therefore maintained title to these lands. The lower provincial courts ruled against the Nisga'a based on the notion that these rights had been extinguished by historical events, i.e., the creation of the British colony and/or by provincial laws that affected their Aboriginal title but that did not specifically extinguish them. The Nisga'a took their case to the Supreme Court of Canada. In 1973, the Supreme Court ruled as follows: six of the seven judges sitting on the case declared that the Nisga'a had held title originally. Three stated that they retained title. Although the Nisga'a did not win their case, the decision did give Aboriginal title legal credibility, causing Ottawa to rethink its position on Aboriginal title.

As discussed before, the actual definition of Aboriginal title was left to negotiations between Ottawa and First Nations. However, a series of legal cases expanded the notion of Aboriginal title, with the 1997 **Delgamuukw** decision of the Supreme Court of Canada establishing the procedures for defining Aboriginal title (Table 3.3).

The Media Turning Point

From 1974 to 1976, Canadian society became more aware of the hidden costs of northern development through the extensive newspaper, radio, and television coverage of the Mackenzie Valley Pipeline Inquiry. This inquiry, conducted by Thomas Berger, considered the potential social, environmental, and economic impacts of constructing a gas pipeline across the north coast of Yukon and then southward through the Mackenzie Valley. The purpose of this pipeline was to ship Alaskan gas from Prudhoe Bay to the United States. The proponents of this megaproject argued that it would stimulate the economy of the Northwest Territories, add to the national economic well-being, and create a modern transportation corridor in the Mackenzie Valley.[13] In public meetings held in northern communities, Aboriginal residents presented a much less favourable interpretation of this proposed construction project. Local Dene and Métis spokespersons expressed strong concerns about impacts on their environment and their social well-being. Three concerns were the focus of their opposition. One was the potential negative impact of this pipeline construction project on the environment. Second, Dene and Métis leaders saw few economic benefits in this proposed project for their people and they worried about possible social costs that might cause great harm to individuals and to Aboriginal culture. Their third and primary concern, however, involved the settling of their land claim. The initial position of the Dene Nation and the Métis Association of the Northwest Territories was that land claim agreements must precede construction on the project.

Table 3.3 Supreme Court of Canada Decisions on Aboriginal Title

Case	Date	Outcome
Nowegijick	1983	Treaties must be liberally interpreted.
Guerin	1984	Ottawa must recognize the existence of inherent Aboriginal title and a fiduciary (trust) relationship based on title.
Sioui	1990	Provincial laws cannot overrule rights contained in treaties.
Sparrow	1990	Section 35(1) of the Constitution Act, 1982, containing the term 'existing rights', was defined as anything not 'extinguished'.
Delgamuukw	1997	Defined how Aboriginal title is proved and accepted oral history of Indian people in land claim cases.
Marshall	1999	Mi'kmaq have the right to catch and sell fish (lobster) to earn a 'moderate living'.

In southern Canada, the extensive media coverage of the Berger Inquiry had a profound impact on public opinion. Night after night, television coverage brought the message into the living rooms of southern Canadians that this proposed industrial project would have a negative impact on Aboriginal communities and permanently damage the fragile northern environment. And for what purpose, other than to supply natural gas to American consumers? Such determined and unyielding opposition from northerners not only challenged the pipeline proposals but cast a shadow over all future industrial developments. By exposing these hidden environmental and social costs, the Berger Inquiry meant that future northern resource projects could be appraised in a more realistic framework. The Inquiry and its subsequent Report (Berger, 1977) changed the way Canadians looked at northern development.

Arctic Sovereignty

Since Europeans first came to North America, sovereignty was demonstrated first by exploration and later by settlement (Table 3.4). In Canada's early history, the fur trade played a key role in both exploration and settlement. With the addition of the Mounted Police and the Anglican and/or Roman Catholic missions, fur trade posts became administrative centres. During and after World War II, military bases and remote radar sites were established. At this time, Aboriginal peoples were relocated to existing settlements that were essentially fur-trading posts. In 1947, Resolute Bay was selected as a military air base because of the frequency of clear, calm weather. The relocation of 17 families from Baffin Island and northern Quebec in the early 1950s to places in the Arctic Archipelago has been the subject of considerable controversy and much study. Were federal officials trying to deal with overhunting in the newly formed settlements and, at the same time, providing an opportunity, through relocation to areas rich in game, for Inuit to maintain a more traditional lifestyle (Rowley, 1998; Dama, 2002)? Or, as others have argued (Dussault and Erasmus, 1994; Tester and Kulchyski, 1994; Marcus, 1991, 1995), was Ottawa attempting to demonstrate Canadian sovereignty through occupancy? Resolute Bay and Grise Fiord received Inuit from Port Harrison, Quebec, and Pond Inlet, NWT (Baffin Island). In 2010, Ottawa formally apologized.

Table 3.4   Key Historic Events Associated with Arctic Sovereignty, 1578–1942

Date	Event
1578	Sir Martin Frobisher searches for the Northwest Passage but discovers Baffin Island.
1610	Henry Hudson's final attempt to find the elusive Northwest Passage ends somewhere in Hudson Bay with his crew setting him, his son, and seven others adrift in a small boat.
1845	Sir John Franklin set sail with two naval ships, a crew of 134 sailors, and supplies for three years. The search for Franklin and his ships took over 11 years when in 1859 ship relics and bodies were found on King William Island. As a result of the search efforts, much of the uncharted waters and islands in the area around the Northwest Passage were mapped.
1898–1902	The Arctic Expedition, privately funded and led by Norwegian Otto Sverdrup, resulted in the discovery of three large islands—Axel Heiberg, Amund Ringnes, and Ellef Ringnes—in Canada's Arctic Archipelago. These islands are also known as the Sverdrup Islands. Otto Sverdrup claimed the islands for Norway, but in 1930 Ottawa obtained title to these islands upon the purchase of Sverdrup's maps, diaries, and records for $67,000.
1903–5	Roald Amundsen, a famous Norwegian explorer, became the first person to sail the length of the Northwest Passage aboard a small sloop, the *Gjöa*, with a crew of only six men.
1913–18	Canadian Arctic Expedition, funded by Ottawa and commanded by Vilhjalmur Stefansson, led to the discovery of four islands in Canada's Arctic Archipelago: Lougheed, Borden, Meighen, and Brock.
1940–2	Under the command of Sergeant Henry Larsen, the *St Roch* sailed 23 June 1940 from Vancouver to traverse the Northwest Passage. Built for the Royal Canadian Mounted Police to serve as a supply ship for isolated, far-flung Arctic RCMP detachments, *St Roch* was frozen in for two winters before reaching Halifax on 11 October 1942. In 1944, *St Roch* returned to Vancouver via the more northerly route of the Northwest Passage in only 86 days.

Indian Affairs Minister John Duncan stated that '[t]he government of Canada deeply regrets the mistakes and broken promises of this dark chapter of our history and apologizes for the High Arctic relocation having taken place' (Curry, 2010).

Since there was no apparent urgency in dealing with Arctic sovereignty, this file remained on Ottawa's back burner. This lassez-faire policy was given a jolt in 1969 when Exxon's ss *Manhattan* traversed the Northwest Passage and then repeated the voyage the following year. The Canadian public was outraged, causing the federal government to react. Clearly, if the oil companies developing Prudhoe Bay in Alaska decided that the most economical manner of transporting the vast oil reserves was by tanker rather than by pipeline, then the US claim that the Northwest Passage was an international waterway would become more insistent. With Exxon deciding to ship its oil by pipeline, however, Ottawa again placed the matter on the back burner, allowing the issue of the Northwest Passage and the much broader question of sovereignty over Arctic waters to remain dormant. All that changed in 1985 when the US Coast Guard icebreaker, *Polar Sea*, crossed these waters without Washington notifying Canada.

Ottawa felt that the US was challenging its control over the Northwest Passage. This topic is discussed more fully in Chapters 8 and 9 and is outlined in Tables 8.1 and 9.1. In the next chapter, demography provides the focus. With the resource boom of the 1950s and 1960s, Canadians moved into northern resource towns and government centres, and this migration greatly increased the northern population and created a number of resource towns.

Challenge Questions

1. During the last Ice Age, ocean levels dropped, exposing more of BC's coastline. Why did sea levels drop and did this facilitate a southward migration of Old World hunters to the heart of North America and beyond prior to the ice-free corridor?
2. Could you make a case for adding a fourth historic phase to reflect advances made by northern Aboriginal peoples since the 1970s? If so, what would you call it?
3. Why did early explorers want to take Aboriginal people back to Europe?
4. How did the Hudson's Bay Company damage the economy of New France?
5. Why did the Hudson's Bay Company and private fur traders take so long to establish posts in the Arctic?
6. Explain why Sir John Franklin would have had a better chance of making it through the Northwest Passage today than in the mid-nineteenth century?
7. Could a case be made that the relocation of Inuit into settlements was a key factor in the socio-political development of the Inuit, which led to the creation of Nunavut?
8. Prime Minister Diefenbaker's Northern Vision focused on extending Canada's transportation system northward, while Richer Rohmer proposed an east–west railway line in the Canadian Subarctic. What were the advantages and disadvantages of Diefenbaker's Northern Vision? What geographical and political problems were inherent in Rohmer's proposal? Which proposal came into reality?
9. What changed between 1950 and 1970 in Canadian society that led to strong Aboriginal and environmentalist voices challenging megaprojects?
10. Of all the images of Canada's North, why do two visions—homeland and hinterland—govern Canadian thought and writings over the past centuries and beyond?

Notes

1. Who the first North Americans were remains a puzzling question for archaeologists. A growing number of archaeologists suspect that early humans arrived in North America well before 13,000 years ago. Their suspicions are partly based on the Bluefish Caves site, where stone tools and bones from several prehistoric animals, including mammoth, antelope, bison, and large cat, and dated as 25,000 to 40,000 years old were found. Unfortunately, the site does not provide geological support that the stone tools can be associated with the smashed bones, i.e., there is a possibility that the stone tools found their way to this site at a more recent time. Other support for the earlier peopling of the Americas comes from the tools dated as 12,500 years old found at Monte Verde in Chile. The Chile find gives support to the arrival of humans some 25,000 to 30,000 years ago. Ancient stone tools found near Grimshaw, Alberta, may have been left at this site some 20,000 years ago—before the site

was covered by the Cordillera ice sheet (Chlachula and Leslie, 1998). The distinguished archaeologist J.V. Wright supports the notion of humans arriving before the last ice advance. Wright (1999: 18) wrote that:

Archaeological, physical anthropological . . . , linguistic . . . , and genetic evidence . . . has been used to argue for three major migrations from Asia into the Western Hemisphere. While controversy exists regarding the specific timing of these events, one sequence has been suggested as follows: the first migration between 30,000 and 15,000 years ago involved the Paleo-Indians whose descendants would constitute the vast majority of native peoples of the Western Hemisphere; a second migration between 15,000 and 10,000 [years ago] gave rise to the members of the Eyak-Athabascan linguistic family and probably some of the current linguistic isolates such as Haida and Tlingit; and a third migration between 9,000 and 6,000 years ago resulted in the historic members of the Eskimo-Aleut . . . .

2. From the time of contact to the late nineteenth century, population estimates were based on accounts of explorers, fur traders, and government officials. The Royal Commission on Aboriginal Peoples suggests a pre-contact population of 500,000. Heidenreich and Galois (Harris, 1987: Plate 69) argue that, in the early seventeenth century, there were about 250,000 Aboriginal people living in what is now known as Canada, suggesting a drop of 50 per cent. By the early 1820s, the Aboriginal population in Canada was estimated at 175,000, indicating a substantial decline from both the pre-contact and the early seventeenth-century figures. The Aboriginal population continued to fall well into the nineteenth century. With the formation of Canada, regular census-taking began. In 1871, the first census recorded just over 102,000 Aboriginal people, which may mark the low point in their population size.

Some tribes, such as the Beothuks, the Mackenzie Delta Inuit, the Sadlermiut, and the Yellowknives, disappeared and their lands were occupied by Europeans or by other tribes. The popular view is that the Beothuks were victims of intentional genocide who were forced to move from the coast by the English and French settlers. Along the coast, food was more plentiful than in the interior, where food sources were more restricted, caribou being the main food source. According to Pastore (1992), the Beothuks did not want to live near English and French settlers, so they decided to remain in the interior of Newfoundland. Prior to the arrival of Europeans, these hunters and gatherers had complex travel and settlement patterns that mixed inland and coastal locations into a successful nomadic lifestyle. Their life depended on being in the right place at the right time to obtain food from the sea and land. This geographic arrangement was interrupted by the arrival of Europeans who occupied coastal sites. A full account is found in Pastore (1992).

3. Transportation is indispensable for resource exports, whether fur pelts or gold ingots. In the days of the fur trade, rivers provided natural highways. Later, resource development depended on expensively constructed roads and railways. Provinces were able to incorporate their northern hinterlands into their economies by building roads and railways. The British Columbia government provides one example. In the early post-World War II period, BC extended its provincial highway system to the Peace River district to provide the rest of the province with a link to that region and to the Alaska Highway and to give Peace River settlers their long-awaited direct outlet to the Pacific coast. The John Hart Highway, built from Prince George through the Pine Pass to a junction with the Peace River district's road system after 1945, pioneered the route to be followed by the Pacific Great Eastern Railway.

4. The Hudson's Bay Company's royal charter gave the company 'sole rights to trade and commerce' in Canada. Two trade goods, guns and blankets, were of special importance. Indians preferred blankets of exceptionally pure, bright colours. Each blanket was graded with a point system corresponding to weight and size. The number of points represented the number of beaver skins required for obtaining a blanket.

5. The HBC agreed to surrender Rupert's Land to the British Crown, which transferred it to Canada. Canada paid £300,000 in compensation to the HBC and allowed the company to

keep its 120 trading posts. Canada also agreed that the company could claim one-twentieth of the land in the Canadian Prairies.

6. In the decades immediately following Confederation, little attention was paid to 'developing' the North; rather, it was a matter of holding on to this vast territory with a minimum of effort and cost. Canada's first Prime Minister, Sir John A. Macdonald, was concerned that American homesteaders might migrate northward into the 'unoccupied' Canadian Prairies, thereby leading to their annexation to the United States. For this reason, his primary concern was to exert political control by building a railway from Ontario to the Pacific Ocean and then to settle these newly acquired fertile lands. The railway was completed in 1885, and the settling of the Prairies continued into the early part of the twentieth century. During all this time, most of Rupert's Land and all of the North-Western Territory and the Arctic Archipelago were left to the fur economy.

7. Resource development also took place in the late nineteenth century along the southern fringe of the North in Ontario and Quebec—forest products and minerals being the main attractions. The building of the two national railways provided access to these northern forests and minerals. In the case of forest resources, the American timber supplies in New England were no longer adequate to meet growing US demands. Soon, Americans looked to their northern neighbour and the vast forests in Ontario and Quebec. In a classic core/periphery relationship, the Subarctic forests attracted capital from the giant newspaper firms in major American cities; in turn, the Canadian forests supplied these firms with pulp that was processed into paper at plants near American metropolitan centres.

8. Canadian Indians, such as the Quebec Cree, fall under the Indian Act (1876). By 1930, most Indian tribes had signed treaties. The main exceptions resided in British Columbia, Quebec, Yukon, and Newfoundland. Until 1939, Ottawa did not acknowledge a responsibility towards the Inuit, but in that year a Supreme Court decision determined that the Inuit of northern Quebec were a federal rather than provincial responsibility.

9. The emergence of the Chipewyan community of Black Lake in northern Saskatchewan demonstrates the swiftness of change (Bone et al., 1973: 22). Prior to 1950, these Chipewyans hunted caribou and trapped fur-bearing animals. Once or twice a year they would visit the Hudson's Bay trading post at Stony Rapids; the rest of the year they spent on the land, moving from one hunting area to another. In their economy, hunting took precedence over trapping. During the early 1950s, a Roman Catholic church was built on the shores of Black Lake and a few Chipewyan families built log cabins near the church. By 1956, over half of the Chipewyan families had established houses at Black Lake and their children were attending the newly constructed school. Family hunting parties were becoming less common because mothers and school-aged children remained in Black Lake. The decision to become permanent residents of Black Lake was influenced by several factors—proximity to their church and the local Hudson's Bay store, as well as the availability of federal health services and programs such as Family Allowances.

10. Aboriginal homelands form another goal. Aboriginal peoples have made some progress in this direction. The creation of the territory of Nunavut is one example. While less well defined, most Aboriginal leaders see this goal best achieved by the combination of land claim agreements and the establishment of a series of political homelands for Aboriginal peoples. This approach requires the settling of many outstanding land claims and negotiation of a series of unique political arrangements with Ottawa or the provinces. Efforts by the Quebec Inuit have met with some success, while the Dene have seen their political dream of Denendeh evaporate. The Lubicon Lake band of Cree in northern Alberta, on the other hand, has not yet achieved a land claim settlement. The Lubicon band claims 10,000 km^2 where oil production is occurring and where the Daishowa pulp mill has a timber lease. These developments have also led to the extension of highways that facilitate further petroleum and logging activities, but such roads have had a detrimental effect on the wildlife.

Without title to land, the Lubicon cannot prevent these changes to the natural environment, nor do they receive any of the resource revenues. Nearly 20 years ago the band claimed around $170 million in compensation for oil and gas taken (*Saskatoon Star-Phoenix*, 1989). The amount today would be staggering. Perhaps as a result, progress towards a land claim settlement has been slow. In reaction, the Lubicon Cree have tried to block further developments. For example, the band negotiated an agreement with Daishowa in March 1988 that no logging would take place on lands claimed by the band until a settlement was reached. Two years later, however, a logging firm that had a contract with Daishowa wanted to begin logging on lands claimed by the Lubicon (MacDonald, 1990). While this kind of conflict slows down economic development, forces local firms to lay off employees, and may result in legal action, the costs are due to the unfinished business of land claims; another cost, of course, is the diversion of Aboriginal energies away from focusing on their own economic development projects. Clearly, unsettled land claims remain a fundamental problem facing Canadian society.

11. The basis of Aboriginal title is found in British justice and was first expressed as a legal document in the Royal Proclamation of 1763. Why did King George III make such a proclamation? The Seven Years War between the British and French was a struggle for military supremacy in North America. Each European nation allied itself with Indians living on lands that it controlled or claimed. The British allied themselves with the Iroquois. At the end of this war, George III issued the Royal Proclamation declaring that lands west of the Appalachians were to remain Indian lands. The British needed the support of the Iroquois to control these lands, which formerly fell under French control. The problem facing Britain was that British subjects in the American colonies wanted to settle on these lands, but under the Proclamation they were not to cross a 'line' following the Appalachian Divide from Maine to Georgia and then to the St Marys River in Florida. Efforts to prevent settlers from moving west proved futile, however, and just before the American Revolution as many as 100,000 colonists may have been living west of this imaginary line (Hilliard, 1987: 149). After the American Revolution, the newly formed United States of America declared the lands extending from the Appalachians to the Mississippi River open for settlement. Thousands of Americans poured over the divide. Within several decades, virtually all of the land east of the Mississippi had passed into American ownership and the original inhabitants were exterminated, assimilated, or living on reservations (ibid., 163).

12. Modern treaties began in 1975 with the signing of the James Bay and Northern Quebec Agreement. In exchange for land, cash, and a form of regional government, the Quebec Cree and Inuit allowed the James Bay Project, which was already well underway, to proceed. Recognizing the need for land claim agreements in other areas not covered by treaty, the federal government established a more orderly process for settling land claims known as comprehensive land claim negotiations. In 1984, the first comprehensive land claim agreement, the Inuvialuit Final Agreement, came into force.

13. The Berger Report attempted to balance the concerns of environmentalists and Aboriginal peoples with the need for economic development in the North. Berger concluded that a pipeline from the Mackenzie Delta up the Mackenzie Valley to Alberta was feasible. But Justice Berger determined that such a pipeline should proceed only after the major obstacles had been overcome, specifically, after further study of how a buried and chilled gas pipeline would avoid frost heave in areas of discontinuous permafrost, and after Native land claims were settled. Berger ruled against building a pipeline across the Yukon coastal plain because of its fragile nature and because of its importance as a calving ground for the Porcupine caribou herd. He felt that a gas pipeline through this calving ground on the Yukon coastal plain would threaten the survival of the caribou herd and the hunting economy of the local people.

References and Selected Reading

Berger, Thomas R. 1977. *Northern Frontier, Northern Homeland: The Report of the Mackenzie Valley Pipeline Inquiry*, 2 vols. Ottawa: Minister of Supply and Services.

Berton, Pierre. 1988. *The Arctic Grail: The Quest for the North West Passage and the North Pole, 1818–1909*. Toronto: McClelland & Stewart.

Bone, Robert M. 2002. 'Colonialism to Post-Colonialism in Canada's Western Interior: The Case of the Lac La Ronge Indian Band', *Historical Geography* 30: 59–73.

———, Earl Shannon, and Stuart Raby. 1973. *The Chipewyan of the Stony Rapids Region*. Mawdsley Memoir 1. Saskatoon: Institute for Northern Studies, University of Saskatchewan.

Bonnichsen, Robson, and Karen Turnmire. 1999. 'An Introduction to the Peopling of the Americas', in Bonnichsen and Turnmire, eds, *Ice Age People of North America: Environments, Origins, and Adaptations*. Corvallis: Oregon State University Press, 1–27.

Canada. 1884. *Census of Canada, 1880–81*. Ottawa: Department of Agriculture.

Canada, Royal Commission on Aboriginal Peoples. 1996. *Report of the Royal Commission on Aboriginal Peoples: Looking Forward, Looking Back*, vol. 1. Ottawa: Minister of Supply and Services Canada.

Chlachala, Jiri, and Louise Leslie. 1998. 'Preglacial Archaeological Evidence at Grimshaw, the Peace River Area, Alberta', *Canadian Journal of Earth Sciences* 35, 8: 871–84.

Cinq-Mars, Jacques, and Richard E. Morlan. 1999. 'Bluefish Caves and Old Crow Basin: A New Rapport', in Bonnichsen and Turnmire, eds, *Ice Age People of North America*, 200–12.

Coates, Kenneth. 1985. *Canada's Colonies: A History of the Yukon and Northwest Territories*. Toronto: Lorimer.

———. 1988. 'On the Outside in Their Homeland: Native People and the Evolution of the Yukon Economy', *Northern Review* 1: 73–89.

Cooke, Alan, and Clive Holland. 1978. *The Exploration of Northern Canada, 500 to 1920: A Chronology*. Toronto: Arctic History Press.

Crowe, Keith J. 1974. *A History of Original Peoples of Northern Canada*. Montreal and Kingston: McGill-Queen's University Press.

Curry, Bill. 2010. 'Ottawa apologizes for 1950s Inuit Relocations', *The Globe and Mail*, 18 Aug. At: <www.theglobeandmail.com/news/politics/ottawa-apologizes-for-1950s-inuit-relocations/article1677179/>.

Damas, David. 1968. 'The Eskimo', in C.S. Beals, ed., *Science, History and Hudson Bay*, vol. 1. Ottawa: Queen's Printer.

———. 2002. *Arctic Migrants/Arctic Villagers: The Transformation of Inuit Settlement in the Central Arctic*. Montreal and Kingston: McGill-Queen's University Press.

Denevan, William M., ed. 1976. *The Native Population of the Americas in 1492*. Madison: University of Wisconsin Press.

———. 1992. 'The Pristine Myth: The Landscape of the Americas in 1492', *Annals, Association of American Geographers* 82, 3: 369–85.

Dickason, Olive Patricia, with David T. McNab. 2009. *Canada's First Nations: A History of Founding Peoples from Earliest Times*, 4th edn. Toronto: Oxford University Press.

Donaldson, Yarmey. 1989. 'Alberta's First Fort', *Western People*, 8 June, 10.

Driver, Harold E. 1961. *Indians of North America*. Chicago: University of Chicago Press.

Duffy, Patrick. 1981. *Norman Wells Oilfield Development and Pipeline Project: Report of the Environmental Assessment Panel*. Ottawa: Federal Environmental Assessment Review Office.

Dussault, René, and George Erasmus. 1994. *The High Arctic Relocation: A Report on the 1953–55 Relocation*. Ottawa: Canadian Government Publishing.

Feit, Harvey A. 1981. 'Negotiating Recognition of Aboriginal Rights: History, Strategies and Reactions to the James Bay and Northern Quebec Agreement', *Canadian Journal of Anthropology* 2: 159–72.

Gern, Richard. 1990. *Cain's Legacy: The Building of Iron Ore Company of Canada*. Sept-Îles: Iron Ore Company of Canada.

Harris, R. Cole. 1987. *Historical Atlas of Canada, vol. 1, From the Beginning to 1800*. Toronto: University of Toronto Press.

———. 1997. *The Resettlement of British Columbia: Essays on Colonialism and Geographical Change*. Vancouver: University of British Columbia Press.

Hearne, Samuel. 1958. *A Journey from Prince of Wale's Fort in Hudson's Bay to the Northern Ocean: 1769–1772*. Toronto: Macmillan.

Hicks, Jack, and Graham White. 2000. 'Nunavut: Inuit Self-Determination through a Land Claim and Public Government?', in Jens Dahl, Jack Hicks, and Peter Jull, eds, *Nunavut: Inuit Regain Control of Their Lands and Their Lives*. IWGIA Document No. 102. Copenhagen: Centraltrykkeriet Skive A/S, 30–115.

Hilliard, Sam B. 1987. 'A Robust New Nation, 1783–1820', in Robert D. Mitchell and Paul A. Groves, eds, *North America: The Historical Geography of a Changing Continent*. Totowa, NJ: Rowman & Littlefield, 149–71.

Kroeber, Alfred L. 1939. *Cultural and Natural Areas of Native North America*. Berkeley: University of California Press.

Lux, Maureen K. 2001. *Medicine That Walks: Disease, Medicine and Canadian Plains Aboriginal People, 1880–1940*. Toronto: University of Toronto Press.

MacDonald, Jack. 1990. 'Daishowa Firm Seeks Police Protection', *Edmonton Journal*, 28 Sept., A7.

McGhee, Robert. 1996. *Ancient People of the Arctic*. Vancouver: University of British Columbia Press.

McMillan, Alan D. 1995. *Aboriginal Peoples and Cultures of Canada: An Anthropological Overview*. Vancouver: Douglas & McIntyre.

Marcus, Alan R. 1991. 'Out in the Cold: Canada's Experimental Inuit Relocation to Grise Fiord and Resolute Bay', *Polar Record* 27, 163: 285–96.

———. 1995. *Relocating Eden: The Image and Politics of Inuit Exile in the Canadian Arctic*. Dartmouth, NH: University Press of New England.

Martin, Horace T. 1892. *Castorologia: or, The History and Traditions of the Canadian Beaver*. Montreal: Drysdale, 1892.

Miller, J.R. 1989. *Skyscrapers Hide the Heavens: A History of Indian–White Relations in Canada*. Toronto: University of Toronto Press.

Mooney, James. 1928. *The Aboriginal Population of America North of Mexico*. Washington: Smithsonian Institution.

Morton, Arthur S. 1973. *A History of the Canadian West to 1870–71*. Toronto: University of Toronto Press.

Newhouse, David R. 2001. 'Modern Aboriginal Economies: Capitalism with a Red Face', *Journal of Aboriginal Economic Development* 1, 2: 55–61.

Nuttall, Mark. 2000. 'Indigenous Peoples, Self-Determination and the Arctic Environment', in Nuttall and Terry V. Callaghan, eds, *The Arctic: Environment, People, Policy*. Singapore: Harwood Academic Publishers.

Page, Robert. 1986. *Northern Development: The Canadian Dilemma*. Toronto: McClelland & Stewart.

Pastore, Ralph T. 1992. *Shanawdithit's People: The Archaeology of the Beothuks*. St John's: Breakwater Books.

Peters, Evelyn J. 1999. 'Native People and the Environmental Regime in the James Bay and Northern Quebec Agreement', *Arctic* 52, 4: 395–410.

Ray, Arthur J. 1974. *Indians in the Fur Trade: Their Role as Trappers, Hunters, and Middlemen in the Lands Southwest of Hudson Bay, 1660–1870*. Toronto: University of Toronto Press.

———. 1976. 'Diffusion of Diseases in the Western Interior of Canada', *Geographical Review* 66: 139–57.

———. 1984. 'Periodic Shortages, Aboriginal Welfare, and the Hudson's Bay Company, 1670–1930', in Shepard Krech III, ed., *The Subarctic Fur Trade: Aboriginal Social and Economic Adaptations*. Vancouver: University of British Columbia Press, 1–20.

———. 1990. *The Canadian Fur Trade in the Industrial Age*. Toronto: University of Toronto Press.

Rea, K.J. 1968. *The Political Economy of the Canadian North: An Interpretation of the Course of Development in the Northern Territories of Canada to the Early 1960s*. Toronto: University of Toronto Press.

Rich, E.E. 1967. *History of the Hudson's Bay Company, 1670–1870*, vol. 2. London: Hudson's Bay Record Society.

Rowley, Graham. 1978. 'Canada: The Slow Retreat of "the North"', in Terence Armstrong, George Rogers, and Rowley, eds, *The Circumpolar North*. London: Methuen, 71–123.

———. 1998. *Cold Comfort: My Love Affair with the Arctic*. Montreal and Kingston: McGill-Queen's University Press.

Saku, James C., and Robert M. Bone. 2000. 'Looking for Solutions in the Canadian North: Modern Treaties as a New Strategy', *Canadian Geographer* 44, 3: 259–70.

Saskatoon Star-Phoenix. 1989. 'Lubicon Band Patrols Oilfields on Disputed Land', 2 Dec., D14.

Scott, Colin H., ed. 2001. *Aboriginal Autonomy and Development in Northern Quebec and Labrador*. Vancouver: University of British Columbia Press.

Smith, Derek G. 1984. 'Mackenzie Delta Eskimos', in William C. Sturtevant, ed., *Handbook of North American Indians, vol. 15, Arctic*. Washington: Smithsonian Institution.

Tester, F.J., and Peter Kulchyski. 1994. *Tammarniit (Mistakes): Inuit Relocation in the Eastern Arctic 1939–63*. Vancouver: University of British Columbia Press.

Thomas, David Hurst. 1999. *Exploring Ancient Native America: An Archaeological Guide*. New York: Routledge.

Thompson, Niobe. 2010. 'Inuit Odyssey', The Nature of Things, CBC-TV. 12 Aug.. At: <www.cbc.ca/documentaries/natureofthings/2009/inuitodyssey/>.

Trudel, Marcel. 1988. 'Jacques Cartier', in James H. Marsh, ed., *The Canadian Encylopedia*, vol. 1. Edmonton: Hurtig, 368.

Watson, J.W. 1964. *North America: Its Countries and Regions*. London: Longmans.

Williamson, Robert G. 1974. Eskimo *Underground: Socio-cultural Change in the Canadian Central Arctic*. Uppsala: Institutionen for Allman.

Wolfe, Eric R. 1982. *Europe and the People without History*. Berkeley: University of California Press.

Wright, J.V. 1995. *A History of the Native People of Canada, vol. 1, 10,000–1,000 B.C.* Mercury Series, Archaeological Survey of Canada, Paper 152. Ottawa: Museum of Civilization.

———. 1999. *A History of the Native People of Canada, vol. 2, 1000 B.C–A.D. 500*. Mercury Series, Archaeological Survey of Canada, Paper 152. Ottawa: Museum of Civilization.

Zaslow, Morris. 1984. *The Northwest Territories, 1905–1980*. Canadian Historical Association Historical Booklet No. 38. Ottawa: Canadian Historical Society.

———. 1988. *The Northward Expansion of Canada, 1914–1967*. Toronto: McClelland & Stewart.

4

Population Size and
Its Geographic Expression

Beyond Canada's **ecumene** lies the thinly populated land of the Canadian North. Fewer than 5 per cent of Canada's population—less than 1.5 million people—inhabit Canada's North, making it the most sparsely populated region in the country. A growing percentage of these northerners are Aboriginal peoples. In fact, just over half of the population of the Territorial North is comprised of inhabitants of Indian, Inuit, or Métis ancestry.

Since the two visions of the North discussed in Chapter 1 divide along ethnic lines, ethnicity is crucial to understanding the region's population geography. While non-Aboriginal northerners comprise approximately 80 per cent of the North's population, they reside in a relatively small number of cities and resource towns. Outside of these urban centres, non-Aboriginals are a minority. In addition, Aboriginal peoples continue to use the seemingly unoccupied lands for hunting and trapping. This population dichotomy has an economic and cultural spatial pattern. For example, most economic opportunities are found in the cities, regional centres, and resource towns, while Aboriginal culture and language are most easily expressed in First Nation reserves, Métis settlements, and Inuit communities.

The North's population has hovered around 1.5 million inhabitants for the past 25 years (Table 4.1). During this time, however, the distribution of people within the North has changed, with fewer inhabitants in the Provincial North and a growing number in the Territorial North. Based on the 2006 census figures shown in Table 4.1, the vast majority—nearly 1.4 million or 93 per cent—resided in the Provincial North while 7 per cent or just over 101,000 lived in the Territorial North.

For next decade, this trend is likely to continue. The key demographic factor in this equation is high Aboriginal birth rates, especially among Inuit. A secondary factor is the economy. If the forest industry remains stagnant, out-migration is likely to continue in the Subarctic, especially in Quebec and Ontario. Another economic possibility lies with the retreat of the Arctic ice cap, which would encourage resource companies to exploit mineral resources in the Arctic because they could ship their bulky commodities to market through the Arctic Ocean. Partial evidence is provided by 2010 population estimates for the Territorial North, which were calculated as 111,504 for 1 July and indicated a nearly 10 per cent increase in Nunavut's population from 2005 (30,3280) to 2010 (33,220) (Statistics Canada, 2011: Table 3.11).

Table 4.1 Population of the Canadian North, 1981, 2001, and 2006

Political Region	Population 1981	Population 2001	Population 2006	% Change 2001–6
Territorial North	68,894	92,779	101,310	9.2
Provincial North	1,402,289	1,376,755	1,368,204	-0.1
Total	1,471,883	1,469,534	1,469,514	0.0

Source: Statistics Canada (1987, 2002, 2007a).

Since the physical geography of the North has a very low carrying capacity, i.e., the ability of the land to support a population, why is the population of the Territorial North increasing? For many, geographic size is equivalent to carrying capacity, but this is not correct. At first, carrying capacity in the North was related to the region's resources—wildlife and, later, mineral resources. By the late twentieth century, however, another rung of financial support emerged, namely Canada's system of transfer payments to less prosperous regions. For the three territorial governments, transfer payments provide the bulk of their revenue and these funds account for most of the salaries of government employees and expenditures (Vignette 4.2). Thus, while the geographic size of the North is enormous and its physical carrying capacity is extremely limited, the role of governments but especially the federal government is of paramount importance. Nonetheless, such a low carrying capacity in the Arctic coupled with a rapidly growing Inuit population raises a serious socio-political question: Does the Arctic face a modern version of the Malthusian population trap?

Geography, but particularly climate and distance, has a negative impact on northern economic development and, therefore, on the level of the North's carrying capacity, which translates into its population size. Four examples demonstrate this negative relationship.

- A short growing season prohibits commercial agriculture.
- The frozen Arctic Ocean cannot serve as a main transportation artery.
- The cold environment makes the cost of construction and the operation of businesses much higher in the North, but especially in the Arctic, than in more temperate areas of Canada.
- The great distance to markets means that its export-oriented resource industry has very high transport costs. Similarly, the cost of importing building materials, foodstuffs, and diesel fuel for community power and heating is extremely expensive and greatly increases the cost of living.

Population Density

Population density figures provide a rough indicator of the capacity of a country or region to support a given population at a particular standard of living. The population density of northern Canada is among the lowest in the world and is far below the Canadian average (Figure 4.1). Expressed as the number of people per unit of land area, Canada's 2006 population density figure was 3.5 persons per km^2

while the figure for the North was less than 0.1 person per km². The population density of Canada is roughly 35 times greater than that of the North. But does the low population density figure for northern Canada imply that this polar region is underpopulated or that its carrying capacity is limited? Not all land is equal. For example, the Food and Agriculture Organization of the United Nations devised a modified version of carrying capacity called physiological density that measures not the total land area but only the amount of arable land per person. Applying this measure to the Canadian North would result in a much higher physiological density than a population density. But given the northern resource base, a carrying capacity measure would have to be adjusted to recognize its non-agricultural land use. Even so, such a measure might well indicate that, given its high unemployment rate, the North is overpopulated!

Within the North, population densities vary widely between the Arctic and Subarctic, as we have seen. Averages, however, can be misleading. For instance, most of the Subarctic contains few people and the density figure is closer to 0.05 persons per km2, but in the southern fringe of the Subarctic most of the 1.4 million people are found, resulting in a much higher figure that reaches nearly two inhabitants per km².

Figure 4.1 Canada's Population Zones

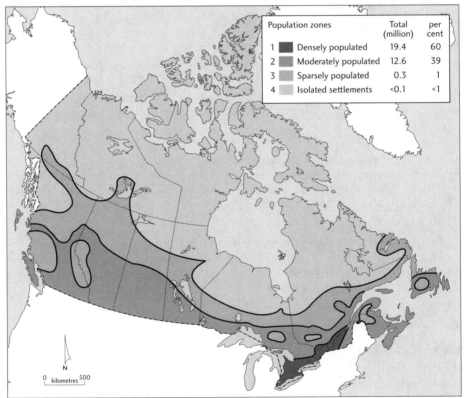

Population Change Since 1867

Since Confederation in 1867, the northern population has increased from 60,000 to just under 1.5 million. However, the growth rate in the North has varied over this time period and, for the last 25 years, population growth has stalled (Figure 4.2 and Table 4.1). Change in the size and character of the northern population has been brought about by the interplay of three demographic variables: fertility, mortality, and migration. While a precise comparison from past censuses is difficult because of changes to census boundaries that approximate the North's southern limit, Figure 4.2 provides a graphic account of population change over time.[1] These demographic changes take the shape of an S-curve, suggesting a connection to the biological concept of carrying capacity. The four phases of population change are:

- slow but steady increase from 1871 to 1941;
- rapid increase from 1941 to 1981;
- population size constant from 1981 to 2006;
- possible population decline, perhaps reflected in the 2011 census, for which figures will be available in February 2012.

Phase 1. In 1871, the population of the North was approximately 60,000 (Figure 4.2). Over the next 20 years, the settlement of the southern fringe of the Subarctic by non-Aboriginal Canadians plus the slow but steady increase in the

Figure 4.2 Population Change in Northern Canada, 1871–2011

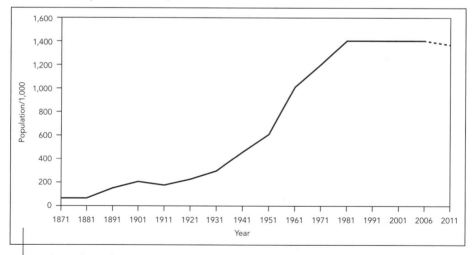

Population change from 1871 to 2011 illustrates the irregularity of population from decade to decade. This irregularity is best understood as three distinct demographic phases, with the last phase demonstrating a halt to population growth. Note that the downward-sloping broken line between 2006 and 2011 is an estimate based on a continuation of declining natural rate of increase for the Aboriginal population and growing weaknesses in the northern economy.

Source: Estimates based on Census of Canada, various years.

Aboriginal population accounted for population growth. By 1901 the North's population had reached 100,000, and by 1931 it had more than doubled to 250,000.

Phase 2. During the 1941–81 period the population in Canada's North jumped from 350,000 to nearly 1.5 million. The four main factors accounting for the enormous increase were: (1) a resource boom that began in the late 1940s and triggered a massive in-migration of southern Canadians and the creation of new resource towns; (2) a high rate of natural increase among the Aboriginal population that saw it more than triple; (3) a policy change in Ottawa that began in 1957 with Prime Minister John Diefenbaker's 'Northern Vision'[2] promoting development in the North; and (4) increased public spending by both the provinces and Ottawa that led to a northern bureaucracy and facilitated the relocation of government officials to the North through a series of northern benefits.

Phase 3. From about 1981 to 2006, the North's population has remained relatively stable at just under 1.5 million. Since the 1980s, annual out-migration has sometimes exceeded in-migration. In 1981 the North had a population of 1.47 million (Table 4.1). Twenty-five years later, that figure was unchanged. The reasons for this demographic reversal from the fast-growth phase prior to the 1980s are complex, but they are related to a net out-migration and a high rate of natural increase of Aboriginal peoples.

Phase 4. While Phase 3 could continue into the near future, changes in two demographic factors—a falling rate of natural increase among the Aboriginal population and a decrease in net migration—could lead to a slow but steady downward slide in the North's population size. The collapse of the forest industry was and remains a primary economic reason for the population decline in the Provincial North. The failure of new resource projects, such as the Mackenzie Gas Project, to take hold and the slowdown of older mining operations, such as diamond mining in the Northwest Territories, may dampen population growth in the Territorial North. Coupled with these two factors, the extremely limited economic opportunities in Native communities and very high unemployment rates may cause greater numbers of younger Aboriginal adults to relocate to southern cities. This potential decline represents a variation of the 'push–pull hypothesis', in this case a struggle between culture (pull to stay in the North) and economy (push to find employment in southern Canada). Yet, until the 2011 population figures are released by Statistics Canada, the true nature of the northern population will not be known.

Stalled Population Growth

The North's population ceased to grow after 1981. The principal reason was the sharp downturn in world demand for natural resources in the 1980s, which reversed the previous in-migration from southern Canada and marked the start of an exodus from the North of many workers and their families.[3] Yet, not all parts of the North experienced population decline. How do we make sense of these conflicting demographic events?

The main point is clear. The North's population remains stalled at just under 1.5 million. Yet, the Territorial North continues to grow, as do a few areas of the Provincial North, especially northern Alberta where oil sands development projects continue to attract workers from across Canada. One of the fastest growing regions is Nunavut, where the population nearly reached 33,000 by 1 April 2010 (Vignette 4.1). On the other hand, much of the Provincial North's economy is based on the forest industry, causing massive out-migration. The worst-hit areas remain northern Ontario and Quebec due to the drop in demand for softwood lumber and pulp from the United States.

Vignette 4.1   **Population Explosion in Nunavut**

While the rate of natural increase has diminished in most areas of the North, Nunavut's high fertility rate continues to fuel its rapidly expanding population. Of Canada's three Aboriginal peoples, the Inuit have the highest fertility rate—more than double the national average. From 1996 to 2006, Nunavut's population increased from 24,730 to 29,474, an increase of just over 19 per cent. By 1 October 2010, an estimate by Statistics Canada recorded Nunavut's population at 33,300 (Statistics Canada, 2010a).

Other factors affecting population growth in the post-1981 era are:

- The era of resource towns is over. Instead, resource companies are turning to a fly-in and fly-out commuting system. Workers and their families continue to live in southern Canada.
- By the early twenty-first century, industry had adjusted its balance of labour and capital so that fewer workers are required.
- The natural rate of increase of the Aboriginal population, especially among Indian communities, continues to decline (though it remains well above the national average).
- The number of Aboriginal families relocating to cities in southern Canada, while small, is increasing. For example, in 2006 nearly 20 per cent of Canada's Inuit population did not live in its four land claim regions and most of these Inuit resided in southern cities (Bone, 2006).

Population Distribution

The uneven distribution of the northern population can be examined from three geographic perspectives, but all reveal a similar spatial pattern—a sharp north/south division with most people living in the southern half of the North.

The first perspective views northern population in terms of the two biomes. The Arctic, with a population of approximately 50,000, contains only a small fraction of the North's total population, perhaps 3 per cent. The Subarctic, on the other hand, is home to over 1.4 million people or nearly 97 per cent of the northern population. The chief explanation for this difference lies in their respective resource bases and their historic development. The Arctic, home for the Inuit and their Thule ancestors for over a thousand years, provided a most challenging environment for a hunting economy and, not surprisingly, its carrying capacity under that economic system could only support a small number of people, perhaps less than half the size of the current population. The most recent figures for the Inuit population come from the 2006 census (Table 4.3). With nearly 40,000 Inuit living in the Arctic, as defined by the four Inuit

Table 4.2 Percentage of Population of the Territorial and Provincial Norths, 1981, 2001, and 2006

Political Region	1981	2001	2006
Territorial North	4.7	6.3	6.9
Provincial North	95.3	93.7	93.1

Source: Statistics Canada (1987, 2002, 2007a).

land claim regions—Nunatsiavut (northern Labrador), Nunavik (northern Quebec), Nunavut, and the Inuvialuit settlement area in the western Arctic (Figure 4.3)—plus fewer than 10,000 non-Inuit, the Arctic is the most sparsely populated natural region in Canada. Unlike the Subarctic, few southern Canadians have relocated to the Arctic. The principal reason is the weak economy, which offers few business/job opportunities. Other factors are an unattractive climate for most Canadians, limited urban amenities and public services, and, except for the Dempster Highway from Dawson, Yukon, to Inuvik, NWT, the absence of a highway system. On the other hand, the vast majority of people residing in the Subarctic are Canadians who relocated from southern Canada to the Subarctic largely because of economic opportunities related to the resource economy and the public service sector.

The second perspective looks at population distribution in terms of the two political norths. Here, too, comparison of the Territorial North and the Provincial North reveals an extreme population dichotomy (Table 4.4). In 2006, 101,310 people or 6.9 per cent of the North's population lived in the three territories while 1.4 million or 93.1 per cent inhabited the northern areas of provinces. Within the Provincial North, Table 4.4 illustrates two geo-demographic factors: (1) that the northern areas of eastern

Table 4.3 Size and Growth of the Inuit Population, Canada and Regions, 1996 and 2006

Regions	2006	Percentage change from 1996 to 2006
Canada	50,485	26
Total: Inuit in Arctic	39,475	18
Nunatsiavut	2,160	3
Nunavik	9,565	25
Nunavut	24,635	20
Inuvialuit region	3,115	−3
Total: Inuit outside Arctic	11,005	62
Rural	2,610	67
Total urban	8,395	60
Census metropolitan area	4,220	97
Urban non-census metropolitan area	4,175	35

Note: A census metropolitan area (CMA) has a total population of at least 100,000, of which 50,000 or more live in the urban core.
Source: Statistics Canada, censuses of population, 1996 and 2006.

Figure 4.3   Inuit Settlements in the Arctic, 2006

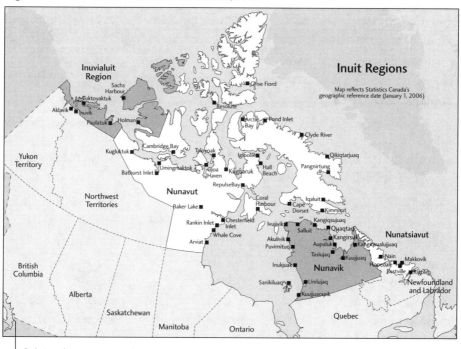

Only Inuvik, a town created under the Diefenbaker government, has a majority of non-Inuit residents.

Source: 2006 Census of Canada. Produced by the Geography Division, Statistics Canada, 2007, at: <geodepot.statcan.ca/Diss2006/Maps/ThematicMaps/OMA_CSD_Maps/InuitRegionsAboriginal_Reference_ec.pdf>.

Canada (Ontario and Quebec) account for the bulk (65.2 per cent) of the population, and (2) that the same areas (northern Ontario and Quebec) suffered a substantial population decline of 21,000, dropping from 979,000 in 2001 to 958,000 in 2006.

The third perspective allocates the northern population into the four nordicity zones of the Near North, the Middle North, the Far North, and the Extreme North (Figure 1.4). The Near North is the most southerly area of the Canadian North, where approximately one million people are located. Most live in cities and towns. Sprinkled among them are northern gateway cities that function as the core within a micro core/periphery model serving sparsely populated hinterlands. Thunder Bay is an important gateway city for northern Ontario. In 2006, some 149,000 people lived within the large, sprawling Thunder Bay Census Division that extends from Lake Superior to well beyond the Canadian National Railway line. This microcosm of the North is divided into the Near North (Thunder Bay and vicinity), where approximately 98 per cent of these people live, and the Middle North (north of the Canadian National Railway tracks) that contains over half of the land base of this census division but has fewer than 2 per cent of the total population. The explanation lies in its physical geography. Dominated by the Canadian Shield, this rough, rocky landscape is ill-suited to agriculture but is suitable for logging, hunting, trapping, and fishing. These economic activities mark a fundamental characteristic of northern land use: most land is not

Table 4.4 Northern Population by Territorial and Provincial Norths, 2001 and 2006

Province/Territory	2001	2001	2006	2006	% Change
Yukon	28,674		30,372		5.9
Northwest Territories	37,360		41,464		11.0
Nunavut	26,745		29,474		10.2
Territorial North		92,779		101,310	9.2
British Columbia	198,289		196,027		−1.1
Alberta	100,479		112,362		11.8
Saskatchewan	32,029		33,919		5.9
Manitoba	66,622		68,279		2.5
Western Canada		397,419		410,587	3.3
Ontario	416,475		413,446		−0.1
Quebec	515,126		499,503		−3.0
Newfoundland/Labrador	47,735		44,668		−6.4
Eastern Canada		979,336		957,617	−2.2
Provincial North		**1,376,755**		**1,368,204**	**−0.1**
Canadian North		**1,469,534**		**1,469,514**	**0.0**

Source: Adapted from Statistics Canada (2007a).

occupied but is used, particularly by resource companies and Aboriginal hunters and trappers. In sharp contrast, the city of Thunder Bay serves as a transport and processing centre as well as a northern gateway. Its geographic location on Lake Superior and its position on the main east–west rail and water transportation route have played a central role in its development.

Canada's Middle North has a population of approximately 400,000. Most of its towns are closely involved in resource industries, especially energy, forestry, and mining. Prince George in British Columbia is the centre of that province's interior forest industry. Alberta's heavy oil centre, Fort McMurray, with a population of more than 52,000 in 2006, is the fastest-growing city in the Middle North. Other resource towns in the Middle North include Thompson, Manitoba, Gagnon, Quebec, and Labrador City.

The Far North lies in the Arctic biome. It contains approximately 50,000 inhabitants. Most are Inuit. Iqaluit, the capital of Nunavut, is located on Baffin Island, and is by far the largest centre in the Far North, with a 2006 population of 6,184.

The Extreme North is limited to the northernmost islands of the Arctic Archipelago. Less than 300 people live in the Extreme North. Resolute is the largest centre, with a 2006 population of 229.

The Number and Distribution of Aboriginal Peoples

The number of Aboriginal peoples in the Canadian North and their percentage of the total northern population are increasing. The pattern of population increase is expected to continue. For instance, in 2006, approximately 320,000 Aboriginal Canadians lived in northern Canada[4] and Aboriginal peoples constituted 21 per cent of the northern population. Most significantly, all three territories witnessed an increase in the proportion of their Aboriginal populations from 1996 to 2006. Over this 10-year period, the total Aboriginal population reached just over 53,000, or 53 per cent of the total population (Table 4.5). In the 2011 census this pattern of high population increase among Aboriginal peoples, but especially the Inuit, is expected to continue.

As we have seen, two demographic factors explain the increasing proportion of Aboriginal peoples within the total population of the North: (1) the higher number of out-migrants among the non-Aboriginal population; and (2) the higher rate of natural increase among Aboriginal populations, especially among the Inuit, than among the non-Aboriginal population. These two factors explain the increasing proportion of the Aboriginal population. Take, for example, the Territorial North, where Aboriginal peoples formed 51.7 per cent in 2001 compared to 52.5 per cent in 2006 (Table 4.5). At the same time, the Aboriginal population is moving through its version of the **epidemiological transition**, which includes a reduction in the fertility rate, a constant crude death rate, a falling infant mortality rate, and an increase in **life expectancy**. The implication is that, if the Aboriginal fertility rates continue to decline and eventually match the national average, then the territorial population growth will slow.

Migration

Migration into the North is no longer the dominating demographic factor. During the rapid expansion of the resource industry from 1950 to 1980, in-migration was large enough and persistent enough to accelerate population growth, alter the demographic structure of the population, and affect its ethnic composition. Viewed in terms of the **push–pull migration theory**, attractive economic opportunities pulled southerners into the North. However, in a variation on this theory, which generally posits 'push' factors in the home country or region and 'pull' factors in the receiving country or region as acting together to spur migration, since the 1980s the North's economy slowed and workers and their families began to return to southern towns and cities: in many instances, even for those with jobs, environmental factors and living conditions in the North 'pushed' them back to southern Canada. In some years, regions have suffered a net loss. In fact, for the last decade, part of the North, i.e., the northern areas

Table 4.5 Aboriginal Populations for Canada's Three Territories, 2001 and 2006

Territory	2001 Aboriginal Population	2001 % Aboriginal	2006 Aboriginal Population	2006 % Aboriginal
Yukon	6,540	22.8	7,580	25.0
NWT	18,730	50.1	20,635	49.8
Nunavut	22,720	84.9	24,920	84.6
Territorial North	47,990	51.7	53,135	52.5

Source: Statistics Canada (2004, 2008b).

of Ontario, Quebec, and Newfoundland and Labrador, have lost people, especially young adults, because of limited economic opportunities. With the modernization of the forest industry, especially in its logging sector, the size of the labour force has been reduced. As a result, many families left for southern Canada. More recently, the depressed forest industry is now the leading factor causing out-migration. Many mills in single-industry towns have closed because of the lack of demand from the United States, and the volume of logging has greatly decreased, causing layoffs in both sectors. On the other hand, Alberta and, to a lesser degree, British Columbia and the Northwest Territories are gaining population largely because of the expanding oil and gas industry. Moving to the North has its attractions for those in the public sector due to high wages, cost-of-living allowances, and subsidized housing. As well, prospects for rapid advancement (often a result of workers replacing those who have opted to return south) are another attraction. These 'extra' benefits have helped southern workers to overcome their concerns about the cold climate and lack of amenities in the North.

Northern migrants, however, often have been 'temporary' residents. Many young people go north to 'make money' for five years or so and then return home with a 'nest egg' for a new start (Petrovich, 1990). Some southern transplants, on the other hand, do become enamoured with the North and make it their permanent home. In resource towns and regional centres along the southern fringe or in the three territorial capital cities, more and more newcomers are laying down roots. The recent mine closures of Faro in Yukon and Leaf Rapids in Manitoba caused many long-term residents to lament the loss of their community and to complain about uprooting their families to relocate elsewhere. Regional loyalties are readily acknowledged as a key geographic theme. For southern migrants, such loyalties are more easily formed in the Subarctic. First, the climate in the Subarctic is not so extreme. Second, the Subarctic contains a number of urban centres, especially resource towns, where community design and structure are similar to that found in southern Canada and where the pace and pattern of life are also similar to that found in southern centres. Third, the Arctic, with 85 per cent of the population Inuit, represents a different cultural setting from that familiar to southern Canadians.

Nevertheless, many southern Canadians are 'economic' migrants attracted by high-paying jobs. For these transplanted Canadians, the North is a frontier for economic opportunity but not a place of permanent settlement. As newcomers, they remain highly mobile and, if their employment ends or an attractive job opens up elsewhere, they are likely to move. Few stay longer than five years. Based on records from 1995 to 2008 for Yukon, the median length of residency for out-migrants (a resident who moved from Yukon in 2008) was three years, while for non-migrants (a resident who remained in Yukon in 2008) the length of residency was seven years (Yukon Bureau of Statistics, 2007: 7; 2010: 8). This tendency to move back home so quickly is reinforced by a desire to be closer to family and friends, to enjoy a wider range of urban amenities, and to return to a more temperate climate. For those who stay longer than three years and then leave Yukon, two other factors often motivate them to return to the south:

- A high degree of job uncertainty common to all resource industries, whether mining, forestry, or oil exploration. During times of expansion there is an in-migration of workers, but during economic contraction the same workers may seek employment outside the North.

- A tendency for young couples to leave the North when their children reach school age. The general feeling is that schools in the North do not provide the same level of education as schools in the south.

Natural Increase

The rate of **natural increase** of a population consists of differences between births and deaths for a given time period divided by its total population and expressed as a percentage. In 2009, the rate for Canada was 0.38 per cent compared to the Territorial North at 1.5 per cent (Statistics Canada, 2010b). The estimated rate for the Provincial North is close to zero. Within the Canadian North, a demographic border exists—much higher rates of natural increase occur in the Territorial North than in the Provincial North. For example, in 2009, the natural rates of increase for Yukon, the Northwest Territories, and Nunavut were 0.52, 1.12, and 2.06 per cent, respectively (Table 4.8). These high rates result from the Aboriginal fertility rate, which is estimated at over two times the national figure. Nunavut, with 85 per cent of its population Aboriginal, has a much higher fertility rate than Yukon, with Aboriginal peoples making up 25 per cent of the population (Table 4.5).

In a modern society, fears of a Malthusian trap and widespread starvation in Nunavut and other areas of the North where Aboriginal Canadians form a majority of the population are far-fetched. Still, Nunavut's carrying capacity has been strained, resulting in the emergence of a population dependence trap, i.e., its residents and government are heavily dependent on financial support from the federal government (Vignette 4.2).

Table 4.6   Crude Birth Rates per 1,000 Persons, Territories and Canada, 2001–2, 2004–5, and 2008–9

Territory	2001–2	2004–5	2008–9
Yukon	11.4	11.7	11.6
Northwest Territories	15.8	16.3	13.9
Nunavut	25.4	25.3	26.7
Canada	10.5	10.5	10.6

Source: Statistics Canada (2007b, 2010b, 2010c: Tables 3.12, 3.13, 3.14).

Table 4.7   Crude Death Rates per 1,000 Persons, Territories and Canada, 2001–2, 2004–5, and 2008–9

Territory	2001–2	2004–5	2008–9
Yukon	5.0	4.5	6.2
Northwest Territories	7.0	5.0	4.7
Nunavut	4.5	4.4	5.1
Canada	7.1	7.3	7.3

Source: Statistics Canada (2007c, 2010b, 2010c: Tables 3.12, 3.13, 3.14).

Table 4.8 Natural Rate of Increase per 1,000 Persons, Territories and Canada, 2001–2, 2004–5, and 2009

Territory	2001–2	2004–5	2009
Yukon	6.4	7.2	5.4
Northwest Territories	8.8	11.3	9.2
Nunavut	20.9	20.9	21.6
Canada	3.4	3.2	3.3

Source: Based on Tables 4.6 and 4.7.

In the broad scheme of demographic change, the demographic transition theory provides a historical explanation linked to socio-economic changes in industrial societies (Vignette 4.3). Is this theory useful for explaining the lag between Canada's low crude birth rate and the much higher northern Aboriginal birth rate, as revealed in Table 4.6? Since a reduction in family size (and therefore the birth rate) depends on the attitudes, needs, and values of a particular population, did the relocation of Aboriginal northerners to settlements occur long enough ago to have had an impact on their birth rate and, therefore, on family size? Historic evidence indicates that family size rose sharply at first, perhaps encouraged by Family Allowance payments, access to modern health care, and public housing. In the last decade, however, a downward trend is noticeable, though the birth rate remains well above the national figure. One likely factor in this downward trend is that Aboriginal women are remaining in the education system for a longer period of time and then entering the workforce, thus

---

Vignette 4.2   A Modern Version of the Malthusian Population Trap

In 1971, Milton Freeman expressed concern about the imbalance between Inuit population size and their resource base. Freeman was particularly concerned that 'a rapid rate of population growth generally prevents any successful attempt to remedy the prevailing unfortunate economic situation.' In 1988, Colin Irwin repeated this concern. Since then, the population has continued to increase at a greater rate than the economy, and a fear has arisen of an Arctic version of the **Malthusian population trap**. In classical Malthusian terms, this trap eventually results in severe food shortage and starvation until a balance between population size and food supply returns. In Nunavut, a modern version of the Malthusian trap leads not to starvation and a reduction in numbers, but to a greater dependency on the federal government through transfer payments. In 2010–11, support through Territorial Formula Financing (equivalent to equalization payments to provinces) for the three governments was: Nunavut, $1,091 million; Northwest Territories, $920 million; and Yukon, $653 million. Beyond these federal transfers, other funds are sent to the territories and provinces, and, on a per capita basis, they reveal the degree of dependency of each jurisdiction on Ottawa: Nunavut receives $35,985 per person; Northwest Territories, $24,221 per person; Yukon, $21,727. Prince Edward Island, at $4,193 per person, represents the highest figure for Canada's 10 provinces while Alberta, at $1,118, receives the lowest amount per person (Department of Finance, 2009).

Vignette 4.3   Demographic Transition Theory

The **demographic transition theory** provides a basis for interpreting major chan-ges in population growth at a global level. It is based on demographic, social, and economic events that took place in Europe over several centuries. These events are linked to the process of industrialization, which originated in Britain and then Western Europe in the mid-eighteenth to early nineteenth centuries. Its essential argument is that factors associated with a rise in the economic well-being of a soci-ety lead to a reduction of mortality and, after a short time lag, a drop in fertility. During these demographic changes, there is a large increase in population but the final phase culminates in little or no natural increase, i.e., zero population growth.

In northern Canada, improved access to medical services certainly accounts for the drop in the mortality rate in the post-World War II decades. Similar declines in mortality have taken place in most developing countries. Yet the second vital decline, that of birth rates, is just beginning in many developing countries and in the Territorial North. In both cases, rates of population increase remain very high. Since the demographic transition theory calls for a drop in birth rates to occur when incomes rise substantially, the low incomes for Indian, Inuit, and Métis families may have the opposite effect. It stands to reason that, if the theory is correct, Aboriginal birth rates are unlikely to decline to levels approximating Aboriginal death rates until well after their income levels rise sharply.

Sources: Weinstein (1976); Newman and Matzke (1984).

delaying family formation. Another factor may be overcrowding in public housing for Aboriginal families and a general shortage of housing for newly formed families. Thus, housing may cause young families to limit the number of children to avoid crowded living conditions. A third factor may be that family planning is now more accepted within the Aboriginal community.

The rate of natural increase varies across the North. The highest rates are found in areas where indigenous people form most of the population, as in Nunavut (Table 4.8). Over the last decade, high rates of natural increase also existed in the Northwest Territories and the provincial norths of Manitoba and Saskatchewan. Low rates were found in the provincial norths of Alberta, British Columbia, Ontario, and Quebec because of the predominance of the non-Aboriginal population.

Fertility in the North is declining. Within the Territorial North, Nunavut's crude birth rate declined the least, dropping from over 30 births per 1,000 persons in the 1990s to 25.2 in 2008–9. Yukon and Northwest Territories in 2008–9 had crude birth rates of 10.7 and 16.7 per 1,000 persons (Table 4.6). While official records do not record the ethnicity/Aboriginality of births and deaths, indirect evidence suggests that the fertility rate for the Aboriginal population is declining as a result of the changing nature of the social fabric of Aboriginal peoples. Demography stresses three points: (1) as the education level of women increases, fertility declines; (2) as the participation rate in the wage economy for women increases, fertility declines; and (3) as family income increases, fertility levels decline. All demographic theories acknowledge that couples in industrial societies decide whether or not to have an additional child on the basis of their personal situation (a kind of cost–benefit analysis within the context of the family's economic, religious, and social situation) and the conventions and customs affecting women in their society/religion. Martel and Bélanger (1999: 164–5) support this proposition based on family planning concepts:

> reproduction has become a matter of choice, because of a major revolution in the history of human populations: the control of fertility through contraception. With the development of effective birth control methods, couples were able to choose relatively accurately the maximum number of children that they wanted and the timing of the births, giving them, to a large degree, control over their fertility.

Demographic Structure

The demographic structure of a population—its age and sex composition—is measured in age cohorts, usually based on five-year intervals. Since the population processes of fertility, mortality, and migration shape a population over time, the demographic structure or population pyramid (Figure 4.4) provides a profile of the particular shape of the age and sex of that population for a particular instant in time, as well as clues regarding its population future and probable implications for its economy and society.

Demographers pay particular attention to the age composition of a given population. A youthful population has a high proportion of its members under the age of 15. Most developing countries fit into that demographic age picture while industrial countries do not. The 2006 census revealed that 17.7 per cent of Canada's population is under 15 years of age, while in the Canadian North those under 15 likely comprise just under 20 per cent (Statistics Canada, 2007d). Taking the Territorial

Figure 4.4 Age/Sex Structure of Nunavut, 2006

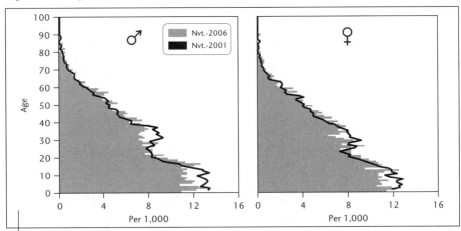

Age is a critical demographic feature. Demographers consider that a younger population means that fewer members fall into the so-called productive sector (ages 15 to 64). Nunavut has an extremely youthful population with 33.9 per cent of its population under the age of 15. In comparison, Canada as a whole has only 17.7 per cent in that age bracket.

Source: Statistics Canada (2007d).

North as an example of the range of 'youthful' populations, Yukon has 18.8 per cent under 15, the Northwest Territories 23.9, and Nunavut 33.9. While an older population exists in most areas of the Provincial North but especially northern Ontario and Quebec, the rest of the Provincial North has a much younger population. For example, Saskatchewan's North, as measured by Census Division 18, had 34.0 per cent of its population under the age of 15, with most of these young northerners of Cree, Dene, and Métis ancestry (Statistics Canada, 2007e). What these age figures demonstrate is that Canada has a dual population in terms of demography, and that duality signifies a parallel duality in economic development.

Demographers also consider those between the ages of 15 and 64 to be the 'productive' population. The implications of a young population are considerable. First, a young population requires more investment in social infrastructure and social programs, ranging from daycare facilities to additional classroom space for kindergarten and primary students. Second, as this young population bulge moves through the life cycle, public investment must be directed to new high schools and post-secondary facilities and then to job training and job creation programs.

Canada's male/female population ratio indicates a slight predominance of females. The reverse is true in the North. A preponderance of males is characteristic of resource frontiers. The demographic gender structure found in Yukon exhibits this 'frontier' characteristic. In 1986, Yukon's sex ratio indicated a majority of males, i.e., 110.4 males for every 100 females. By 2001, Yukon's sex ratio had dropped to 105.2. Five years later, the slide continued, reaching 101.3. Why did the Yukon's sex ratio change? The main reasons are that Yukon, especially Whitehorse, is no longer viewed as a male-oriented 'rough and tough' territory; and that the role of women has changed, with a higher proportion of women now in the Yukon labour force. Not surprisingly, then, Whitehorse has more females than males, with 96.8 males per 100 females (Statistics Canada, 2007f).

The age of Canada's population is increasing. This demographic fact is reversed in the North. A unique variation to the North's age pyramid is due to the in-migration of more young males than females. For example, Yukon has a population 'bulge' in its age structure between the ages of 20 and 45. This bulge is more prominent among the male population. Other areas of the North, such as Nunavut, have been less affected by economic development and therefore by in-migration. Consequently, their age structures do not exhibit such a bulge. In other northern areas, but especially northern Ontario, Quebec, and Newfoundland and Labrador, economic retrenchment has seen an outflow of surplus populations to cities in southern Canada. Since the migrants tend to fall in the 20–45 age range, these areas have lost a segment of their labour force, causing a retraction in age structure.

Urban Population

The North's physical geography encourages people to live in urban centres. With agriculture limited to the southern fringe of the Canadian North, such as in the Peace River country of Alberta and British Columbia and in the Lac St Jean lowlands of Quebec, and with the relocation of Indian, Inuit, and Métis from the land to Native settlements, virtually all of the northern population resides in small villages and larger cities. Northern urban centres have the following characteristics:

- Cities with populations of 25,000 or more are located in the Provincial North.
- The vast majority of northern centres have a population under 5,000 inhabitants.
- Most Aboriginal urbanities reside in villages that were the former sites of fur-trading posts and that, today, have very weak local economies.
- The cost of living and doing business in these centres tends to increase with distance from southern cities that supply most foodstuffs and building materials.
- Cost of construction and operating public facilities is well above similar construction and operations in southern cities, largely because of the cold weather and the presence of permafrost.
- The vast majority of Arctic centres have less than 1,000 residents. Only one centre—Iqaluit—has more than 6,000 inhabitants (Table 4.10). In spite of their small size, almost all Arctic communities, unlike most Subarctic ones, have recorded high rates of population increase from 2001 to 2006. Three Arctic centres, Inuvik, Igloolik, and Repulse Bay, achieved rates over 20 per cent while Iqaluit reached 18 per cent.

Many urban centres in the Canadian North have lost population over the last several decades. The four largest cities in the North are Saguenay (formerly Chicoutimi–Jonquière) at 151,643, down 2.1 per cent from 2001; Thunder Bay, 122,907, up slightly from 2001; Prince George, 83,225, down 2.1 per cent from 2001; and Sault Ste Marie, 80,098, up by 1.5 per cent from 2001. However, the 10 largest centres in the Territorial North have bucked this trend (Table 4.10).

From 2001 to 2006, the larger cities in the Provincial North are, with a few exceptions, either contracting in size or showing little increase (Table 4.9). Two cities in northern Alberta—Grande Prairie and Wood Buffalo (Fort McMurray)—are the exception and they are growing at very rapid rates. Grande Prairie, with its diversified economy based on agriculture, forestry, and natural gas, saw its population jump from 37,000 in 2001 to 47,000 in 2006, an increase of 27 per cent. Wood Buffalo, a single-industry town in the heart of the oil sands developments, grew from 42,500 in 2001 to 52,600 in 2006, an increase of 24 per cent. In general, the weak economy of the Provincial North (with the exception of Alberta) lies behind the urban population decline. On the other hand, the Territorial North continues to see its major towns and cities expand. The basic reason for this growth is due to the high rate of natural increase among its Aboriginal residents, most of whom are reluctant to move to southern centres because of their close ties to their settlement and its people.[5]

Classification of Urban Centres

Economic and cultural elements allow for the division of northern population centres into three categories: regional centres, resource towns, and Native settlements. Regional centres range from small service centres like The Pas in northern Manitoba to capital cities like Yellowknife in the Northwest Territories. In all instances, the capital cities are the major service centres in the territories. These regional centres are the source of certain goods and services for those living within their catchment areas. Central place theory seeks to explain the relative size and geographic spacing of urban centres as a function of shopping behaviour. Its central argument is that basic or lower-order goods and services are found in all urban centres while specialized or higher-order ones are found only in larger urban centres. Successively larger centres offer a greater variety of goods and services and so command a broader market territory. For example, Yellowknife provides a wider range of goods and services than do smaller centres, such as the nearby village of Behchokò (Rae-Edzo). Residents of Behchokò have to travel to Yellowknife to obtain a higher order of good or service—e.g., dining, shopping—not available in their home community. For a firm to offer a set of goods at a particular place, it must sell enough to meet operating costs and, presumably, to turn a profit. The concept that a minimum level of demand is necessary to allow a firm to stay in business is called the threshold. The same argument can be made for public services, i.e., a hospital and high school are located in regional centres while nursing stations and primary schools are found in smaller centres. Fort Simpson and Wrigley provide a simple example of the urban hierarchy. Fort Simpson, with a population in 2006 of 1,216, is the regional centre in the Deh Cho region of the Northwest Territories. Fort Simpson has both a hospital and a high school. Wrigley, with only 122 residents in 2006, has neither a hospital nor a high school. Students from Wrigley who wish to attend high school must travel nearly 200 km or move to Fort Simpson. Similarly, residents of Wrigley requiring medical care have to go to the Fort Simpson hospital. Not surprisingly, Wrigley's population declined from 2001 to 2006 by 26 per cent.

Resource towns or single-industry towns are products of the resource economy. Many were created during the resource boom from 1950 to 1980. Resource towns are vulnerable to the vicissitudes of global demand for their natural resources. Tumbler

Table 4.9 Top 10 Provincial North Urban Centres by Population Size, 2001–6

Urban Centre	2006	2001	% Change 2001
Saguenay, Quebec	151,643	154,938	−2.1
Thunder Bay, Ontario	122,907	121,986	0.8
Prince George, BC	83,225	85,035	−2.1
Sault Ste Marie, Ontario	80,098	78,908	1.5
Wood Buffalo, Alberta*	52,643	42,581	23.6
Grande Prairie, Alberta	47,076	36,983	27.3
Timmins, Ontario	42,997	43,686	−1.6
Rouyn–Noranda, Quebec	39,924	39,621	0.8
Alma, Quebec	32,603	32,930	−1.0
Val-d'Or, Quebec	32,288	32,423	−0.4
Baie-Comeau, Quebec	29,808	30,401	−2.0
Sept-Îles, Quebec	27,623	27,623	0.7
Total	687,589	699,492	−0.3
Provincial North	1,368,204	1,376,755	−0.1
Percentage of Provincial North	50.3	50.8	

*Fort McMurray.
Source: Statistics Canada (2007a).

Table 4.10   Top 10 Territorial Urban Centres by Population Size, 2001–6

Urban Centre	2006	2001	% Change
Whitehorse, Yukon	20,481	19,058	7.4
Yellowknife, NWT	18,700	16,541	13.1
Iqaluit, Nunavut	6,184	5,236	18.1
Hay River, NWT	3,648	3,510	3.9
Inuvik, NWT	3,484	2,894	20.4
Fort Smith, NWT	2,364	2,185	8.2
Rankin Inlet, Nunavut	2,358	2,177	8.3
Arviat, Nunavut	2,060	1,899	8.5
Behchokò (Rae-Edzo), NWT	1,894	1,552	22.0
Baker Lake, Nunavut	1,728	1,507	14.7
Total	62,901	56,559	11.2
Territorial North	101,310	92,779	9.2
Percentage of Territorial Population	62.1	61.0	

Source: Statistics Canada (2007a).

Vignette 4.4   Dominance of Capital Cities in the Territories

Whitehorse is the capital of Yukon, and is also the transportation, business, and ser-vice centre for the territory. In 2006, its population was 20,461, making it the largest city in the three territories. In the same year, Whitehorse made up 67 per cent of the territory's population of 30,372. This modern city is located in the Yukon River Valley and along the surrounding terraces. In the past, the Yukon River represented a major transportation link between Whitehorse and Dawson. Since 1947, the Alaska Highway not only connected the urban centres of Watson Lake, Whitehorse, and Dawson, but it also has provided a link with Alaska and the provinces of British Columbia and Alberta.

Yellowknife is the capital of the Northwest Territories. Located on the north shore of Great Slave Lake, Yellowknife was originally a gold-mining town but its two mines, Con and Giant, closed in 2003 and 2004, respectively. With the opening of several diamond mines in the Territorial North, Yellowknife has added diamond process-ing to its business functions. In 1967, Yellowknife was selected as the capital of the Northwest Territories. Its population in 2006 was 18,700, making it the second largest city in the three territories. In 2006, Yellowknife formed 45 per cent of the population of the Northwest Territories (41,464).

Iqaluit is the third largest city in the territories. Located on Baffin Island, Iqaluit had a population of 6,184 in 2006. Its rapid growth (18.1 per cent from 2001 to 2006) is largely due to its role as the political capital of Nunavut. Unlike the other two cap-ital cities in the Territorial North, Iqaluit formed only 21 per cent of Nunavut's 2006 population of 29,325.

Ridge, for instance, has spanned two global business cycles and thus lived, died, and was reborn as a centre for coal production in the northeastern area of BC's Cordillera.[6]

Resource towns also are susceptible to plant closures, especially those based on non-renewable resources. While forest industry towns are founded on a renewable resource, mining towns are not so fortunate. Many mines cease operations after about 30 years. During that time, these towns have gone through rapid population growth, population stability, and population collapse. Pine Point in the Northwest Territories, Uranium City in northern Saskatchewan, Leaf Rapids in northern Manitoba, and Schefferville in northern Quebec provide examples of mining towns that have gone through such a 30-year cycle of population expansion and decline. All lost their primary mining function, causing the miners and their families to move. By 1993, Pine Point was abandoned, while Schefferville and Uranium City had shrunk to less than 200 inhabitants each, most of whom were either Métis or Indians who lived in the area prior to the mining developments. Leaf Rapids faced a similar fate with the closure of its mine in 2002. In 2001 its population totalled 1,309, but by 2006 Leaf Rapids had lost over half its population.

These four communities went through a classic resource town life cycle ending in closure (Table 4.11). As resource towns go through this cycle they undergo five phases of expansion and contraction, each marked by dramatic changes in population. The first phase precedes a company announcement about developing a local resource. At that time, the site is uninhabited (though it may form part of a traditional hunting territory). The second phase occurs as workers and their families arrive to occupy the newly constructed town. As production reaches its peak during phase three, the demand for additional workers ceases and the town's population stabilizes and then begins to decline as the ore body becomes more difficult to mine and technological advances reduce the number of workers. The fourth phase is linked to the company's announcement that the mine will soon close. At this point, there is a sharp decrease in the town's population. With phase five, the closure of the mine, the town is abandoned and its population drops precipitously.

The mining town of Pine Point illustrates the full boom/bust cycle (Bone, 1998: 252–3). In the late 1950s, Cominco decided to develop a lead/zinc mine just south

Table 4.11   Classic Population Life-Cycle Model for Resource Towns

Phase	Population Characteristics	Associated Events
1	Uninhabited site	Company announces plans to build a resource town.
2	Sudden increase in population size	Workers and their families arrive in recently completed company town.
3	Population size reaches peak and then remains stable	Resource production attains its maximum and the need for additional workers ceases.
4	Sharp decline in population size	Company decides to close operations. Workers and their families depart.
5	Return to an uninhabited site	Mine closed and its buildings and housing demolished.

Source: Bone (1998: 250). Reprinted by permission of the publisher.

of Great Slave Lake. The problem of access to outside markets was solved by the construction of the Great Slave Lake Railway by the Canadian National Railway. With considerable help from Ottawa, construction of the new town of Pine Point and its lead/zinc mine began in 1962. Three years later, the mine was in full production. From 1962 to 1976, this resource town witnessed extremely rapid growth, reaching nearly 2,000 inhabitants by 1976. By that time, the company had extracted most of the more accessible and higher-quality ore and the cost of mining began to increase. In 1983, the mine ceased production and most work was focused on milling the ore for shipment to a smelter at Trail, British Columbia. From 1976 to 1986, the size of the community began to decrease, dropping to 1,500 by 1986. With the final export of processed ore in 1988, the mill was closed and preparations were underway to turn the townsite back to nature. All buildings were destroyed or removed. By 1991, the town of Pine Point had disappeared from the landscape. As one former Pine Point resident observed:

> Don't think people realize how hard it is to leave a town where all three children were born. . . . Seems unreal to see your home town slowly disappear. It's too bad that some other industry couldn't have been brought in to keep the town alive. . . . It's a shame to see buildings (arenas, school, etc.) abandoned when so many other settlements have nothing of the sort and could use the facilities. (Kendall, 1992: 134)

Some resource towns, such as Fort McMurray, are still growing. Situated in the heavy oil sands of northern Alberta, Wood Buffalo (Fort McMurray) is undergoing a spectacular expansion. Its population increased from 35,213 in 1996 to 41,466 in 2001, an increase of 17.8 per cent. By 2006, Fort McMurray had reached 52,643,

Figure 4.5   Whitehorse, Yukon

Located in the boreal forest, the city of Whitehorse lies along the river terrace of the deeply entrenched Yukon River.

Source: City of Whitehorse. Reprinted by permission.

gaining nearly 24 per cent over its 2001 figure. With several huge oil sands construction projects underway plus the expansion of existing heavy oil plants, the demand for workers exceeds the supply, driving wages higher and higher. High wages and steady employment attract many from across Canada and beyond, especially from economically depressed areas troubled by high rates of unemployment.[7] Not surprisingly, Newfoundland has supplied so many workers that Newfoundlanders make up around one-third of Fort McMurray's population (Mahoney, 2002: A6; Storey, 2009).

Native settlements in the North are a product of the post-colonial era in Canada. Many have fewer than 1,000 inhabitants; most are situated in remote locations; access to some is possible only by air transportation (most have water transportation for part of the year); cost of living is extremely high because store food and other goods are transported

Figure 4.6   Fort McMurray, Alberta

This aerial view of Fort McMurray (popularly referred to as Fort Mac) is from the north looking south. Fort McMurray lies at the confluence of the Athabasca and Clearwater rivers. The Grant MacEwan Bridge over the Athabasca River connects the downtown with the new Thickwood subdivision (in the foreground). Fort McMurray, although its 2006 population exceeded 50,000, is not an incorporated city but rather is the largest community in the Regional Municipality of Wood Buffalo.

Source: Gord McKenna. Reprinted by permission.

by truck or air from southern cities; and, without a local tax base, the federal government, directly or indirectly, contributes most funds for public services and housing. Yet, Native settlements continue to exist because they represent a small piece of their traditional homeland. The Cree of Kashechewan, a remote First Nations community of about 1,900 people on the shores of the Albany River in northern Ontario, when faced with the option to relocate near Timmins following repeated pollution and degradation of their water system, chose to remain in their flood-prone community (CBC News, 2007). Culture, then, trumps economic reality (for more on this subject, see Chapter 9).

The operating cost of Native settlements is high because of their remote location (high cost of delivering public services) and small size (lack of economies of scale). With a negligible local tax base, local councils depend on transfer payments from territorial and provincial governments. Given rising costs of delivering existing services and the costs of building additional facilities, a major question facing governments is whether they can properly support these communities. As the 1999 draft report on Clyde River, Nunavut, indicated, there is no easy answer to this question (Vignette 4.5). Over 10 years later, Clyde River is still a functioning community.

Vignette 4.5 Are 'Have-Not' Native Settlements Doomed or Cultural Survivors?

A draft consultant's report on Clyde River in Nunavut suggested that the existence of this community might have to be 'reviewed'. According to the report, the problem in Clyde River is not unique and is found in many other Arctic communities that have no economic activities other than hunting and public construction projects. Local tax revenues fall far short of the funds needed to provide local services. With high unemployment rates, a critical shortage of housing, and an insufficient tax base to pay for public services, increasing transfer payments from other levels of government are necessary to maintain a minimum quality of life in these communities. Unfortunately, the territorial government is hard-pressed to keep such funding at its current level and has little room to increase funding. Capital projects, such as sewer and water systems, provide temporary relief, but the cost of maintaining these systems becomes an additional burden on the community. A local Clyde River official, Mr Palluq, bitterly said of the consultant's report: 'The guy writing this was getting $200 a day. And he says we should cut back employee benefits. Lots of these people have been working for the municipality for $12 an hour for 15 years on half-days.' But cuts are coming and retiring hamlet employees may no longer receive a payment for unused sick leave. Clyde River's mayor, James Qillaq, unhappily said, 'it seems like it's a slope and we're still sliding down.'

Can Clyde River expand its economic base? In southern Canada, single-industry towns, notably Chemainus, that faced closure turned to the arts and tourism (Barnes and Hayter, 1992). In the case of Clyde River, this isolated Inuit community represents an ideal location for an Inuit culture and language centre; hence such a centre, Piqqusilirivvik, was formed in 2009. Here, young Inuit from across Nunavut will learn from Elders about culture, language, and the land/ice, thanks to grants from the federal and territorial governments (O'Neill, 2009).

Source: McKibbon (1999).

Figure 4.7   Clyde River, Nunavut—Going, Going, Gone?

The challenge facing the Inuit community of Clyde River is typical of isolated Native communities. Clyde River, located on a fjord on Baffin Island, represents a cultural homeland for its residents, but the community lacks an economic base other than its traditional seal hunting. Yet, its residents remain committed to the community, indicating the strong pull of culture. Its population increased to 820 by 2006—a jump of 4.5 per cent over its 2001 population. The population is young, with a median age of 20.8 years in 2006, just over half of Canada's median age of 39.5 years. Without a tax base, however, Clyde River depends heavily on funds from the Nunavut government. Yet, Clyde River survives and grows—its population in 2009 was 850 inhabitants.

Source: Photo taken August, 1997. Courtesy Dr Ansgar Walk.

Labour Force

Since labour force statistics for northern Canada are organized by provinces and territories, a single set of figures is not readily available. To throw some light on this subject, the labour force for the Northwest Territories serves as an approximation of the northern workforce. As in other resource frontiers, the NWT labour force, based on data from 2001 to 2009, has the following characteristics (NWT Bureau of Statistics. 2010):

- Fluctuations in the size of the labour force correspond with the health of the resource economy, especially the diamond industry. In 2001, the size of the labour force was 22,000; in 2006, it jumped to 24,000; and then it decreased to 22,200 in 2009.
- Labour force increases largely reflect an inflow of skilled workers from southern Canada when high-paying jobs are readily available. In a sense, the Northwest Territories has a labour force adjustment mechanism, that is, a portion of its workforce is of a 'temporary' nature—southern workers who come to fill employment opportunities and then leave when the jobs cease, as happens, for example, with the closing of a mine.

- Beyond the active labour force, a large pool of underskilled and therefore unemploy-able northern adults exists. Because they are not actively seeking employment, such workers are not included in the official unemployment figures. If they were included, the unemployment rate would be at least three times greater than the official one.
- The relative lack of skilled northern tradespeople can be attributed to various fac-tors, including a lack of trade schools in the North, the length of time involved in obtaining trades certification, the absence of role models in the Aboriginal com-munity, and the registration and hiring of union workers for northern jobs in southern labour centres.
- Compared to Canada's national unemployment rate, the Northwest Territories' unemployment rate appears remarkably low—8.6 per cent in 2001 and a low point in 2006 at 5.4 per cent, with an increase to 6.3 per cent in 2009. The explanation is that the high level of underemployment in the Territorial North, especially among the Aboriginal labour force, is not included in the calculations of unemployment rates. The potential NWT labour force is defined as those between 15–64 years of age while the active labour force contains those who are employed or are seeking employment. For the Northwest Territories, an unusually high number of people are classified as 'not in the labour force'. Some are students, housewives, and disabled persons, but many have simply given up looking for work and therefore are not counted in the employment and unemployment data. For example, in 2009, the 'not in the labour force' figure was 9,200, or 29 per cent of the working-age population. Of that group, approximately one-third are not seeking employment because they believe no jobs are available in their communities, and most of these communities are small, isolated Native communities (NWT Bureau of Statistics, 2006).

Another way to view the northern labour force is by the four industrial sectors—primary, secondary, tertiary, and quaternary. Primary activities include agriculture, mining, quarrying, logging, fishing, hunting, and trapping. Secondary activities involve the processing or manufacturing of primary products, as well as construction and utilities. Tertiary activities are services provided by public agencies and private business, in other words, the service sector that includes most government workers, teachers, computer programmers, hospital workers, store clerks, restaurant workers, and the like. Quaternary activities, which are hardly relevant to our discussion, are best described as policy decision-making activities taken by senior government offi-cials and by the management sector of corporations.

By dividing the economy of the Northwest Territories into the primary, secondary, and tertiary sectors we can gain a better understanding of the structure of the northern economy. The very small quaternary sector, which comprises less than 1 per cent the labour force, becomes part of the tertiary sector.

On first glance, given the dominating role of the resource industry in terms of value of production, it would seem logical that primary activities would be the major employer. This is not the case. Table 4.12 demonstrates the relatively weak position of the primary sector, comprising only 6.4 per cent of the total employed workers, compared to the tertiary sector at 83.9 per cent. The largest single subsector is public administration at 21.8 per cent.

Table 4.12 Employment by Economic Sectors, Northwest Territories, 2006

Economic Sector	Subsector	No. of Employees	Percentage
Primary		1,515	7.1
	agriculture, forestry, fishing, hunting	150	0.7
	mining, oil and gas	1,365	6.4
Secondary		1,921	9.0
	construction	1.295	6.0˙
	manufacturing	336	1.6
	utilities	290	1.4
Tertiary		17,915	83.9
	education	1,605	7.5
	health care and social assistance	2,000	9.4
	retail trade	2,045	9.6
	public administration	4,650	21.8
	transport and warehousing	1,805	8.5
	other	5,810	27.1
Total		21,350	100.0

Source: Adapted from NWT Bureau of Statistics (2008: Table 3).

The prominence of the tertiary sector is characteristic of hinterlands in industrial countries. Such a structural pattern is due to the funding from the industrial state to provide a similar level of public services in each region of the country. This public policy is a cornerstone of Canada's social contract. In the Canadian North, geography and jurisdictional complexity exacerbate the need for a large public sector. Geography, with its combination of a vast space and few people, makes the delivery of public services expensive. In Nunavut, most residents live in very small communities where administration, education, and health services operate without the benefit of economies of scale found in larger centres. The principle of access to public services regardless of the size of the community is necessary but expensive. For this reason, the ratio of government employees to residents is much higher than in the rest of Canada. Another reason is that administrative bodies representing different levels of government and Aboriginal organizations often exist in the same community. While each has its own mandate, this jurisdictional arrangement provides a series of hierarchical layers of government for relatively few people living in urban centres. In Nunavut, for example, Nunavut Tunngavik Inc. (NTI) is the Inuit organization responsible for the implementation of the Nunavut Land Claim Agreement and, according to the Clyde River Protocol that formally outlines the relationship between NTI and the government of Nunavut, that responsibility involves regular consultation with government officials (Légaré, 2000). NTI's goal is to ensure that Inuit interests are expressed in legislation.

The Tungavik Federation of Nunavut (later renamed Nunavut Tunngavik Inc.), for example, took the lead in a harvester support program (Wenzel, 2000).

Summing Up

The North's population remains just under 1.5 million. Although it is difficult to know if the population will change in the coming decade, two factors—a falling rate of natural increase among the Aboriginal population and an increase in net out-migration (including Aboriginal migrants)—may cause a population decline. Another factor is that resource towns have lost favour and companies are using air-commuting systems to supply labour to their remote mines and mega-construction jobs. Aboriginal culture, Native settlements, and the future for a predominately Aboriginal northern population are significant issues that reflect the struggle between culture and economics. This raises the question: *How important is human diversity?* The popular anthropologist, Wade Davis, sees it as our greatest legacy. Davis coined the term **ethnosphere** to capture the notion of a global social web of human life:

> You might think of this social web of life as an 'ethnosphere', a term perhaps best defined as the sum total of all thoughts and intuitions, myths and beliefs, ideas and inspirations brought into being by the human imagination since the dawn of consciousness. (Davis, 2009: 2)

At the centre of this web are the many languages and cultures, which, unfortunately, have dwindled in our economically and socially globalized world. In the North, Aboriginal languages and cultures have already been lost and more are on the edge of extinction.

Native communities lie on the margins of economic well-being. Clyde River, for instance, typifies the dilemma facing such communities. Funds necessary to supply modern services, ranging from health care to education, to provide economic support for the unemployed, and to operate and maintain public services such as housing are far beyond the local tax base. Fiscal dependency on the government of Nunavut and ultimately the federal government is essential and is growing as the populations of such communities increases.

Within the North, the population is unevenly distributed. For example, over 92 per cent of the northern people reside in the Provincial North. The same spatial pattern is even more acute for the North's two biomes, with the Subarctic accounting for nearly 97 per cent of the people in Canada's North. Also, some areas are losing population while others are gaining. Major resource projects, such as the oil sands developments in northern Alberta, account for population growth, yet, as we have seen, this growth is neither constant nor permanent because resource extraction is finite in nature and dependent on world markets. Nonetheless, from 2001 to 2006, northern Alberta had the fastest rate of increase, at nearly 12 per cent, while northern Newfoundland and Labrador had the highest rate of population decline, at 6.4 per cent. In absolute numbers, the greatest population loss took place in northern Quebec—a drop of nearly 16,000 persons. On the other hand, high rates of natural increase among the Aboriginal population play a central role in the maintenance and increase of population in places where the Aboriginal population forms a majority, such as Nunavut and northern Saskatchewan. In places where the economy is sluggish and where the percentage of Aboriginal peoples is low, out-migration accounts for population declines.

From a demographic perspective, the Aboriginal population differs from the rest of the northern population in two ways. First, the Aboriginal population is growing at a much faster rate. Second, the Aboriginal population is much younger. Both of these factors have consequences. For example, Native settlements have a chronic shortage of housing and a growing mismatch between employment opportunities and the number just entering the labour market. From a geographic perspective, most of the Aboriginal population resides in Native settlements, often located in remote areas of the North. This means that many urban amenities are limited to lower-order goods and services, i.e., basic foodstuffs and services such as elementary schools and nursing stations.

Thus, the population geography of Canada's North provides an insight into the area's cultural and economic duality. As the homeland of a number of Aboriginal peoples, the North provides a cultural refuge allowing Indian, Inuit, and Métis to maintain a geographic connection with their traditional lifestyles and, in some cases, to establish different forms of self-government. At the same time, the North represents a last wilderness for some southern Canadians and is often understood as a resource frontier. Most southerners who have relocated to this frontier have done so for economic opportunities. What is unfolding is a struggle to find a compromise between the two visions of the North. This struggle forms a central theme in the remaining chapters in this book.

Challenge Questions

1. Why has the population of the Territorial North increased while that of the Provincial North has declined, thus preventing the North's population to increase?
2. What demographic conditions signal the start of 'phase 4', which would mark a declining population in the North?
3. Do you agree that physiological density is a more meaningful measure of the North's population density?
4. Why are Yukon birth rates (Table 4.6) similar to those for Canada as a whole while the Northwest Territories and Nunavut have much higher rates?
5. What are the economic implications of the youthful nature of the population in the North, but especially Nunavut, compared to that found in the rest of Canada?
6. Why did Yukon's sex ratio change from a preponderance of males to a more balanced ratio? Is this demographic adjustment likely to occur in other parts of the North?
7. Suggest some reasons why the North's population is so concentrated in small urban places.
8. Why do so many resource towns follow a boom/bust life-cycle? See Table 4.11.
9. Should Canadians take measures to ensure the well-being of Aboriginal languages?
10. Clyde River, Nunavut, remains a vibrant community. Did a preference by local residents for culture trump economic reality? Explain.

Notes

1. Within territories and provinces, the statistical units employed by Statistics Canada may change over time. Territories and provinces have the right to ask for internal boundary changes to meet their particular statistical needs. Within territories and provinces, the next geographic level of a statistical unit after provinces and territories is the census subdivision.

2. A huge immigration was associated with the resource boom of the 1950s and 1960s. Ottawa played a role in that boom. In 1957, Prime Minister John Diefenbaker announced his 'Northern Vision'. Abandoning its earlier laissez-faire policy, Ottawa began to aggressively promote development, particularly in the Provincial North: highways were built to resource sites, and new administrative and resource towns were created.

3. The out-migration of northerners varies across the North by ethnicity and by region. Southerners who settle in the North often relocate to southern Canada within five years. This high turnover is partly due to the uncertainty within mining towns. Downsizing or even mine closing usually translates into most miners and their families leaving the North. Aboriginal residents have shown little interest in relocating to southern cities and towns, although some (usually the more educated ones) are gravitating to regional centres or capital cities (but not to resource towns).

4. The census enumeration asked Canadians to state their ethnicity. This approach means that ethnicity is based on the respondent's perception of his or her ancestral background. The number of Aboriginal people by ethnicity in 1986 was 711,725, with some 210,000 living in the North. According to the 2006 census, the Aboriginal population in Canada had reached 1.2 million, with about 300,000 residing in northern Canada (Statistics Canada, 2008b).

5. In the 1950s, Aboriginal Canadians relocated to small settlements where they traditionally traded their furs. These tiny villages are classified as Native settlements. Few relocated to regional centres and virtually none settled in resource towns. Within a decade, some moved to regional centres for a variety of reasons, including specialized health services and employment opportunities. Most migrants were drawn from the growing number of young, more educated Aboriginal Canadians. In regional centres, many are now employed by Aboriginal organizations and government agencies.

6. With huge coal reserves in the nearby mountains, Tumbler Ridge was created as a planned resource community in 1981. Situated in the eastern flank of the Rocky Mountains, the town's future seemed secure because the coal companies had long-term contracts to supply coal to Japanese steel mines. Yet, the global economy slid into a recession in the 1980s, causing the demand for coal to drop dramatically and appearing to spell an end to Tumbler Ridge (Halseth and Sullivan, 2002). China, however, needed coal, and by the early twenty-first century, coal mines in northeast BC were in full production, giving Tumbler Ridge a second chance. Tumbler Ridge's population changes reflect its ups and downs as a resource town. In 1986, Tumbler Ridge had 4,566 residents and the town reached a population peak in 1991 with 4,794 people. Over the next 10 years, Tumbler Ridge saw its population decline, reaching a low of 1,851 people in 2001. Its population rebounded to 2,454 in 2006 due largely to the restart of coal mining (Statistics Canada, 2007g).

7. In terms of its potential labour force, the North has very few professional and skilled workers, especially among the Aboriginal population. For that reason, most skilled workers are hired in southern Canada and brought to northern worksites. Since these workers face much higher living costs and a lack of urban amenities found in southern centres, they will not move to the North unless the employer (companies and governments) pays high wages and offers incentives. Often, these incentives include subsidized housing and an allowance to travel to southern Canada. The federal government provides several additional incentives, including a northern isolation allowance for its employees and a northern personal

tax deduction for all northern employees and business people. Since northern communities vary widely in terms of their isolation, determining who should get these incentives is difficult. As well, the degree of isolation could change with a highway reaching a community. Northern isolation allowance payments provide cash compensation. Revenue Canada (now the Canada Customs and Revenue Agency) has used Hamelin's concept of nordicity to create two northern zones. Places in Zone A are more isolated than locations in Zone B and, therefore, individuals in Zone A filing personal income tax receive more tax relief than those in Zone B. All places in Yukon, the Northwest Territories, and Nunavut fall into Zone A. Zone B corresponds to places in the Provincial North, including resource towns. Examples include Chibougamau in Quebec, Red Lake in Ontario, The Pas in Manitoba, La Ronge in Saskatchewan, Fort McMurray in Alberta, and Tumbler Ridge in British Columbia.

References and Selected Reading

Barnes, T., and R. Hayter. 1992. '"The Little Town That Did": Flexible Accumulation and the Community Response in Chemainus, British Columbia', *Regional Studies* 26:647–67.

Bone, Robert M. 1998. 'Resource Towns in the Mackenzie Basin', *Cahiers de Géographie du Québec* 42, 116: 249–56.

———. 2006. 'Inuit Research Comes to the Fore', in Jerry P. White, Susan Wingert, Dan Beavon, and Paul Maxim, eds, *Aboriginal Policy Research: Moving Forward, Making a Difference*, vol. 3. Toronto: Thompson Educational Publishing.

CBC News. 2006. 'Kashechewan: Water Crisis in Northern Ontario', 9 Nov. At: <www.cbc.ca/news/background/aboriginals/kashechewan.html>.

———. 2007. 'Ottawa Nixes Relocation for Flood-Prone Kashechewan', 30 July. At: <www.cbc.ca/canada/ottawa/story/2007/07/29/kashechewan-deal.html#skip300x250>.

Davis, Wade. 2009. *The Wayfinders: Why Ancient Wisdom Matters in the Modern World*. Toronto: House of Anansi Press.

Department of Finance. 2009. *Federal Transfers to Provinces and Territories*. At: <www.fin.gc.ca/access/fedprov-eng.asp>.

Freeman, Milton M.R. 1971. 'The Significance of Demographic Changes Occurring in the Canadian East Arctic', *Anthropologica* 13, 1 and 2: 215–37.

Halseth, Greg, and Lana Sullivan. 2002. *Building Community in an Instant Town: A Social Geography of Mackenzie and Tumbler Ridge, British Columbia*. Prince George: University of Northern British Columbia Press.

Irwin, Colin. 1988. *Lords of the Arctic: Wards of the State*. Special Edition of Inungnut. Rankin Inlet: Keewatin Inuit Association.

Kendall, Glen. 1992. 'Mine Closures and Worker Adjustment: The Case of Pine Point', in Cecily Neil, Markku Tykklainen, and John Bradbury, eds, *Coping with Closure: An International Comparison of Mine Town Experiences*. London: Routledge, ch. 6.

McKibbon, Sean. 1999. 'Report: Clyde River's Very Existence Is Questionable', *Nunatsiaq News*. At: <www.nunatsiaq.com/nunavut/nvt91029_01html>.

Mahoney, Jill. 2002. 'Where the Jobs Are', *The Globe and Mail*, 12 Mar., A6.

Martel, L., and Alain Bélanger. 1999. 'An Analysis of the Change in Dependency-Free Life Expectancy in Canada between 1986 and 1996', in Bélanger, *Report on the Demographic Situation in Canada 1998–1999*. Statistics Canada Catalogue no. 91–209. Ottawa: Minister of Industry, 164–86.

Newman, James L., and Gordon E. Matzke. 1984. *Population: Patterns, Dynamics, and Prospects*. Englewood Cliffs, NJ: Prentice-Hall.

Northwest Territories (NWT) Bureau of Statistics. 2006. NWT Labour Supply Presentation, 5 July. At: <www.stats.gov.nt.ca/labour-income/labour-supply/>.

———. 2008. *Newstats: Labour Market Activities, 2006 Census*. At: <www.stats.gov.nt.ca/census/2006/Labour%20Force_2006.pdf>.

———. 2010. *Statistical Quarterly* (June). At: <www.stats.gov.nt.ca/publications/statistics-quarterly/sqjune2010.pdf>.

O'Neil, Katherine. 2009. 'Traditional Skills To Be Taught at Nunavut's New Cultural School', *Globalcampus*, 9 Aug. At: <www.globecampus.ca/in-the-news/article/traditional-skills-to-be-taught-at-nunavuts-new-cultural-school/>.

Petrovich, Curt. 1990. 'The Best Reason to Come North . . . Or Is It?', *Arctic Circle* 1, 1: 36–41.

Pleizier, Christina. 2006. Personal communication, 13 Apr., Indian and Northern Affairs Canada.

Statistics Canada. 2004. *Aboriginal Identity Population, 2001 Counts, for Canada, Provinces and Territories—20% Sample Data*. At: <www12.statcan.ca/English/census01/products/highlight/Aboriginal/Page.cfm?Lang=E&Geo=PR&View=1a&Table=1&StartRec=1&Sort=2&B1=Counts01&B2=Total>.

———. 2007a. *Population and Dwelling Counts, for Canada, Provinces and Territories, Census Divisions and Census Subdivisions (Municipalities), 2006 and 2001 Censuses—100% data*. At: <www12.statcan.ca/english/census06/data/popdwell/Filter.cfm?T=302&S=1&O=A>.

———. 2007b. *Births and Birth Rate by Province and Territory*. At: <www40.statcan.gc.ca/l01/cst01/demo04b.htm>.

———. 2007c. *Deaths and Death Rate by Province and Territory*. At: <www40.statcan.gc.ca/l01/cst01/demo07b.htm>.

———. 2007d. *Portrait of the Canadian Population in 2006, by Age and Sex: Provincial/Territorial Populations by Age and Sex*. At: <www12.statcan.ca/english/census06/analysis/agesex/ProvTerr7.cfm>.

———. 2007e. *Age Groups and Sex for the Population of Canada, Provinces, Territories, Census Divisions and Census Subdivisions, 2006 Census, 100% Data*. At: <www.statcan.ca/bsolc/english/bsolc?catno=97-551-XWE2006013>.

———. 2007f. 'Whitehorse City', *Community Profiles 2006*. At: <www12.statcan.ca/english/census06/data/profiles/community/Details/Page.cfm?Lang=E&Geo1=CSD&Code1=6001009&Geo2=PR&Code2=60&Data=Count&SearchText=whitehorse&SearchType=Begins&SearchPR=60&B1=All&Custom>.

———. 2007g. 'Tumbler Ridge', *Community Profiles 2006*. At: <www12.statcan.ca/english/census06/data/profiles/community/Search/SearchForm_Results.cfm?Lang=E>.

———. 2008a. 'Table 8: Size and Growth of the Inuit Population, Canada and Regions, 1996 and 2006'. At: <www12.statcan.ca/english/census06/analysis/aboriginal/tables.cfm>.

———. 2008b. 'Aboriginal Peoples: Highlight Tables, 2006 Census'. At: <www12.statcan.ca/english/census06/data/highlights/Aboriginal/index.cfm?Lang=E>.

———. 2010a. 'Canada's Population Estimates: Third Quarter 2010', *The Daily*, 22 Dec. At: <www.statcan.gc.ca/daily-quotidien/101222/dq101222a-eng.htm>.

———. 2010b. 'Deaths', *The Daily*, 23 Feb. At: <www.statcan.gc.ca/daily-quotidien/100223/dq100223a-eng.htm>.

———. 2010c. 'Quarterly Demographic Estimates—June to March 2010', *Canada's Population Estimates*. At: <www.statcan.gc.ca/pub/91-002-x/91-002-x2010001-eng.htm>.

———. 2010d. *Canada Yearbook*: Table 24.9, 'Birth Rate, by Province and Territory, 2003/2004 to 2008/2009'. At: <www.statcan.gc.ca/pub/11-402-x/2010000/chap/pop/tbl/tbl09-eng.htm>.

———. 2011. *Annual Demographic Estimates: Subprovincial Areas, 2005–2010*, 3 Feb. At: <www.statcan.gc.ca/pub/91-214-x/2009000/tablelist-listetableaux3-eng.htm>.

Storey, Keith. 2009. 'Help Wanted: Demographics, Labor Supply and Economic Change in Newfoundland and Labrador', presentation at 'Challenged by Demography: A NORA Conference on the Demographic Challenges of the North Atlantic Region', Alta, Norway, 20 Oct.

Weinstein, Jay A. 1976. *Demographic Transition and Social Change*. Morristown, NJ: General Learning Press.

Yukon Bureau of Statistics. 2010. *Yukon Migration Patterns, 1999–2008*. At: <www.eco.gov.yk.ca/stats/pdf/migration2008.pdf>.

5

Resource Development, Megaprojects, and Northern Benefits

The Canadian North is part of the global economy. Resource development is critical to its economic growth. Most projects are large-scale undertakings known as megaprojects that require vast capital investments and long-term horizons. Not surprisingly, then, multinational companies and Crown corporations are the leading firms involved in the northern economy. Megaprojects, some think, are the engine of northern economic growth, but these huge industrial efforts create a very narrow economic base that is also subject to a construction version of the boom-and-bust cycle. Without a doubt, megaprojects accelerate northern industrial growth, but such projects can do little to broaden the frontier economy. Since the nature of regional development has a much broader agenda than simply increasing economic output, it follows that megaprojects are best described as 'engines of economic growth' rather than 'engines of regional development'.

Unlike southern Canada, the North has several structural weaknesses that severely limit a widening of the northern economy as a result of spinoff effects from megaprojects. These weaknesses include:

- excessive economic leakages;
- the boom-and-bust cycle of non-renewable resources and resource projects;
- a northern labour force unable to meet the needs of huge construction projects, thus necessitating the importing of skilled workers from southern Canada.

Can these weaknesses be overcome or at least modified? Key to this question is a shift in corporate behaviour to the North and its people so that it becomes closer to stewardship rather than exploitation. A brief synopsis of the past century provides some positive findings and, if the trend continues, hope for a brighter future. To simplify matters, this synopsis is divided into two time periods.

The early pattern of resource development in Canada's northern hinterland was associated with a laissez-faire approach by Ottawa, allowing the classical lines of the core/periphery model to dominate the direction of northern development and ignoring the needs and wants of the Aboriginal population: capital flowed to the North from large companies and Crown corporations to construct megaprojects that served the interests of the provinces, the nation, and the international business community. While all regional economies interact with each other, the level of dependency of the North on the rest of the world was extreme. During this period, both private and public companies ignored local Aboriginal peoples and paid little attention to the environmental and social consequences of their projects. An excellent example began in 1947 when the first American megaproject in Canada's North was announced by US iron and steel interests.[1] Their plan called for the exploitation of the vast iron deposits in northern Quebec and Labrador and the building of a rail–sea transportation system capable of supplying iron ore to American steel plants. Two years later, financing was secured for the project and construction began. Until the James Bay Project, this iron mining project represented the largest single capital investment in the Canadian North. By the early 1950s, a massive iron mining and transportation system was completed, but with no regard for the Aboriginal people and the Subarctic environment found in northern Quebec. At that time, few rules and regulations were in place because Canadian society and its governments saw megaprojects as an effective way of opening the North.

In the late twentieth century, the pattern of resource development shifted from a laissez-faire approach by government to a more active one, from denying Aboriginal rights to entering into land claim negotiations, and from ignoring environmental damage caused by industrial projects to demanding environmental assessments and post-project cleanup. By the twenty-first century, proposed resource projects worked within a framework of federal, provincial, and territorial rules and regulations.

Several factors were instrumental in moving governments and companies to a more progressive approach to resource development, where rules and regulations govern corporate actions. First, Canadians recognized that industrial growth had its dark side, namely, that resource projects had hidden costs associated with the disposal of toxic wastes into the air and water. As noted in Chapters 1 and 3, the Berger Inquiry in the 1970s played an important role in changing public opinion by holding hearings in local communities. These hearings, covered by the national media, exposed on a daily basis specific environmental threats to the fragile North and its Aboriginal peoples. Furthermore, the public felt that companies, rather than taxpayers, should take responsibility for these hidden costs. Second, the environmental movement, through various non-governmental organizations (NGOs) such as the Sierra Club, Pollution Probe, and Greenpeace, exposed the worst examples of industrial pollution and, with this evidence, pestered governments to force companies to take responsibility for their impact on the environment. Third, federal and provincial governments enacted legislation to ensure that projects undergo environmental impact assessments with the goal of minimizing damage to the environment.

At the same time, decisions by the Supreme Court of Canada enabled a land claims settlement system to unfold both for Aboriginal groups under treaty and for those groups that had never signed treaties with the government. This system has enabled Aboriginal peoples to gain control of a portion of their traditional land base and to share in the decision-making concerning the approval/rejection of industrial proposals in their territory. As well, Aboriginal leaders pressed for a share of the profits from megaprojects that occurred beyond their land claim territory but within their traditional land base. A recent example is the Meadowbank gold mine in Nunavut, which lies within the traditional Inuit lands but outside the territory claimed under the Nunavut Land Claims Agreement. These lands are defined as Crown lands, where mineral resources fall under the jurisdiction of the federal government. On 20 June 2010, officials of the company, Agnico-Eagle, pledged to respect the land and its people at a ceremony at Baker Lake. Leading Inuit dignitaries participated in this ceremony. Even though the life of this mine is estimated at 19 years, Jose Kusugak, president of the Kivalliq Inuit Association, declared that Baker Lake is now 'the happiest community in Nunavut because of the hope generated by Agnico-Eagle's Meadowbank gold mine', while the mayor of Baker Lake, David Aksawnee said, 'I see a lot of people working on this site. They're well-fed now and I'm happy to see that' (Bell, 2010). This 'happiness' was made possible by three factors:

- the Nunavut Land Claims Agreement (NLCA), which includes a provision for a 5 per cent share of subsurface royalties to flow to the government of Nunavut;
- by an Impact Benefit Agreement between the company and Kivalliq Inuit Association that spells out the company's contractual obligation to provide employment, training, business opportunities, and funding arrangements;
- the recognition by the company that mining on Inuit traditional land has both economic and social obligations so that it must demonstrate respect for the land and its people.

Resource Development: Theory and Reality

The economic history of Canada is largely based on the exploitation of resources in opening Canadian frontiers to a market economy. In southern regions of Canada, resource development initiated a process of economic growth by exploiting a major resource that, in some cases, led to further economic activities and eventually to the diversification of regional economies. Canadian scholars have recognized and analyzed this historic pattern of regional development in Canada. In 1930, Harold Innis first observed that Canada's regional development was triggered by resource development that, over time, matured into a more diversified economy. Known as the staple thesis, Innis's approach presented a fresh interpretation of regional economic history and geography—from a Canadian perspective.

Mel Watkins (1963) transformed Innis's thesis into a more conventional theory of regional development. Regional development is not an automatic process of the market economy. In fact, the most positive outcome of staple projects is a broadening and maturing of the resource economy, but, in the worst-case scenario the resource economy flounders and eventually slides into a staple trap where, instead of development, underdevelopment takes hold (Watkins, 1977). Hayter and Barnes (2001: 37) put it more gently: 'But as Innisians emphasized, there is nothing automatic about such diversification.'

While the staple theory provides a broad historic framework to explain regional development, what about current reality? For example, the price of oil—now above US$90 a barrel—is unlikely to return to levels experienced in the last decade of the twentieth century when a barrel of oil was worth less than US$20. As Vignette 5.1 points out, commodity prices have risen to new levels and the range of prices within this new level appears firm. For the resource-oriented North, higher commodity prices are good news, making existing operations more profitable and encouraging more investment. On the other hand, five factors distinguish the North from other regions of Canada and these factors hamper the process of regional growth in the form of megaprojects from gaining a grip and thus leading to a mature, diversified economy.

1. Physical factors provide special challenges to developers. These challenges include permafrost, a fragile environment, and sea ice. Focusing on sea ice, two former Arctic mines provide an insight into the transportation challenges posed and the need for co-operation between mining firms. Polaris (on Little Cornwallis Island) and Nanisivik (on the northern tip of Baffin Island) were lead/zinc mines that shipped their ore through the eastern end of the Northwest Passage (Lancaster Sound) to Germany. But because the Arctic ice pack blocked Lancaster Sound for most of the year prior to 2007, both mines had to store their ore for 10 months and then ship it by a specially designed ship, the MV Arctic. Even so, floating ice was often present and the double steel-reinforced hull of the MV Arctic proved most effective to deal with floating ice but not with the pack ice. In 2002, both mines closed when their commercial ore bodies were exhausted.

2. High transportation costs are the Achilles heel for northern projects. One reason is a very limited transportation infrastructure; another is the great distance required to haul supplies and equipment to construction sites and then to ship resources to world markets. In Nunavut, for example, few people—less than 40,000—occupy this largest territory in Canada, yet no part of Nunavut is connected to the rest of Canada by a highway or railway, forcing companies like Agnico-Eagle to rely on air and sea transportation. In the northern reaches of provinces, provincial governments have tried to overcome this distance barrier by building modern transportation routes into their hinterlands.[2] In the 1920s, the British Columbia government sought to stimulate resource development by building the Pacific Great Eastern Railway (PGE), but the high cost of construction in the Cordillera and limited funds in the provincial treasury prevented its completion until 1952. Hence, the PGE was known as 'the railway that begins nowhere and ends nowhere'. By connecting Prince George to the port of North Vancouver, the railway opened the vast forests of British Columbia's interior to global markets, and in time the PGE reached further north to Fort St John and further east to Dawson Creek and the Peace River country. This railway was renamed British Columbia Railroad in 1972 and BC Rail in 1984. Canadian National Railway purchased BC Rail in 2004.

3. Demographic factors lead to extreme economic leakage. First, the urban and business network is small and limited, and thus incapable of responding to the needs of large-scale industrial projects. Consequently, the demand for equipment and supplies is satisfied by firms outside of the North. Second, the northern labour force is largely unskilled and inexperienced. Companies turn to southern labour markets for skilled workers. These two structural weaknesses lead to a high level of economic leakage of possible northern benefits. For example, the wages earned by workers from the south are often spent in the south rather than circulated in the northern economy.

4. Economic geography reveals that, while the construction phase of large-scale industrial projects results in a high level of economic activity, such activity is short-lived and does not lead to economic stability and diversification. In fact, such a sharp burst of intense economic activity, by focusing on a small geographic area, overheats

the local economy, places heavy demands on the public infrastructure, and draws the local Aboriginal workforce into temporary jobs and business contracts. The Norman Wells Oil Expansion and Pipeline Project exemplifies this process. During the construction phase from 1982 to 1985, the largely non-Aboriginal residents of the small community of Norman Wells (about 650 inhabitants) enjoyed the economic boom, but they suffered from the construction noise and traffic and saw their public services stretched to the limit. By the end of the construction, local businesses that had expanded during the boom had to readjust to the reality of a much smaller local market, causing some to close and others to scale back their operations.

5. The Aboriginal component of the northern labour force and business community is just beginning to gain the skills and capital necessary to compete in this northern version of the market economy. As Poelzer (2009: 448) observed:

> The supply-versus-demand problem has two dimensions: First, the inadequate supply of northern residents qualified to meet the labour demand in the public and private sectors; and second, the insufficient supply of diverse educational opportunities, particularly at the bachelor's-degree level.

Looking back at the early pattern of resource development, few northern Aboriginal people were prepared for the market economy, and consequently they could gain little from development and not much more from job and business opportunities. This issue was certainly a major theme in the 1977 **Berger Report**. In some developments, such as the James Bay Project in northern Quebec, the Cree saw their traditional land-based economy negatively affected. In the twenty-first century, however, resource projects are taking place after most land claims have been settled, and companies are negotiating 'private' impact and benefit agreements with locally affected Aboriginal groups (Slocombe, 2000; Bradshaw and Prno, 2007). With these agreements in place, coupled with recognition by companies of a need to provide northern benefits, resource projects are now more likely to have a positive impact and such impacts may lead to economic diversification within the local area and even beyond.

Two questions posed in this chapter are:

- Can resource development in the North lead to economic stability and diversification?
- Is sustainable development possible in an economy based on non-renewable resources?

Given the substantial Aboriginal population, fundamental to shaping the answers to these two questions was the 1997 Supreme Court decision in *Delgamuukw*, which provided a firm legal basis for establishing Aboriginal title to lands traditionally occupied by Aboriginal peoples (see Chapter 3). The resolution of landownership and comprehensive land claim agreements have opened the door for greater Aboriginal participation in resource development (see Chapter 7). On the other hand, the Aboriginal labour force remains mismatched with the demands of industry. Underemployment is widespread and jobs go begging for skilled workers. As Berger (2006: 18) pointed out in the case of Nunavut, 'there are in the vicinity of 1,500 jobs that could be claimed by Inuit had they the necessary skills.'

Vignette 5.1   A New Dance Floor for Commodity Prices?

In the first years of the twenty-first century, global prices for energy and resources have jumped significantly. Is this part of the normal boom-and-bust commodity price system whereby prices rise when the world economy is expanding but fall when the world economy contracts? Or is something else happening? In short, has the world's economy reached a higher plateau where the demand for energy and resources will fluctuate, but within a much higher range of prices? The argument goes like this: global demand for resources has increased due to the rapidly expanding economies in Asia, especially in China and India. The impact of increased demand has seen record price jumps. Suddenly, resource-producing regions—at the same level of production—are benefiting from higher prices. While the world business cycle will no doubt continue to expand and contract, some believe that the price levels for resources will be maintained at a higher range of commodity prices. As Brian Oleson, a University of Manitoba agribusiness professor has said: 'We are entering a new era. It's almost as if the platform [for energy, grain, and mineral prices] has been raised and we're all dancing on a new dance floor' (Greenwood, 2007).

Resource Base

Canada's northern resource base—except for forestry and wildlife—is best understood through its geology. As discussed in Chapter 2, the geomorphic regions of Canada are associated with particular energy and mineral wealth. The Canadian Shield, for instance, is associated with 'hard rock' minerals such as diamonds, gold, iron, nickel, and uranium while the Interior Plains and Arctic Lands are associated with vast sedimentary basins that contain huge quantities of oil and natural gas.

The resource base divides into non-renewable and renewable resources. Non-renewable resources are mineral and petroleum deposits. These resources have a fixed life—once a mine opens, for example, the first shovelful presages its death. Two examples illustrate the variation in longevity of gold mines. On the one hand, the Giant gold mine near Yellowknife operated for 56 years (1948 to 2004). On the other hand, the gold deposit at the Meadowbank mine near Baker Lake, which opened in 2010, is estimated to last less than 20 years. While the extraction of these resources stimulates the northern economy, the long-term impact is limited because of their finite nature. This raises the questions of the sustainability of non-renewable resources and the extent to which the exploitation of these resources offers the potential for economic diversification. In both instances, the historical record provides a largely negative answer.

The principal renewable resource, the boreal forest, lies in the Subarctic where the climate permits tree growth in its warm summers. Properly managed, the harvesting of the forest resource can lead to sustainable development, but without processing it will not lead to a diversified economy. Processing renewable resources, by achieving a higher level of value added, moves the economy towards greater diversification. Then, too, sustainable logging practices are imperative. Unfortunately, the desire for short-term gains has led to destructive resource exploitation in the past (Clapp, 1999).

The forest industry remains in a depressed state. For the past 10 years, the forest industry of the Subarctic has suffered with the closure of many sawmills and pulp and paper mills. Its demise is not due to a decline in the forest resource but to a sharp drop in demand for its products. For the forest industry to regain its previous level of production, US imports of Canadian forest products, especially softwood lumber, must return to their early 1990s level. Such a recovery is linked to a strong rebound in the US housing market from its current record low.

Our discussion of the resource base now turns to specific cases, beginning with renewable resources—forests and water—and ending with two types of non-renewable resources—diamonds and petroleum.

The Forest Resource

Canada has a portion of the circumpolar forest that forms a 'green halo' near the top of the globe. As the dominant natural vegetation found in the Subarctic, the boreal forest extends from Newfoundland and Labrador to Yukon. The forest represents both a key logging area and a rich zone of biodiversity, as described in Chapter 2.

Only the southern section of the boreal forest has mature stands, where logging takes place in British Columbia, Alberta, Saskatchewan, Manitoba, Ontario, Quebec, and Newfoundland and Labrador. In the territories, Yukon and the Northwest Territories have small logging operations in their portion of the boreal forest, but no logging exists in Nunavut, which is north of the treeline. With few exceptions, access to forest lands requires a company to obtain a timber lease from a provincial or territorial government. First Nations have obtained timber leases on lands where they claim Aboriginal title

---

Vignette 5.2 The Resource Economy and Boom-and-Bust Cycles

With its narrow resource base, the northern economy is vulnerable to wide fluctuations in its economy, caused partly because of the nature of non-renewable resources but also because of shifts in global demand. These world shifts are part of the global business cycle. This cycle—a normal part of the market economy—affects all parts of the world but the downturn is especially troubling to resource hinterlands that have few other economic activities to fall back on. If the downturn lasts for an extended period, workers and their families will depart, thus aggravating the economic situation.

The impact of the **boom-and-bust cycle** occurs at three levels. At the primary level, a small mine begins with boom-like conditions during the construction phase, followed by an operations phase when a more modest level of economic activity takes place. Eventually, the resource is exhausted and the operation closes, marking the bust phase of non-renewable development. A second type of cycle is associated with megaprojects. Their impacts follow a similar pattern with two exceptions. First, the economic and social impacts spread over a larger area—a regional impact rather than a local one. Second, the impacts last for a much longer period of time. The third type of boom–bust cycle that affects the North is due to global economic downturns that cause resource companies to halt production or to close their operations, at least until commodity prices increase again.

and, by forming joint ventures with established forestry firms, have become involved in logging and lumber production. Not all First Nations have had success. Located in the heart of the boreal forest of Alberta, the Lubicon Cree band has yet to achieve a land claim agreement with Ottawa. In the 1980s, the Alberta government was eager to diversify its oil-based economy and awarded timber leases for most of the boreal forest in Alberta to five forest companies. Yet, the awarding of these leases before the Lubicon First Nation settled its land claim has created a bitter and smouldering dispute. Meanwhile, the Japanese firm, Daishowa, whose pulp plant is located in Peace River, Alberta, continues to harvest timber from Lubicon traditional land.

For the past 10 years, the forest industry has been contracting, causing high levels of unemployment in the communities. This has led to out-migration, which compounds the economic difficulties facing forest communities. A measure of the decline is provided by the downward trend in the annual timber harvest. In 2004, Canada's forests hit a peak annual harvest of about 200 million cubic metres of timber, but since then the depressed state of the industry has seen the annual figure fall well below 200 million cubic metres and shows no signs of recovery (Natural Resources Canada, 2010a: 31). In 2008, for example, the annual harvest had fallen to 137 million cubic metres (ibid., 9). Approximately 70 per cent of this harvest is in softwood lumber, with the remainder going to the pulp and paper industry. This decline was due to a drop in exports to the United States, where traditionally nearly three-quarters of Canada's forestry output was sold. In 2002, exporters had to accept a 27 per cent tariff on Canadian lumber imposed by the US. Once the tariff was removed in 2006, however, the US housing market had collapsed, thus greatly reducing US demand for softwood lumber.

Like many small towns along the southern edge of the northern coniferous forest, Meadow Lake, Saskatchewan, provides a microcosm of the impact of low prices and falling demand on boreal forest communities. Meadow Lake has struggled to maintain its forest industry, including seeking public funds for its forest companies to survive, but the Saskatchewan government was unwilling to provide more subsidies, causing Millar Western to close its pulp mill in 2007. In addition, the town's sawmill suffered from the high US duties on softwood lumber from 2002 to 2006, followed by the catastrophic drop in US demand for softwood lumber. One unique feature of this forest complex is the involvement of the Meadow Lake Tribal Council, which holds timber rights to a 3.3 million-hectare forest management licence area. The timber from this forest supplies the sawmill, owned and operated by the Meadow Lake Tribal Council. In 1992, the Tribal Council formed a partnership with Millar Western Forest Products to supply wood pulp to the Millar Western pulp mill. With the disappearance of the US market, the pulp and sawmills fell on hard times. Both ceased production in 2007, marking the end to this promising partnership.

While the forest industry is 'on the ropes' due to the sharp drop in demand from its traditional customer, the United States, efforts to find other markets are underway with federal and provincial support as well as an initiative to tap into 'green' markets in Europe. One example of a turnaround is taking place in the single-industry town of Mackenzie, British Columbia. First, after a shutdown of two years, its pulp plant was reopened in August 2010 with 80 per cent of its high-quality bleached kraft paper being shipped to Chinese markets (Macdonald. 2010). Second, its two sawmills were reopened in November 2010 with sales going primarily to Canada and

---

Vignette 5.3   **Resource Differences in the Arctic and Subarctic**

The resource geographies of the Arctic and Subarctic differ sharply. The Subarctic has several advantages over the Arctic:

- Its resource base is much broader and includes a wider range of resources—major energy, mineral, timber, and water resources.
- Its geographic location makes it more accessible to the American market.
- Its natural environment presents less of a barrier to industrial projects because of a milder climate, a longer river/lake navigation season, and a land less affected by permafrost.

All of these advantages translate into lower costs of resource development in the Subarctic than in the Arctic. In addition, sustainable development prospects are more promising in the Subarctic than in the Arctic because of the greater array of renewable resources in the Subarctic, especially the boreal forest.

United States (its traditional markets) but also to new markets in China and Japan (ForestTalk.com, 2010).

Mineral Resources

Canada is one of the world's leading mineral producers and exporters. Nearly half of this production comes from northern mines. Since 2000, the annual value of Canadian mineral exports climbed from $20 billion in 2000 to $47 billion in 2008. Annual demand and prices rose significantly as the world economy expanded until the severe downturn in late 2008. Then, except for gold, demand for commodities dropped sharply. In 2009, the total value of Canadian mineral output fell to $32 billion (Natural Resources Canada, 2010b: 6). As the global economy picks up speed, demand and prices for resource commodities, especially coal, iron, and oil, are likely to increase once again, perhaps reaching or surpassing the 2008 level.

Within the world of mineral exploration, the discovery of commercial mineral deposits has an element of suspense and surprise—a prospector never knows when or if he/she will hit pay dirt. Diamonds are a case in point. Until diamonds were discovered in the Canadian Shield of the Northwest Territories, most geologists did not believe that such deposits existed in Canada (see Vignette 5.4). The timing was perfect for the Northwest Territories because, in the 1990s, the central pillar of its resource economy—gold—was coming to an end. Equally unexpected, the recent discovery of a large gold deposit near the famous Klondike placer goldfields[3] has amazed geologists, and if a mine is developed it will certainly kick-start Yukon's sluggish economy (Figure 5.1). The boom-and-bust economic model based on the rich and varied geology of the Cordillera north of 60 degrees has become a fact of life in Yukon, where many producing mines based on gold, silver, copper, tungsten, asbestos, cadmium, lead, zinc, nickel, and platinum have come and gone. The prospects for another mineral boom in the second decade of the twenty-first century are bright due to two factors. First, high gold and other mineral prices have accelerated exploration efforts

Figure 5.1   Yukon's White Gold Discovery: A New Klondike?

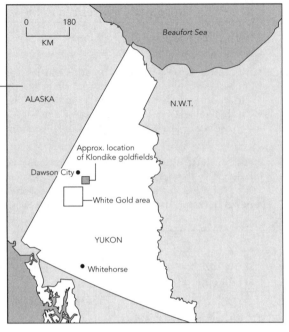

After just more than a hundred years, gold fever is back in the Klondike following a discovery in 2004 by Shawn Ryan, a local prospector, who found what may be the source of the Klondike placer gold near the confluence of the White and Klondike rivers just south of Dawson City. Ryan's discovery, coupled with record high prices for gold, means that an underground mine is almost assured in the White Gold area by 2015 while further north, Kinross Gold Producer plans to open its Victoria gold mine by 2013 subject to approval of its environmental statement and the signing of an Impact and Benefits Agreement by the Na-Cho Nyak Dun First Nation.

Sources: Bouw (2010); Thompson (2011).

in the last few years. Second, strong demand now and for the foreseeable future has encouraged mining companies, and a series of new mines are on the brink of production, pushing Yukon's mining industry towards another potential boom (Figure 5.1).

In the Canadian North, production ranges from highly precious minerals, such as diamonds, gold, silver, and platinum, to bulky, lower-valued minerals, such as lead, zinc, copper, and nickel. Given the high cost of transportation to world markets, highly valued minerals are often extracted first. Diamonds, gold, and silver are examples of high-valued minerals while coal, iron, and nickel fall into the lower-valued mineral category. The implication of minerals of high and low value centres on transportation costs. High-valued minerals can stand high transportation costs while low-valued minerals cannot.

Major mining operations occur across the North (Figure 5.2). In the Territorial North, diamond production is the leading mineral by value. Three diamond operations (Ekati, Diavik, and Snap Lake) are found in the Northwest Territories. In 2010, gold mining began at the Meadowbank mine in Nunavut. In the Provincial North, iron ore mining takes place at the Labrador City and Wabush mines in Labrador; nickel at Thompson, Manitoba; Raglan, Nunavik, and Voisey's Bay, Labrador; one diamond mine at Victor, Ontario; and uranium at three mines (McArthur River, McClean Lake, and Rabbit Lake in northern Saskatchewan (Table 5.1 and Figure 5.2).

Uranium Mining

Northern Saskatchewan's Canadian Shield contains some of the world's richest uranium ore deposits and its mines account for 20 per cent of world production (see Figure 5.3 and Table 5.2). The large size and high grade of these ore bodies make them

Figure 5.2 Major Mines, 2010

more economical to mine than deposits in other countries. Some of these deposits are mined by open-pit techniques while others are mined using hard-rock mining techniques. At Cigar Lake, which has a particularly rich uranium deposit, remote-controlled underground mining techniques will be used to reduce the threat of radiation to the workers.

Diamond Mining

The discovery of diamonds in the 1990s was the most exciting and unexpected mining story in the Northwest Territories (Vignette 5.4). Until the late twentieth century,

Table 5.1 Mineral and Petroleum Production in the Territorial North, 2009 ($ millions)

Mineral Type	Yukon	Northwest Territories	Nunavut	Territorial North
Metallic	245	50	0	295
Non-metallic	6	1,459	0	1.465
Petroleum	10 (est)	700 (est)	0	710
Total	261	2,209	0	2,470

Sources: Natural Resources Canada. 2010; Bureau of Statistics, NWT. 2010; and Yukon Statistics Review 2008.

Table 5.2 Uranium Mines and Mills in Northern Saskatchewan

Facility	Licencee	Licence	Type
Cigar Lake Mine	Cameco Corporation	Construction	Under Construction
Cluff Lake Mine Site	AREVA Resources Canada Inc.	Decommision	Under Decommisioning
Key Lake Mill	Cameco Corporation	Operation	Licensed to produce up to 7,200,000 kg of uranium per year; licensed to receive ore slurry from McArthur mine
McArthur River Mine	Cameco Corporation	Operation	Licensed to mine up to 7,200,000 kg of uranium per year
McClean Lake Mine/Mill	AREVA Resources Canada Inc.	Operation	Licensed to produce up to 3,629,300 kg of uranium per year
MidWest Joint Venture Mine	AREVA Resources Canada Inc.	Site preparation	Site activities suspended indefinitely pending environmental assessment
Rabbit Lake Mine	Cameco Corporation	Operation	Licensed to produce up to 6,500,000 kg of uranium per year

Source: Canada Nuclear Energy Commission. (2009).

geologists did not believe that the Canadian Shield contained diamonds. In 1991, two prospectors, Charles Fipke and Stewart Blusson, discovered diamonds near Lac de Gras in the mineral-rich **Slave Geological Province** of the Canadian Shield some 300 km northeast of Yellowknife. Other diamond deposits were found in the same geological zone, consisting of ancient kimberlite pipes within volcanic formation, but only three deposits (Ekati, Diavik, and Snap Lake) had commercial potential. Ekati was the first mine to open, in 1998; Diavik followed in 2003; and the mine at Snap Lake opened in 2008. Each mine took about 10 years from discovery to production, reflecting another feature of megaprojects, that only very large companies have the financial strength to tackle such projects. Ekati and Diavik are open-pit mines and Snap Lake is an underground mine. Each deposit has an estimated 20- to 30-year supply of diamonds, with peak production estimated at 2013 and shut down of the last of the three mines around 2030 (Hoefer, 2009: 408). Within a year of Ekati opening, diamonds joined oil as the backbone of the resource economy of the Northwest Territories. By value of production, number of employees, and spinoff effects, these three mines constitute the heart of the northern economy. Ekati and Diavik are located about 300 km northeast of Yellowknife while Snap Lake lies some 200 km from Yellowknife. All mines are linked to Yellowknife by a winter ice road. In 1998, the Ekati mine accounted for $500 million worth of diamonds. By 2001, its production was valued at $847 million. Just across the border in Nunavut, another diamond mine—Jericho—began production in 2006, but it abruptly suspended operations on 6 February 2008. In 2009, the annual value of diamond production and exports from the Northwest Territories was $1.5 billion (Table 5.3).

Figure 5.3 Uranium Activities: Past, Present, and Potential

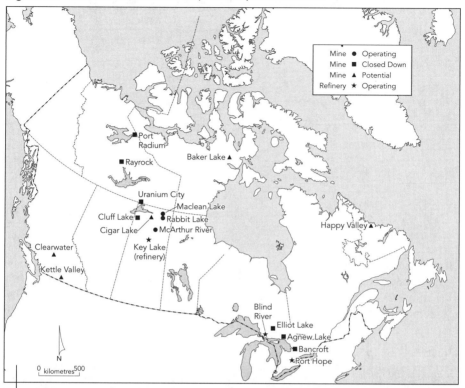

Source: From 'Uranium Discussion Guide', Canadian Coalition for Nuclear Responsibility (www.ccnr.org).

The quality and production of Canadian diamonds are very high. By 2009, Canada's diamond production ranked second in the world by value and third by volume of production (Antwerp Facets, 2010). However, the diamond mines in the Northwest Territories have removed the most accessible ore by open pit mining. By 2010, tunnel mining had begun, suggesting that the diamond boom was over (Vanderklippe, 2010).

Diamond Processing

Processing resources remains an elusive goal. Yet, economics normally calls for processing to take place near major markets. Diamond processing is one exception and represents a small step in the direction of regional development.

While most raw diamonds are exported to the United Kingdom and Belgium for sorting and cutting, a small portion is allocated to the new diamond sorting and cutting industry in Yellowknife. Normally, rough diamonds are shipped to a few well-established diamond sorting, cutting, and polishing centres, but the NWT government insisted that some processing take place in the Northwest Territories. Accordingly, BHP Billiton Diamonds, Diavik Diamond Mine, and now De Beers's Snap Lake operation sell rough diamonds to manufacturers in the Northwest Territories as part of

Table 5.3 Value of Diamond Shipments from Northwest Territories

Year	$ millions
2002	793
2003	1,588
2004	2,097
2005	1,763
2006	1,567
2007	1,765
2008	2,084
2009	1,448*

*Preliminary figure.

Source: Bureau of Statistics, NWT. 2010:37 [2002 to 2008]; and Natural Resources Canada. 2010 [preliminary figure for 2009].

an understanding between producers and the territorial government to promote and support value-added diamond activity in the Northwest Territories (NWT, 2007). Unfortunately, value-added activity did not take hold and, by 2010, only one diamond cutting and polishing company remained open. Even its future is in jeopardy.

Figure 5.4   Snap Lake Diamond Mine

The Snap Lake mine site, located in the lake-strewn landscape of the Canadian Shield, began production in 2008. The mine is owned and operated by De Beers. The site is situated in the natural vegetation transition zone between the boreal forest and tundra vegetation.

Source: De Beers Canada (2007). Reprinted by permission of De Beers Canada Inc.

Vignette 5.4   **Timetable for Diamonds in the Northwest Territories and Nunavut**

The Northwest Territories is the diamond centre for North America. By 2006, two major mines, Ekati and Diavik, accounted for 12 per cent of the world's diamond production. Already, a small proportion of Ekati diamonds are sorted and cut in Yellowknife. Diamonds have been mined in the Northwest Territories only recently, as the following timeline shows. In 2006, Jericho diamond mine, located near the Lupin gold mine in Nunavut, began production, but it operated for only a short time.

Date	Event
1990	Geologist did not believe that the geological structure necessary for kimberlite and hence diamond deposits existed in the Canada Shield.
1991	Charles Fipke and Stewart Blusson discover diamonds near Lac de Gras, Northwest Territories, sparking the largest staking rush in Canadian history.
1993	BHP opens an exploration camp near the original discovery just north of Lac de Gras.
1998	BHP Ekati open pit diamond mine (Panda) opens.; BHP and the NWT government sign an agreement whereby diamond sorting and valuation of some Ekati diamonds will take place in Yellowknife.
1999	Sirius Diamonds Ltd opens a cutting and polishing facility in Yellowknife.
2000	Ottawa approves the Diavik proposal and construction of the Diavik mine begins; Deton'Cho Diamonds Inc. and Arsianian Cutting Works NWT Ltd open cutting and polishing facilities in Yellowknife.
2001	BHP opens a second open-pit mine called Misery.
2003	Diavik mine starts production.
2006	Jericho mine, the first diamond mine in Nunavut, begins operations but closes in 2008.
2008	De Beers's Snap Lake mine begins production.
2010	Only one diamond cutting and polishing firm remains in business in Yellowknife.
2013	Likely peak of diamond mining by the three mines
2030	Likely end of operation by the last of the three mines

Source: Adapted from Hoeler 2009 and NWT Diamonds Timeline, at: <www.gov.nt.ca/RWED/diamond/timeline.htm>.

This quick turnabout raises the question, 'Are efforts to capture value-added activities viable?' Sudbury's diamond processing plant apparently has flourished because the Ontario government insisted that 10 per cent of diamonds produced at De Beers's Victor mine in Ontario's Subarctic go to the Sudbury plant. John Kaiser, a market analyst, was quoted by Gagnon (2010) as saying:

> Diamond processing plants in the Northwest Territories and Ontario is one of these make-work businesses that make the local politicians proud of themselves

that they have some sort of industry . . . in their local area, but it's really outside the economic box. On the other hand, without government intervention in the marketplace, the North will receive few benefits from mega resource projects.

Figure 5.5　Diamond Mines in the Northwest Territories and Nunavut

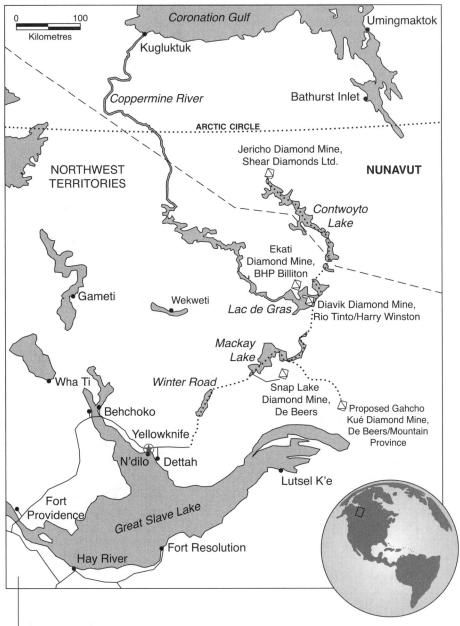

Source: Diavik Diamond Mines Inc.

The North, then, is confronted with a trade-off—companies pay more for diamond processing in the North but northern communities obtain employment and training benefits, plus a few highly skilled diamond cutters and polishers immigrate to Yellowknife from global diamond-processing centres.

Nickel Mining

The Arctic does contain two major mines, both of which produce nickel. One, the Raglan mine owned by Xstrata Nickel, is situated in Nunavik, the Inuit homeland in northern Quebec; the other is at Voisey's Bay along the Labrador coast at latitude 56°N. The Raglan mine is linked by all-weather roads to an airstrip at Donaldson and to the concentrate, storage, and ship-loading facilities at Deception Bay on Hudson Strait. Production began at the Raglan mine in 1997. The current mine life is estimated at more than 30 years (Welch, 2010). The mine is a fly-in/fly-out operation with employees working for 28 days, followed by 14 days off. Its workforce totals approximately 500, with only 1 per cent, or five workers, being Inuit. The ore is crushed, ground, and treated on site at the mill to produce a nickel-copper concentrate that is trucked 100 kilometres to Deception Bay, shipped to Quebec City, and then moved by train

Figure 5.6   Raglan Nickel Mine

The all-weather gravel road that extends 100 km to Deception Bay is shown in the foreground. Since the mine lies within Nunavik, the company has made an Impact Benefit Agreement with Makivik, the Inuit company that manages the financial aspect of the James Bay and Northern Quebec Agreement.

Source: Xstrata Nickel. Reprinted by permission.

Table 5.4 Voisey's Bay Timeline

Year	Activity
1994	Diamond Fields discovers potential 'elephant' nickel deposit.
1996	Diamond Fields sells its nickel site to Inco.
June 2002	Inco signed a $2.9 billion development agreement with the Newfoundland and Labrador government.
2002–3	Preparation work clearing land and building construction facilities at Voisey's Bay and Argentia begins.
2003	Construction commences on the open-pit mine at Voisey's Bay, the Inco Innovation Centre in St John's, and the hydro-metallurgical demonstration plant in Argentia.
2004	Inco's Innovation Centre completed.
2006	Voisey's Bay begins production and Argentia pilot plant receives first concentrates. Vale, a Brazilian mining company, by purchasing Inco, owns and operates Voisey's Bay mines.
2007	Testing of hydromet process at Argentia pilot plant starts.
2008	Vale decides whether to build a commercial hydromet processing facility or to build a matte processing facility that will produce finished nickel from concentrate smelted elsewhere.
2018	Vale decides whether to expand the open-pit mine at Voisey's Bay into an underground mine.
2036	With ore exhausted, Vale closes its Voisey's Bay mining operation.

Source: Adapted from Hasselback (2002: fp7).

to Sudbury for smelting. The final refining process takes place at Xstrata's Norwegian refinery where pure nickel is produced.

At Voisey's Bay, one of the world's richest nickel deposits was discovered by accident—the prospectors were searching for diamonds. The size of this deposit is staggering, with 150 million tons of proven nickel ore plus smaller deposits of cobalt and copper. This deposit has an estimated lifespan of 30 years. The ore is located in three sites: the Ovid, the Eastern Deeps, and the Western Extension. Apart from the claims of the Innu and Inuit of Labrador, the ownership of this deposit has changed hands several times. Vale, a giant Brazilian mining company, purchased Inco in 2006 and therefore the Voisey's Bay nickel deposit. In 2009, the value of nickel produced from Voisey's Bay mine reached nearly $500 million (Natural Resources Canada, 2010b). From open-pit mines, the ore is shipped by freighter and then by rail to its refinery in Sudbury. Beginning in 2013, however, the ore is to be shipped to Vale's nickel processing plant at Long Harbour in Placentia Bay, near St John's on the island of Newfoundland—an outcome resulting not from market conditions but from the government of Newfoundland and Labrador obtaining a commitment for local processing in its 2002 agreement with Inco. In 2005, Inco opened its demonstration hydromet processing facility at Long Harbour, and because it proved successful Vale now will process nickel concentrate at this new plant (CBC News, 2010a).

Figure 5.7 Loading Nickel from Voisey's Bay

The freighter *Umiak* is shown loading nickel concentrate in May 2006 at Edward's Cove, Labrador, as part of the new multi-billion dollar mining operation at Voisey's Bay.

Source: <www.vbnc.com/ShowGallery.asp?GalleryID=4#>. Reprinted with permission from Vale Inco.

Northern Ontario's Ring of Fire

The Canadian Shield contains vast, undiscovered minerals. One recent discovery took place in the remote James Bay Lowland of northern Ontario. Known as the Ring of Fire, this huge deposit consists of chromite, copper, nickel, platinum, and other minerals. The size of the deposit is similar to the Sudbury ore body, suggesting that mining could last for over 100 years. However, the project is stymied by the lack of land-based transportation. Given the bulky nature of the ore bodies, a railway to southern refineries and markets is necessary. Not only is the estimated cost of such a railway high—around $2 billion (Sudol, 2011)—but provincial and federal governments have shown no interest in contributing to a mining railway. Other barriers to the Ring of Fire project exist, including the concerns of the Webequie and Marten Falls First Nations about the lack of consultations to determine their role in the development; and, given the potential risks of such mining to the environment, the proponents face a major hurdle in gaining approval for their yet-to-be-prepared environmental impact report.

Figure 5.8   Labrador and Voisey's Bay Nickel Mine

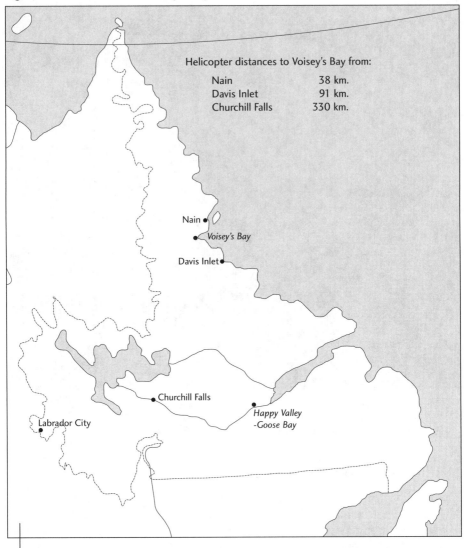

Source: 'Voisey's Bay: An Introduction', at <arcticcircle.uconn.edu/SEEJ/voisey/intro.html>. Reprinted by permission of Arctic Circle.

Petroleum Resources

The North contains most of Canada's petroleum reserves in a variety of sedimentary basins. These basins—with both proven and unproven reserves—are well beyond the edge of commercial exploitation because of their remote geographic location. The most accessible Arctic reserves—those of the Mackenzie Delta and the Beaufort Sea, which were identified by exploratory drilling in the late twentieth century—are still decades from development (Vignette 5.5). The least accessible reserves classified as probable lie deep beneath the Canadian sector of the seabed of the Arctic Ocean,[4] while even larger possible reserves are suspected to exist in the international area or unclaimed zone of the Arctic Ocean. These latter reserves may not reach the exploitation stage this century. As discussed in more detail in Chapter 8, the scramble by nations bordering the Arctic Ocean for ownership of the international section of the Arctic Ocean is due, in large part, to these sedimentary structures that may contain much more oil and natural gas than exists in the rest of Canada (Gautier et al., 2009).

Within the Subarctic, oil and natural gas production takes place in northern British Columbia and Alberta as well as the adjacent areas of Yukon and the Northwest Territories. Of these projects, the Alberta oil sands represent one of the greatest non-conventional oil reserves in the world and it supplies much of Canada's oil to the United States.

Natural Gas: The Energy Source of the Future?

In a world concerned about greenhouse gas emissions, the clean-burning nature of natural gas, compared to coal and oil, greatly reduces the emission of greenhouse gases to the atmosphere. Conversion of coal-burning thermal electric plants continues and even the North American trucking industry is considering switching from diesel fuel to natural gas. The current low price for natural gas makes conversion more attractive. Ontario, for example, is in the process of closing some of its coal-burning thermal plants and replacing them with natural gas plants in order to reduce air pollution.

Natural gas from northern British Columbia and Alberta is already flowing into the energy-starved Chicago market. Huge deposits of natural gas in northern Alberta and BC were discovered but lacked the means to reach US markets. In 2000, the completion of the Alliance Pipeline solved that transportation problem, thus allowing natural gas from this region to reach the Chicago market. With the discovery of the largest deposit of natural gas in Canada at the Horn River Basin shale gas field in northeastern

---

Vignette 5.5   Oil and Gas Reserves in the Beaufort Sea and Mackenzie Delta

Within the sedimentary rocks in the Mackenzie Delta and in the shallow continental shelf waters of the Beaufort Sea, geologists have already found vast quantities of oil and natural gas. Petroleum exploration began in the mid-1960s. Nearly 200 exploration wells have been drilled in the region, resulting in the discovery of nearly 60 oil and gas fields. The largest discoveries in the Mackenzie Delta include the Taglu, Parsons Lake, and Niglintgak natural gas fields. The Amauligak oil field represents the largest offshore discovery. More remains to be found. In 2007, for example, Devon Energy Corporation discovered a huge oil pool in the Beaufort Sea. The petroleum reserves of the Beaufort Sea and Mackenzie Delta amount to over 10 per cent of Canada's petroleum reserves. With the exception of local gas production for the town of Inuvik, these petroleum deposits still await commercial development. The Mackenzie Gas Project (MGP) is one such potential development: a 1,220-km pipeline to markets in the United States would have to be constructed for this project to be feasible—and with gas prices so low, prospects for the next year, and maybe decade, are not bright for this project.

BC, a similar transportation challenge emerged. Pacific Trail Pipelines has proposed a natural gas transmission pipeline system from the Horn River gas deposit to the Pacific coast near Kitimat, where a planned LNG terminal would allow for shipment of lique-fied natural gas by tankers to China and other Pacific Rim countries. The company still has many hurdles to pass, including federal and BC environmental assessment and the opposition of affected First Nations and NGOs.

In the last century, prices of natural gas, along with oil, were rising. But the supply of natural gas has greatly increased due to new technology—the so-called **horizontal drilling** and **hydraulic fracturing** system. As a result, the discoveries of 'elephant-size' shale deposits of natural gas in the provincial norths of British Columbia and Alberta and the adjacent Fort Liard area of the Northwest Territories have turned the supply/demand equation on its head with natural gas prices declining. The size of these deposits, coupled with their relative closeness to the national pipeline system, provides relatively easy access to the American market and possibly to the East Asian market.

This technological breakthrough has delayed the exploitation of Arctic gas and the building of Arctic gas pipelines. With fresh supplies of natural gas readily avail-able in the Subarctic, Arctic gas is no longer necessary to supply southern markets. Consequently, the start of the first Arctic pipeline to North American markets has been back-burnered. Yet, the two contenders for this potentially lucrative pipeline continue to jostle for position. The huge natural gas reserves found in the North Slope of Alaska (which includes Prudhoe Bay) and the Mackenzie Delta may have to wait for the next decade before market conditions are positive. Both need costly pipelines to reach North American markets. The proposed Alaskan gas pipeline route would cross Canadian territory on its way to the Chicago market while the MGP would paral-lel the Mackenzie River and then the route of the Norman Wells Pipeline. Since the Mackenzie Gas Project is less expensive than the Alaskan project (estimated in 2005 at over $30 billion compared to $16 billion for the all-Canadian one), the Mackenzie Delta line is more likely to reach American markets first. Yet, these cost estimates continue to increase, with the MGP coming in at around $20 billion as of April 2010. At such a figure, the project would only go ahead with federal government financial help—something that Ottawa has avoided so far (Selleck, 2009).

How do the two projects compare? The Alaskan project has two very compelling points (Huston and Yu, 2005):

- Alaska's reserves are much larger than those in the Mackenzie Delta—(Alaska holds known reserves of 35 trillion cubic feet and an estimate of an additional 100 trillion cubic feet compared to about 9 trillion cubic feet of known reserves at Mackenzie Delta, plus estimated reserves similar in size to those in Alaska.
- American security concerns call for the development of American petroleum deposits.

On the other hand, the Mackenzie Delta Project has three main advantages:

- The Canadian pipeline promises much lower construction costs because of a shorter distance to market and because the diameter of its pipe is much smaller (and therefore less costly) than that proposed for the Alaskan pipeline.
- The projected construction time (six years) is much shorter compared to the pro-posed Alaskan pipeline (10–12 years).

- Aboriginal participation through the Aboriginal Pipeline Group, which would have a share in the ownership of the pipeline, obviates to some extent the kind of jurisdictional and legal wrangling that can tie up major projects for a long time.

As things now stand, however, with the recent technological capability of exploiting natural gas in the shale deposits further south, the issue may well be moot, at least for the time being, as to which—if any—pipeline route from the Far North wins the day.

Water Resources

One of Canada's most valuable resources is water. Much of that resource lies in the Canadian North. The Mackenzie River, for instance, is one of the world's longest rivers. Consequently, the rivers and lakes of the Canadian North contain enormous potential for hydroelectric power development. However, market conditions have dictated that only those resources in the Subarctic warrant development. Crown corporations have taken the lead and undertaken massive hydroelectric projects in three provinces (Quebec, Manitoba, and British Columbia). The famous and controversial Churchill Falls in Labrador was a private project built by British Newfoundland Development Corporation and now managed by Nalor Energy. In the case of Churchill Falls, the nearby iron ore mines provided the market for much of its energy, which is committed to Hydro-Québec until 2041. In other instances, the power is shipped south to Canadian consumers with the surplus sold to American utilities. At first, these huge projects were seen as a solution to both energy shortages and air pollution problems. As it turned out, hydroelectric development had a negative impact on the natural environment by transforming it into an industrial landscape, not only at the immediate dam sites but by flooding and polluting large areas that had been Native hunting and fishing grounds. In turn, since the natural environment was the basis of the traditional economy of Aboriginal peoples living in these areas, their lives and traditional diets have been irrevocably altered by spending more and more time as village dwellers rather than as hunters on the land (see Chapter 6 for more details on environmental impacts and Chapter 7 for the effect on the Quebec Cree).

Vignette 5.6 An Innovation in Transmission of Electric Power

Development of remote hydroelectric resources depends on the transmission of electrical energy via high-voltage power lines to southern markets. Without access to large markets, large-scale hydroelectric projects would not be viable. For many years, this potential power was unused due to the high cost of transporting electrical energy to major industrial areas. Increased demand and technological advances in electric power transmission have made remote hydroelectric projects commercially attractive. In 1903, for example, the transmission of electrical power some 140 km from Shawinigan Falls to Montreal was considered an engineering feat. Today, Hydro-Québec's high-voltage transmission lines from its power stations in James Bay to the cities and industries in southern Quebec are almost 10 times longer than the line between Shawinigan Falls and Montreal.

Most hydroelectric power is generated in two geomorphic areas, the Cordillera and the Canadian Shield, which have ideal natural conditions for generating electricity: a large volume of water from heavy annual precipitation and terrain with steep drops in elevation. Potential hydroelectric sites still exist in northern British Columbia, Quebec, Manitoba, Newfoundland and Labrador, and Saskatchewan. Ontario, which has the greatest need for more energy, lacks similar sites in its Canadian Shield because of relatively low elevations.

Major hydroelectric projects require large water reservoirs. To smooth out fluctuations in river flows, storage dams and water diversions supplement the water supply to the reservoirs. Ensuring a reliable flow of water throughout the year maximizes the installed generating capacity of the hydroelectric power plant. Even with a complex system of reservoirs, storage dams, and diversions, a long period of below-average precipitation can reduce river flow and hence electrical power generation. In the 1980s, the Subarctic experienced below-average precipitation and hydroelectric power production at installations in northern Manitoba and Quebec was adversely affected. The following decades saw a return to average precipitation and the restoration of water levels. However, these variations in annual precipitation, while unpredictable, are part of a natural hydrological cycle. Such long-term variations should send a warning to those dependent on hydroelectric energy and to provinces enjoying revenues from the sale of such power.

Of all the regions of Canada, Quebec has the greatest potential for the production of hydroelectricity. The reason for this natural advantage lies in the physical geography of Quebec's portion of the Canadian Shield. The James Bay Project occupies a huge area of the Canadian Shield in northern Quebec. In a sense, the Canadian Shield in Quebec acts as a huge reservoir with its many lakes. The river system consists of three major rivers and several smaller ones—all flow into James Bay. The main attraction of these rivers for hydroelectric development is threefold:

- These rivers originate in the uplands of the Canadian Shield in the interior of northern Quebec; the elevation ranges between 500 and 1000 m and thus provides a natural drop before reaching James Bay.
- The Canadian Shield contains large lakes that are easily transformed into huge reservoirs.
- The climate of northern Quebec has a high annual amount of precipitation.

The negative aspects of such huge projects are related to three factors. First, much land is submerged to create reservoirs and river flows are altered from their seasonal rhythm to control the release of water from the reservoirs to meet the high demand for electricity in winter. Second, the hydroelectric landscape harms wildlife, which was the primary source of food for Aboriginal hunting societies and remains an important source. Third, ownership of Crown land is not as clear-cut as once thought. Aboriginal peoples who occupy these lands have claims that must be addressed before construction proceeds. However, until the James Bay and Northern Quebec Agreement in 1975, Aboriginal claims to the land were ignored.

The nature of hydroelectric projects is shown in Figure 5.9. Water is stored upstream in a reservoir, and the water (headrace) flows into the penstock (a tunnel leading to the powerhouse or generating station). Electrical power is created as the water drives the turbines in the powerhouse. After passing through the powerhouse, the water (called tailrace) flows into the river. From the reservoir to the river, the drop in elevation and the amount of water released are critical to generating power.

Nelson River Project

The Nelson River drainage basin extends over nearly 100,000 km², bringing waters from the Rocky Mountains to Hudson Bay via the Saskatchewan River as well as the waters from the Red River basin, much of which lies in the United States (Figure 5.10). The Nelson River itself is 1,600 miles in length, stretching from Lake Winnipeg to Hudson Bay. Much of its journey to the sea is through the rugged Canadian Shield, thus providing many ideal sites for hydroelectric dams. In addition, waters from the Churchill River have been diverted into the Nelson River to increase the volume of water. A series of dams and power stations dot the Nelson River and are known as the Nelson River Project. While much smaller than the James Bay Project, the Manitoba hydroelectric effort was but one hydroelectric development began by provincial Crown corporations in the early 1960s.[5]

The Nelson River Project was initiated in 1961 with the completion of the Kelsey dam and power station, well before the James Bay Project was announced in 1971 and before the Quebec Cree and Inuit signed the James Bay and Northern Quebec Agreement of 1975. One critical difference between the two hydro projects was that the Manitoba First Nations had taken treaty. While the disruption of both aboriginal communities was similar, Manitoba First Nations obtained a much narrower arrangement known as the Northern Flood Agreement of 1977 compared to the James Bay and Northern Quebec Agreement of 1975 which went far beyond compensation for direct damage to the land and way of life and into issues of regional self-government.

The Nelson River hydroelectric development, which now includes five power stations with more in the planning stage (Figure 5.12), represents the major source of electric power in Manitoba. Surplus production, like in northern Quebec, is exported

Figure 5.9 Hydroelectric Generation

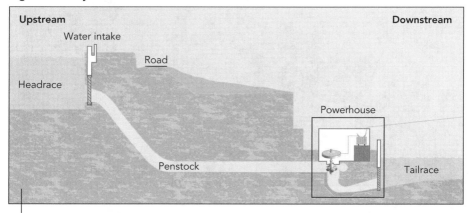

Source: Hydro-Québec (2007b).

to the United States and neighbouring provinces. Ontario's energy shortage may result in an increase in exports. Also, Manitoba Hydro employs the strategy of diverting water from another river (the Churchill River) to augment the flow of water through the power stations on the Nelson River. Long-term power sale contracts were signed with Northern States Power of Minneapolis. Lake Winnipeg, the twelfth largest lake in the world, serves as a giant reservoir for the Nelson River generation system.

With the demand for electricity increasing, Manitoba Hydro began site preparations for the Wuskwatim hydroelectric project in the late 1990s. Construction is scheduled for completion in 2011. Located on Burntwood River 45 km southwest of the nickel-mining town of Thompson, construction should be completed by 2012. However, what is remarkable about this hydro project is the dramatic shift in policy of Manitoba Hydro to engage in partnerships with First Nations. In a remarkable break from past practices, Manitoba Hydro, in 2008, formed a partnership with Nisichawayasihk Cree Nation (Wuskwatim Power Limited Partnership, 2010). This partnership has moved the province beyond the sour relations between Manitoba Hydro and First Nations associated with flooding of traditional lands and destroying fishing grounds. The partnership sees Manitoba Hydro providing construction and management services to Wuskwatim Power

Figure 5.10   Nelson River Drainage Basin

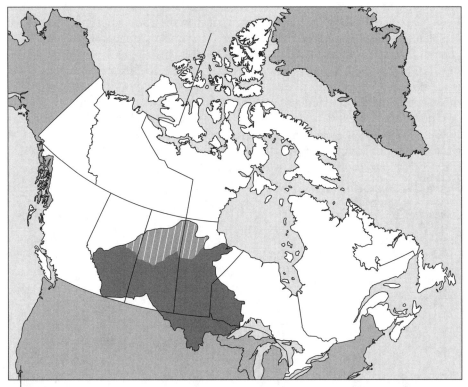

Nelson River's watershed and the watershed of the upper Churchill River diverted to the Nelson.

Source: 'Canada Drainage Basins', *The National Atlas of Canada*, 5th edn (Ottawa: Natural Resources Canada, 1985), at: <atlas.nrcan.gc.ca/site/english/maps/archives/5thedition/environment/water/mcr4055>.

Figure 5.11   Jenpeg Generating Station

Located on the upper reaches of the Nelson River, Jenpeg produces hydroelectric power and is used to control the outflow of water from Lake Winnipeg.

Source: Manitoba Hydro (2007). Courtesy Manitoba Hydro. Photograph by Mario Palumbo CPA.

Figure 5.12   Power Plants on the Nelson River in Northern Manitoba

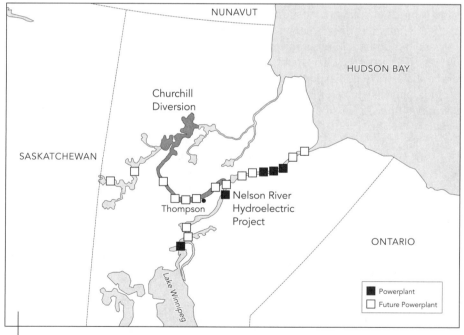

The names of the power plants located on the Nelson River and shown in solid black on the map begin with Jenpeg (1979), just north of Lake Winnipeg; the next four power plants are Kelsey (1961), Kettle Rapids (1974), Long Spruce (1979), and Limestone (1990). While not shown on this map, the first power plant in northern Manitoba, Grand Rapid (1960), was built on the Saskatchewan River as it enters Lake Winnipeg.

Source: Wikipedia (2007).

Limited Partnership in accordance with the Project Development Agreement (PDA) signed in June 2006. This PDA ensures economic benefits to Nisichawayasihk members in jobs and training, with the prospect of becoming Manitoba Hydro employees (ibid.).

Manitoba Hydro's most ambitious undertaking is on the lower Nelson River. The Conawapa Generating Station, if built, would be the largest hydroelectric project in Manitoba. With 2021 as the estimated date for such power needs, the start of construction is some distance off (Manitoba Hydro, 2010). While a long and expensive transmission line would be required, Conawapa could supply energy-poor southern Ontario with much-needed electricity.

The Role of Megaprojects in Northern Development

Megaprojects are huge industrial undertakings that transform small areas of the northern hinterland into industrial nodes with the sole purpose of exporting a commodity to other parts of the world. In one sense, megaprojects hold the key for resource development in the Canadian North. Time, however, is a factor in overcoming the North's unyielding geography, especially the years of construction due to the fragile environment, which may limit construction to the winter when the ground is frozen, the challenge of supplying remote sites, sometimes by ice roads, and extra time needed to deal with permafrost. Given all these challenges, only large corporations, especially multinationals, have the capacity to finance, design, construct, and operate megaprojects. Multinational corporations also have excellent political connections that often lead to various forms of government support, including price supports, tax concessions, and outright grants. Provincial Crown corporations are an exception to this rule. Crown corporations do have a substantial advantage in that they have a monopoly. In the case of hydroelectricity, they have public funding to develop electrical power, sell it at set prices above costs, and export surplus power to other provinces and US states.

Megaprojects, however, are not an economic panacea for the North. They have had disruptive impacts on Aboriginal societies and pose a serious threat to the environment. Past examples in the twentieth century, such as the mercury contamination of the English–Wabigoon River system by a pulp plant at Dryden, Ontario, and the resulting devastating impact on the residents of Grassy Narrows and Whitedog reserves, should alert corporations in the twenty-first century to the dangers of toxic pollution (see Chapter 7 for a more detailed account). Yet, Suncor and Syncrude have been mining the oil sands for decades, and have created a stark industrial landscape with giant tailings ponds. The 'treated' effluence from these ponds is released into the Athabasca River. Indications are that this effluence has negatively affected aquatic life and the health of Aboriginal people at Fort Chipewyan, where the river flows into Lake Athabasca (Kelly et al., 2009; University of Alberta, 2010; CBC News, 2010b).

Another drawback of megaprojects in the North is the high level of economic leakage, thus limiting the impact on local and regional growth. During the construction phase, business opportunities abound but the small northern business sector cannot satisfy the needs of the construction firms, and economic spinoffs generated by the construction are lost to firms in southern Canada. Equipment for the oil sands project is built in southern Canada or in foreign centres.[6] The explanation is simple: the North does not have a manufacturing capacity. Similarly, the North does not have a skilled labour force and companies employ many southern workers. Another form of leakage stems from air commuting, which results in workers and their families spending their wages in their home communities in southern Canada. The net result is that virtually all of the multiplier effects accrue to other regions of Canada. As well, boom towns suffer from social problems—often a combination of single men and high wages (Vignette 5.7)

Megaprojects have two phases: construction and operation. Of the two, the construction phase places the greatest pressure on the impacted community and surrounding area. The construction phrase creates an intense work atmosphere where thousands of workers live in trailers, work long hours, and take home big paycheques. The impacted community is under immense pressure from the constant noise from the equipment and trucks, from the construction dust and air pollution, from the number of new families and their demand for community services, and from the shortage of housing. During construction, the common characteristics of megaprojects are:

- Construction costs alone exceed $1 billion.
- The construction period exceeds two years.
- Substantial short-term employment and business opportunities are generated.
- Outside labour and businesses play a central role because local workers and firms are unable to satisfy demands from the northern megaproject.
- The sheer size of the construction workforce overwhelms existing community services and infrastructure.
- Construction work changes and sometimes damages the local environment.
- The local transportation system is expanded to meet the needs of the megaproject.

A basic question facing the North is that the very nature of the global economy emphasizes corporate profits, not environmental and social responsibilities. Given

---

Vignette 5.7 Booming Fort McMurray

Megaprojects do result in economic growth, but they also have a dark side. Along with the arrival of large numbers of construction workers, social problems arise. Fort McMurray continues to experience 'the price of prosperity'. Back in 2002, Tom Barrett of the *Edmonton Journal* (2002: EJ15) wrote:

> The latest economic boom to sweep through Fort McMurray has brought with it a predictable increase in crime. Crime, crack, and cocaine are problems for a community that plays as hard as its works. Take a lot of disposable income, add a huge population of single men without roots in the community and you can expect some problems, says RCMP Staff Sgt. Scott Stauffer. Money buys cocaine, ecstasy and a host of softer drugs, plus plenty of liquor. Heavy drinking often leads to fights and traffic accidents.

Ten years later, Barrett's comments still hold true. Fort McMurray remains overwhelmed, and the newcomers, by their sheer numbers, place enormous pressure on the community's housing stock, infrastructure, and recreational facilities, thereby forcing radical changes to the community's character and lifestyle.

the fragile nature of the northern environment and the presence of large numbers of Aboriginal peoples who are just finding their footing in the market economy, do these firms have an obligation to encourage social development and to protect the environment? Peter Drucker thought so (Vignette 5.8).

Vignette 5.8 **Corporate Responsibility**

In 1939, Peter Drucker wondered how the vast power held by multinational corporations could be justified within the tradition of Western political philosophy. As a leading social philosopher, Drucker recognized that the corporation was not inherently conservative, such as the state, church, and army. On the contrary, the corporation, but especially the multinational corporation, must constantly transform itself to meet new circumstances. Drucker believed that the corporation was the ideal candidate to lead a rapidly changing modern industrial society. Indeed, if corporations refuse to take responsibilities for solving social and environmental problems besetting the global community, who else will have the power to do so?

Vignette 5.9 **Regional Multiplier**

In neo-classical economic theory, the most common form of economic impact analysis is based on the Keynesian concept of the multiplier effect. The multiplier effect is a measure of the economic impact of a new development, such as a factory or mine, on the local or regional economy. A high regional multiplier translates into a region heading towards economic diversification.

There are three types of impacts: the direct impact of the wages, salaries, and profits of the new development; the indirect impact from payments to regional industries supplying goods and services to the new firm; and the induced impact, which is the increase in payments to retail stores and their regional suppliers brought about by the spending of the new income.

The regional multiplier is expressed mathematically as $1/(1-s)$ where s is the marginal propensity to consume goods and services within the region. Goods and services purchased outside the region represent economic leakage. For example, let us assume that the induced impact is determined by a regional multiplier of 1.5. This multiplier indicates that one-third of every dollar spent on supplies and wages by the owners of the new enterprise occurs within the region. It is calculated from the expression $(1/1-s)$ where $1/(1-0.33) = 1.5$. Arriving at the total annual income impact involves applying the multiplier to the total expenditures from direct and indirect impacts within the region. If this annual amount was $4 million, then by applying the multiplier of 1.5, the indirect impact is $2 million and the total impact is $6 million ($4 million direct impacts and $2 million indirect impacts).

In the case of the Alberta-Pacific pulp project, which began production in 1993, the multiplier was used to calculate the number of anticipated indirect jobs (jobs not directly associated with the project). In this example, the multiplier was assumed to be low—1.2—and the number of jobs in the northern region was calculated as follows: 600 direct jobs X 1.2 = 720 total jobs (120 indirect jobs).

The Era of Megaprojects

After World War II the North's resources became more attractive as supplies elsewhere were either exhausted or insufficient to meet the demand. American firms initiated this new era of resource development in Canada's northern hinterland. Their first huge investment took place in northern Quebec and adjacent areas of Labrador where iron mines, mining towns, a railway, and port facilities were built to supply ore to US steel plants (Table 5.5).

By the twenty-first century, Canada's North had attracted companies from around the world. After American companies, European firms arrived and finally an Asian presence put its stamp on northern resource development. These companies are engaged either directly or indirectly in megaprojects. For example, while Suncor and Syncrude are the dominant players in the oil sands, these huge reserves of oil have attracted the major oil companies, including ExxonMobil (United States); Royal Dutch Shell (Netherlands); BP (United Kingdom); Total SA (France); Chevron Corporation (United States); ConocoPhillips (United States); and Statoil (Norway). These global companies often form Canadian subsidiaries or purchase a share in existing projects, thus spreading risk and sharing profits. A smaller oil field at Norman Wells is owned by Imperial Oil, a subsidiary of Exxon.

The trend towards fewer large commodity companies results in rising concerns by importing companies about supply and prices. In the global mining economy, mergers in 2006 and 2007 left only four major players (Xstrata, BHP, Rio Tinto, and Vale) dominating global coal, iron, and nickel production. To counter this move, China, Japan, and South Korea have encouraged their companies to secure supplies in other countries. For instance, China's bursting economy needs to import increasing amounts of energy to sustain its growth and to satisfy its consumer demand. Accordingly, in 2010, three Chinese companies invested into the oil sands. China also needs raw materials. China imports huge quantities of iron ore through BHP and Vale. One Chinese steel company invested in a new iron mine in northern Quebec to diversify its sources of iron ore. (Vignette 5.10)

Crown Corporations and Hydroelectric Power

With the exception of Nova Scotia, each provincial government has created a Crown corporation designed to develop and market hydroelectric and thermal electrical sites. In this way, provincial Crown corporations play a major role in the development of water resources in northern areas of provinces. Provincial hydroelectric Crown corporations in British Columbia, Manitoba, and Quebec have built huge hydroelectric projects to produce low-cost energy for their consumers and to sell surplus energy to the United States. While the James Bay Project is by far the largest hydroelectric facility in North America and the second largest in the world (after the Three Gorges project in China), other large hydroelectric sites are found in Canada's Subarctic, including Churchill Falls in Labrador, the Manicouagan–Outardes complex on the Quebec North Shore, Sir Adam Beck station on the Niagara River in Ontario, Gordon Shrum Dam on the Peace River in northern British Columbia, and the Columbia River facility in the southern part of British Columbia.

Table 5.5 Megaprojects in the North

Project	Resource	Production Date	Major Market	Ownership	Transportation
Quebec/Labrador iron ore project	iron ore	1954	United States	The Iron Ore Company of Canada	rail and lake carriers
Great Canadian Oil Sands plant	oil sands	1967	United States	Sun Oil Company	pipeline
Syncrude Canada plant	oil sands	1978	United States	Syncrude Canada Ltd	pipeline
Peace River hydro-electric project	electricity	1980	United States	BC Hydro	transmission lines
Northeast Coal Project	coal	1984	Japan	Denison Mines	rail and ocean ship
Norman Wells Oil Project	oil	1985	United States	Esso Resources Canada	pipeline
James Bay hydro-electric project	electricity	1986	United States	Hydro-Québec	transmission lines
Peace River pulp plant	pulp	1990	Japan	Daishowa	rail/ship
Alberta-Pacific forest plant	pulp	1993	Japan	Mitsubishi and Oji Paper Company	rail/ship
Ekati Diamond Project	diamonds	1998	United States	BHP	aircraft
McArthur River Project	uranium	2000	United States	Cameco Corporation	truck
Alliance Gas Pipeline	natural gas	2000	United States	Alliance Pipeline Inc.	pipeline
Suncor Millennium Project	oil sands	2001	United States	Exxon Mobil Corp.; Royal Dutch/ Shell Group; and Petro-Canada	pipeline
Diavik Diamond Project	diamonds	2003	United States	Rio Tinto	aircraft
Voisey's Bay nickel mine	nickel	2005	United States	Inco	ship
Horizon Oil Sands Project	oil sands	2008	United States	Canadian Natural Resources Ltd	pipeline
Snap Lake diamond mine	diamonds	2008	United States	De Beers	aircraft

Vignette 5.10   **China's Strategy of Diversifying Imports**

Wuhan Iron and Steel Corp. (Wisco), China's third-largest steelmaker, seeks to divers-ify its sources of iron ore away from giants such as Vale and Rio Tinto. In 2009, Wisco grabbed the opportunity to invest in the Bloom Lake project, a nearly completed iron ore mine in the Quebec–Labrador Iron Ore Trough. The developer, a relatively small Canadian mining firm called Consolidated Thompson, needed capital to com-plete the mine. As well, Wisco purchased half of Bloom Lake's annual output of eight million tonnes at market rates. The first shipment of iron ore concentrate left Sept Îles on 27 July 2010 for China. With over a 100-year ore supply at Bloom Lake, Wisco has successfully diversified for the long haul.

Sources: Gibben (2009); Consolidated Thompson (2010).

The James Bay Project

In 1971, Quebec Premier Robert Bourassa announced plans to turn the 'wilderness' of northern Quebec into an energy landscape consisting of a series of hydroelectric dams, reservoirs, and power stations. Hydro-Québec was assigned the task of imple-menting this project. The provincial strategy went beyond producing power. Hydro-Québec had two other objectives—to accelerate the growth of Quebec's industrial economy; and, obviously, to make this huge investment financially viable. The first objective called for Hydro-Québec to offer low-cost power to firms locating in Quebec. Second, the financial viability was ensured by the sale of surplus electricity at market prices to utilities in New England, and these profits in turn help pay for the project. In fact, without long-term agreements with American utilities in place, the first phase of the James Bay Project might not have happened. The first site was focused on La Grande River and its lakes. Later, the remaining two large basins in northern Quebec—the Great Whale River Basin and the Nottaway–Eastmain–Rupert River Basin—would be developed.

The first phase, known as La Grande Project, began in 1972, but was halted by legal action launched by the Quebec Cree and Inuit. At that time, Quebec, like other provinces, did not recognize Aboriginal rights to their traditional lands. The Quebec court ruled that the two parties must reach an agreement. In 1975, the James Bay and Northern Quebec Agreement became the law of the land. Construction then moved ahead, and La Grande Project was completed in 1985 at a cost $15 billion.

The Great Whale River Project was to be the second phase. Announced in 1985, the strategy of Hydro-Québec was again to seek long-term contracts for its electricity and to use the initial payments to pay for the construction costs. Maine, New York, and Vermont all negotiated contracts with Hydro-Québec Although hydroelectric power is touted as environmentally friendly energy, environmental organizations, including the Sierra Club and the New York Energy Efficiency Coalition, and the Quebec Cree under the leadership of Chief Matthew Coon Come strongly opposed the development of the Great Whale River Project and launched a public relations campaign in New York claiming that the project would cause great harm to the land and people. In 1992, the state of New York cancelled two contracts with Hydro-Québec totalling $17.6 billion.

These contracts were critical for the project. While the relatively low price of natural gas was the principal factor affecting this decision, environmental concerns related to the flooding of 5,000 km² of land in northern Quebec also played an important role. Ironically, much of the natural gas purchased by thermal electricity plants in New England came from western Canada via the extension of the TransCanada Pipeline network in Ontario to the Duke Pipeline network in New York. In 1994, the Quebec government announced that the Great Whale River Project would not proceed until the demand and price for its electricity improved.

While the Great Whale River Project remains on hold, the Nottaway–Eastmain–Rupert River system to the south came into the spotlight at the beginning of the twenty-first century. A major turning point in relations between the Quebec Cree and the Quebec government sparked interest in the developing all resources in the traditional Cree territory. In 2002, the two parties signed the Paix de Braves Agreement, which called for the development of the natural resources of northern Quebec, a transfer of $3.6 billion in annual payments spread over 50 years to the Cree Regional Authority, and the sharing of the profits from these undertakings. This agreement with the Cree cleared the way for harnessing the Eastmain River and for diverting the Rupert River northward to La Grande Project, where the water will pass through the La Grande power stations (see Vignette 7.9).

Figure 5.13 Eastmain River, Its Reservoir, and the Eastmain-1 Power Station

With the signing of the Paix des Braves in 2002, construction began almost immediately in the Nottaway–Eastmain–Rupert Basin. In 2005 its reservoir was filled, and two years later the Eastmain-1 power station, shown above, was completed.

Source: Hydro-Québec (2006), at: <www.hydroquebec.com/eastmain1/images/batir/images_photos/destination2005_g_6_11.jpg>.

Multinational Corporations and Oil Projects

Development of the vast petroleum deposits in the sedimentary basins of northern Canada has been slow because of the high cost of transporting the product to markets in the United States. As noted earlier, with the rise of China as an economic power, the prospects of marketing petroleum in China have become a possibility. The first proposal focuses on developing a huge natural gas deposit in northeastern BC, transporting it by pipeline to the Pacific coast, and then shipping it by LNG tankers to China. But existing petroleum operations, whether oil or gas, are aimed at the US market. The Alliance gas pipeline, completed in 2000, is the longest gas pipeline in North America and carries natural gas from Fort St John in northeastern BC to Chicago. Other gas pipelines related to the Mackenzie Gas Project and Alaskan reserves, as discussed earlier in this chapter, are also under consideration. Twenty years earlier, oil began flowing from Norman Wells to US markets, primarily in the Midwest and along the Pacific coast.

The Norman Wells Project

With oil prices soaring in the 1970s, Esso Resources Canada saw an opportunity to expand production at its Norman Wells oil field[7] by constructing a pipeline and supplying much of the new production to American markets. Until then, oil from the Norman Wells field supplied the local communities and mines along the Mackenzie River and the two major lakes, Great Bear and Great Slave. This huge construction undertaking, known as the Norman Wells Oil Expansion and Pipeline Project, faced two challenges. First, the project was proposed shortly after the Berger Report and before land claim agreements were settled. Second, the buried pipeline route was in the zone of discontinuous permafrost.

As the first megaproject to take place in the Northwest Territories, this project was caught up in the political undertow caused by the Mackenzie Valley Pipeline Inquiry and its report. While much smaller than the Mackenzie Valley pipeline proposal, the Norman Wells oil pipeline proposal raised many similar concerns to those expressed by opponents to the earlier proposal for a Mackenzie Valley gas pipeline. Ottawa saw the Norman Wells proposal as a first step to creating an energy corridor in the Mackenzie Valley, and as a way to circumvent the Dene resistance to such developments prior to land claim settlements. Rightly or wrongly, the federal government strongly believed that economic development would lead to social improvements in the communities of the Mackenzie Valley.

The Norman Wells Project called for drilling additional wells to increase output from the shallow oil field[6] lying beneath the Mackenzie River and for the construction of a pipeline from Norman Wells to Zama, Alberta, where it would connect with existing lines for transport to southern markets. In 1978, Esso submitted its plans to Ottawa for an environmental review that would lead to the approval of its oil expansion and pipeline proposal. In 1980, the federal Environment and Assessment Review Office recommended that the project proceed, but that construction should be delayed until 1982 to permit further land claim negotiations. Even though no land claim agreements were reached, construction began in 1982. By 1985, the project was completed at a cost of just under a billion dollars. Production from the new wells greatly increased

Figure 5.14 Norman Wells Pipeline Route

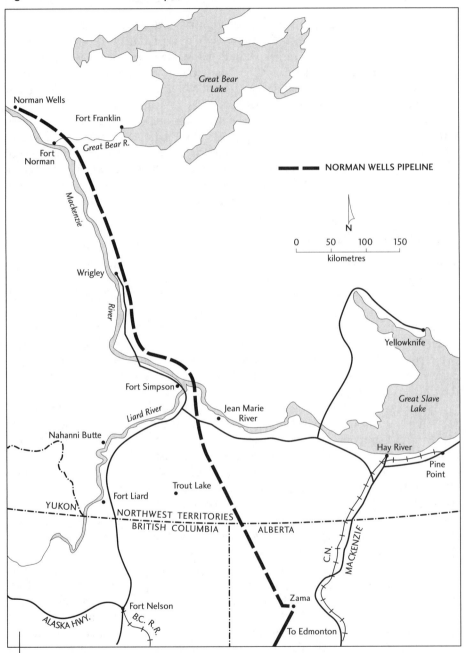

The Norman Wells oil field continues to produce oil, which is carried south by the first completely buried oil pipeline in Canadian permafrost terrain. The pipeline, which began operations in 1985, is viewed as a pilot project for future pipelines in permafrost environments.

Source: Bone (1984: 63). Reproduced with the permission of the Minister of Public Works and Government Services Canada, 2008.

the total Norman Wells output, from 175,000 m³in 1984 to 1.2 million m³ in 1985. Production hit a high in 1989. Since then, production has slowly declined, dropping to 1.0 million m³ in 2008. Prices, of course, have varied from year to year. In 1984, production was worth $20 million while in 2008 it reached $616 million (Table 5.6).

During construction, 200 new oil and water injection wells were drilled, six artificial islands were built to serve as drilling platforms, an oil-gathering system was put in place, and a central processing plant that conditions oil for pipeline transmission was built. The relatively small-diameter (324 mm) pipeline, buried along its entire length, has a capacity of about 5,000 m³/day. It transports oil at near-ground temperature, thereby reducing the potential for frost heave. Three pumping stations are located near Norman Wells, Wrigley, and Fort Simpson. Esso assigned Interprovincial Pipe Lines Ltd (IPL) the task of constructing a buried pipeline (Figure 5.14). The pipe was laid during the winter to minimize damage to the environment. Both Esso and IPL subcontracted the work to other firms, including local companies. Surveying and line cutting took place in the winter months of 1982–3 while the pipe was assembled, welded, and buried in a trench over the course of two winter seasons (1983–4 and 1984–5).

The performance of the Norman Wells oil field is remarkable. Originally seen as a small deposit that might continue to pump oil for 20 years, the field continues to produce relatively large quantities of oil, partly because of the injection of water to ensure a greater percentage of the deposit is captured. How much longer the Norman Wells oil field will remain productive is unknown, but eventually this non-renewable resource will be exhausted.

Alberta Oil Sands

The Alberta oil sands represent one of the largest petroleum deposits in the world (Figures 5.15 and 5.16). In fact, the oil reserves of this deposit rival the size of conventional oil reserves found in the Middle East. According to Alberta Energy (2010):

- Alberta ranks second, after Saudi Arabia, in terms of proven global crude oil reserves.
- In 2008, Alberta's total oil reserves were 171.8 billion barrels, or about 13 per cent of total global oil reserves (1,342 billion barrels). Alberta accounts for an overwhelming majority (more than 95 per cent) of Canada's proven oil reserves.
- The vast majority of Alberta's oil reserves (170.4 billion barrels, or 99 per cent) are found in the oil sands.

Alberta has almost all of Canada's tar sands except for a small area in northwest Saskatchewan. Development of these tar sands involves three steps:

- extracting the bitumen by either open-pit mining or by a steam injection system;
- upgrading the bitumen into a heavy oil;
- refining the heavy oil into commercial products.

Production, whether by open-pit mining or **steam-assisted gravity drainage** (SAGD), takes place in three areas of Alberta: the Athabasca area (40,000 km²) near Fort McMurray, which is the most exploited; the Cold Lake area (22,000 km²); and the

Table 5.6 Production and Value of Norman Wells Petroleum, 1981–2008

Year	Oil Production (000 m³)	$ Value (millions)	$ Value/m³
1984	175	20	114.3
1985	1,148	195	169.9
1990	1,864	248	133.1
1995	1,590	206	129.6
2000	1,536	382	208.3
2005	1,090	425	389.9
2006	1,168	482	412.7
2007	1,110	493	441.1
2008	1,037	639	616.2

Source: NWT Bureau of Statistics, *Statistics Quarterly*. Reprinted by permission of the Government of the Northwest Territories.

Peace River area (8,000 km²). The locations of these three bitumen deposits are shown in Figure 6.4.

Bitumen is a challenge to extract because the oil sands lie beneath an overburden of 1–3 metres of water-logged muskeg, below which is up to 75 metres of clay and barren sand. The underlying oil sands are typically 40–60 metres thick and lie atop relatively flat limestone rock. While some deposits are close to the surface, most lie beneath some 50 metres of overburden (Figure 5.15).

Yet, as a mixture of sand, water, clay, and bitumen buried beneath the ground, the conversion of oil sands into oil is not a simple matter. This conversion takes two tonnes of oil sand to produce one barrel of oil. Clearly, the extraction and processing of oil sands calls for a series of megaprojects, each of which calls for huge investments, a long construction phase, and expensive technology.

Do these oil sands projects represent a class beyond megaprojects? Joseph (2010) makes a case for calling them 'gigaprojects' because of the huge investment and the scale of the projects. With six operating open-pit mines, nine mines under construction, and 17 steam-driven wells, plus more on the drawing board, extracting bitumen provides an economic burst with benefits that spread well beyond Alberta (Alberta, 2011). The economic impact is enormous. With 20 oil sands companies located in Alberta, capital—especially foreign capital—is rushing into the oil sands in record amounts to grab a piece of one of the world's largest oil reserves (Manta, 2011). Canadian labour also benefits. Indeed, as many as 10,000 trades workers from Newfoundland and Labrador regularly commute to Fort McMurray for high-paying jobs. But open-pit mining, the creation of huge tailings ponds, atmospheric pollution, and the release of effluent into the Athabasca River represent an incalculable cost to the environment (see Chapter 6).

In 1967, the world's first oil sands mine was started by Great Canadian Oil Sands (a predecessor company of Suncor). The Syncrude mine followed in 1978 and is now the largest mine by geographic area. In 2003, Shell Canada opened its Albian Sands mine at Muskeg River followed in 2008 by the Horizon mine owned by Canadian Natural

Resources. All open-pit mines have bitumen upgraders that convert the oil sands into synthetic crude oil for shipment to refineries in Canada and the United States.

Back in the 1960s, oil companies were hesitant to tackle the oil sands because they believed that the high cost of converting the oil sands into synthetic crude oil was too great, thus making the viability of oil sands development improbable. Even in 1967, when Great Canadian Oil Sands pioneered extraction, its costs were much higher than the cost of most conventional oil. Over time, mining and processing oil sands resulted in efficiencies and economies of scale that have greatly reduced costs and made Suncor and Syncrude highly profitable. In 2008 Nexen opened its in situ facility at Long Lake, which contains the most advanced technology currently available. As a result, Nexen (2010), by combining steam-assisted gravity drainage with cogeneration and upgrading, expects its operating costs to average about $25/barrel, which is substantially lower than other companies because of the reduced need to purchase natural gas. Even so, producing oil from these projects remains more expensive than conventional oil production in Canada and certainly much more expensive than the cost of producing Middle East oil. As well, the separation process requires large quantities of energy and water, which has both cost and environmental concerns.

In 2008, Alberta's production of bitumen reached an all-time high of 1.3 million barrels per day (bbl/d), and by 2018 bitumen production is expected to soar to 3 million bbl/d (Alberta Energy, 2010).

Approximately 60 per cent of current production is generated by six open-pit mines (Suncor's Millennium and Steepbank mines; Sycrude's Mildred Lake and Aurora mines; Shell's Muskeg River; and Canadian Natural Resources' Horizon mine); three other open-pit mines have been approved. The remainder of oil sands production comes from SAGD operations—10 in the Athabasca oil sands; six in the Cold Lake reserves; and one in the Peace River deposit (Strategy West, 2010). Current upgrading capacity in Alberta is approximately 1,209,000 bbl/d of bitumen with synthetic crude oil output at approximately 1,037,500 bbl/d (Alberta Energy, 2010). Six **upgrader plants** are operating and another seven are proposed (Alberta Energy, 2011). Two operating upgraders are situated in Alberta's **Upgrader Alley**—one in Edmonton (Suncor's Strathcona upgrader) and the other at Fort Saskatchewan (Shell's Scotford upgrader). The remaining four upgraders are near their respective open-pit mines (Syncrude at Mildred Lake; Suncor at Base and Millennium; Nexen at Long Lake; and Canadian Natural Resources at Horizon). Early in the twenty-first century, companies proposing additional oil sands production decided to export diluted bitumen to US heavy oil refineries because of cost advantages. The Alberta government has supported this position and has actively promoted the Keystone XL project in the US. Not wanting to see all new production flow south of the border, however, the Alberta government in 2009 announced its **Bitumen-Royalty-in-Kind program** (BRIK). In doing so, Alberta elected to collect its royalties in kind, i.e., in bitumen, rather than in cash to ensure a sufficient supply of bitumen for proposed upgraders in Alberta (Alberta Treasury Board, 2010).

The oil sands are important to Alberta and Canada for three reasons:

- *Employment.* The oil sands operations draw Canadian workers from across the country. In Atlantic Canada, working in Fort McMurray and leaving the family behind is known as the 'Big Commute' (CBC News, 2007).

- *Capital investment*. Each project requires billions of dollars, with much of this funding coming from foreign sources.
- *Spinoff effects*. Each project requires processing facilities, i.e., upgraders and heavy oil refineries, as well as vast amounts of building supplies, equipment, and specialized products. Companies throughout Canada, especially in western Canada, Ontario, and Quebec, benefit from these spinoff effects.

The driving force for rapid expansion is the ever-growing US demand for a secure supply of oil. The United States imports more oil from Canada than from any other country. Most Canadian oil comes from the oil sands. Declining US oil production and the US concern about politically unstable oil producers in the Middle East drive the American hunger for Canadian oil. More than ever before, America's continental energy policy is based on Alberta, causing US and other foreign multinational firms to invest billions in northern Alberta oil sands projects. The numbers are staggering. In 2008, Suncor allocated $20.6 billion to expand production by 2012 (Suncor, 2008). The National Energy Board (2006: 11) calculated that proposed project investments for the period 2006–15 totalled over Cdn$125 billion. This massive investment will accelerate oil sands production and exports to the US. Currently, oil sands production exceeds one million barrels per day and could reach three million barrels by 2018 (Natural Resources Canada, 2007). Most new production will be exported to the US as diluted bitumen and refined there. Concern has been raised that Alberta is losing the value-added component of its oil sands industry by not doing more refining in the province (see Vignette 5.11).

While the US stranglehold on Alberta oil is unlikely to change in the short run, Canada's oil sands have attracted increasing attention from foreign oil companies, especially Asian companies seeking to satisfy growing demand in their countries and

---

Vignette 5.11 Keystone Pipeline and Its Proposed Expansion

A $12 billion pipeline completed in June 2010 transports Athabasca diluted bitumen to heavy oil refineries in Illinois. Its proposed extension (Keystone XL) to Gulf states will result in more bitumen processed in the United States than in Canada. The source of bitumen for Keystone XL will come from oil sands projects planned or now under construction. However, as of October 2011, Keystone Pipeline XL has not yet been approved by the US State Department. The delay is due to opposition from the US Environmental Protection Agency, which is concerned about Alberta's 'dirty oil' having adverse impacts on the US environment as well as flying in the face of the US policy of clean energy. On the other hand, oil refineries along the Gulf coast of Texas are calling for quick approval so that the supply of Alberta bitumen will keep their refineries operating at full capacity. Traditional sources of heavy oil from Venezuela and Mexico are declining, causing a shortage of supply at these refineries and contributing to the price spread between West Texas Intermediate (the benchmark for North American crude oil pricing) and North Sea Brent (the benchmark for European crude oil pricing).

Source: TransCanada (2010).

Figure 5.15   Oil Sands Development in the Fort McMurray Area

The Syncrude oil sands complex north of Fort McMurray turns oil sands into oil. Huge quantities of water and energy are required; toxic waters from the upgraders are placed in tailings ponds and later released into the Athabasca River, which may be affecting the aquatic life and the health of residents of Fort Chipewyan, many of whom eat fish on a daily basis.

Source: Ron Garnett/AirScapes.ca.

secure equity stakes in oil sands projects. These Asian companies are pursuing a strategy of buying resources as a hedge against inflation and devaluation of the US currency. As of 2011, eight major Asian investments had taken place in the oil sands (Fekete, 2010; US Energy Information Administration, 2010):

- In 2005, Sinopec, China's second largest domestic oil producer, obtained a 9 per cent interest in Syncrude; two years later, it purchased a 50 per cent stake in the Northern Lights oil sands project under development by the French multinational, Total.
- In 2006, Korea National Oil Corporation purchased the Blackgold bitumen deposit from Newmont.
- In 2007, Chinese National Petroleum Company secured exploration rights for a 260-acre tract in the Alberta's oil sands.
- In the same year, China National Offshore Oil Corporation gained a position in the Christina Lake oil sands project.
- In 2010, PetroChina purchased a 60 per cent ownership in the MacKay River and Dover projects of Athabasca Oil Sands Corporation.
- A second investment in 2010 was undertaken by China Investment Corporation (CIC), which joined Penn West Energy Trust to develop the trust's oil sands in the Peace River region.

Figure 5.16   Location of Oil Sands in Geological Cross-Section

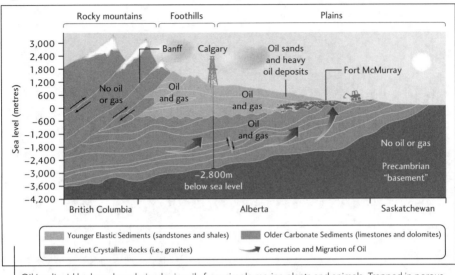

Oil is a liquid hydrocarbon derived primarily from simple marine plants and animals. Trapped in porous sedimentary rocks, oil moved more than 100 km eastward and upward from its origin. Eventually, the oil reached and saturated large areas of sandstone at and just below the surface of what is now northern Alberta and Saskatchewan. Plans to exploit the deeper Saskatchewan deposits call for in-situ extraction techniques known as steam-assisted gravity drainage.

Sources: Oil Sands InfoMine (2008); Centre for Energy (2007). Reprinted by permission of the Canadian Centre for Energy Information, at: <www.centreforenergy.com>.

- A third investment in 2010 was made by the Thai oil and gas company (PTT), which purchased a 40 per cent share in the Norwegian energy giant Statoil's Kai Kos Dehseh Project. Three years earlier, Statoil purchased this oil sands site from North American Oil Sands Corporation.
- In 2011, China National Offshore Oil Company (CNOOC) purchased a minority position in the Long Lake project by replacing Opti Canada for $2.1 billion.

Conclusion

Two questions were posed at the beginning of this chapter:

- Can resource development in the North lead to economic stability and diversification?
- Is sustainable development possible in an economy based on non-renewable resources?

On the surface, given the nature of non-renewable resources and the depressed situation for the forest industry, the answer to both questions is no. Yet, the continued evolution of the resource economy towards greater involvement of Aboriginal peoples and their institutions might change the outcome.

With the North as a resource hinterland, northern economic stability and diversity are unlikely to follow the pattern that emerged in southern Canada. Yet, a northern version is possible. Back in 1939, Peter Drucker wondered how the vast power held by multinational corporations could be justified within the tradition of Western political philosophy (Vignette 5.8). Part of the answer may lie in joint ventures such as those between Aboriginal organizations and multinational corporations. Within this business structure, economic and social development could then coalesce, leaving governments to play their traditional supporting role. As the James Bay and Northern Quebec Agreement illustrates, times change, paradigms shift, and old ways are left behind. The Paix des Braves, a bold step beyond the JBNQA, is but one example where a crisis atmosphere provided an arena for new thinking and an agreed balancing of interests. Perhaps the Report of the Joint Review Panel (2010) examining the Mackenzie Gas Project (see Table 5.7 and Figure 5.17) is an initial step in that direction—even though the MGP appears dead in the water with one of the key partners, Shell, backing away from the project proposal in the summer of 2011. The apparent failure of the MGP

Table 5.7   Mackenzie Gas Project: Summary of Five Key Sustainability Issues

1. *Cumulative Impacts on the Biophysical Environment*
   Impacts on the longer-term resilience of ecosystems and what they provide (as recognized in special conservation areas, protected areas, and land-use plans) and on the wildlife harvesting and other traditional land-based cultural and livelihood activities that they support during Project life and beyond.

2. *Cumulative Impacts on the Human Environment*
   Impacts on community economic and socio-cultural well-being during the stages of the Project life and beyond, including vulnerability to cumulative impacts on community economic and socio-cultural well-being, and vulnerability to boom-and-bust impacts.

3. *Equity Impacts*
   The distribution of positive and negative impacts (especially concerning access to opportunities and resources, revenue flows, and exposure to burdens and risks) within and among communities, and between men and women, youth and Elders, and present and future generations, including the impacts of the anticipated use of hydrocarbon resources (upstream and downstream impacts of product life cycle from gas exploration to end use of gas and greenhouse gas [GHG] loadings).

4. *Legacy and Bridging*
   Impacts from use of the Project and associated revenues and other impacts as a bridge to more sustainable livelihoods and generally more sustainable futures for the Beaufort Delta and Mackenzie Valley regions. They also include use of the Project and associated activities for building capacities of individuals, communities, agencies and other organizations to manage impacts, and to obtain and retain benefits from Project-related opportunities.

5. *Cumulative Impacts Management and Preparedness*
   The preparedness of government agencies and other responsible authorities to manage the cumulative impacts of the Project and associated activities in a way that ensures lasting, multiple, mutually reinforcing gains, including their capacity and preparedness to apply, monitor, enforce, and adjust necessary terms and conditions. They also include carrying out the design and delivery of impact mitigation or enhancement programs, planning and management for acceptable development scale and pacing, and dealing with uncertainties and surprises, positive and negative.

Source: Adapted from Report of the Joint Review Panel (2009: 589, Table 19-2).

Figure 5.17   Proposed Mackenzie Gas Pipeline Route

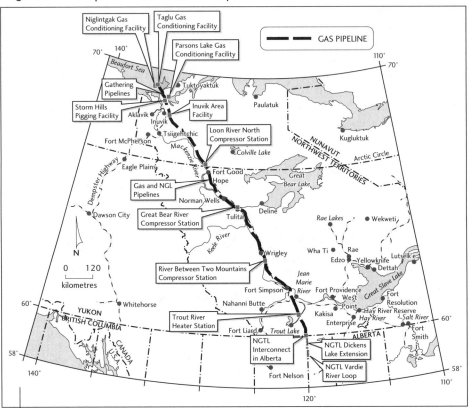

Source: Mackenzie Gas Project (2007). Graphic courtesy of the Mackenzie Gas Project, Imperial Oil.

indicates the danger of relying on megaprojects to move the North towards the goals of economic stability and diversity and sustainable development.

That sustainable development might be possible in a resource-based economy seems oxymoronic. Innis, Watkins, and Hayter and Barnes noted the downside of non-renewable resource-based economies—once the resources are exhausted, the land and people are worse off than before. Yet, if sustainable development meets the needs of the present without compromising future generations to meet their own needs, perhaps even non-renewable resources can lead to a modified version of sustainability. In the case of non-renewable resources such as oil and gas, sustainable development requires conversion of a portion of profits from those resources into other assets that can provide lasting benefit. In other words, future generations may not be able to produce oil and gas, but they should still benefit from today's use of those resources. Public trust funds, as established by Alberta and Norway, are examples of conversion of finite resources into lasting capital assets, although the government of Norway takes a much more active role in the marketplace. Beyond using oil and gas revenues to achieve sustainable development, even more important, especially in the Canadian

North, is the human resource, especially the creation of a better-educated society and a more highly skilled workforce. How critical is the human resource issue to achieving sustainability? Clearly, human resources are a critical variable. As Greg Poelzer (2009: 427) has noted, education that bridges both Aboriginal and non-Aboriginal worlds forms a critical foundation for a sustainable North: 'Education is key to facing the complex dualities, enormous challenges and tremendous opportunities in the contemporary Canadian North.'

For the long term, focusing on the basic social building blocks, such as education, may well be a more reliable route to the goal of sustainable development than the uncertain, piecemeal approach of mega resource projects.

Challenge Questions

1. Given the spread of subsurface resources across the North, can you identify a relationship to the North's geological structure?
2. What structural weakness exists in the northern economy that holds it back from diversifying like other Canadian regions?
3. Does the establishment of a northern diamond processing industry in Yellowknife and Sudbury signal a new approach by multinational companies to ensure northern benefits and economic diversification, or is this processing industry a product of the intervention of the state?
4. Whether you take the oil sands with reserves lasting at least 100 years or the Meadowbank gold mine with reserves lasting less than 20 years as an example, can you make a case that the 'staple trap' represents the ultimate 'underdeveloped' future, or is this 'trap' concept simply a theoretical outcome found in Marxist ideology?
5. In the early 1990s, the economy of the Northwest Territories depended heavily on gold mining. With the closure of its gold mines, diamond mining replaced gold. Do you think that this 'replacement' of one non-renewable resource for another represents a northern version of 'sustainable development'?
6. In the late twentieth century, the James Bay Project was bitterly opposed by Quebec Cree. How did the twenty-first-century version of the James Bay Project take on a new life and win their active support?
7. Is there a danger in putting too much faith in resource megaprojects to move the North to sustainable development?
8. Given the predictions of a longer ice-free navigation season in the Beaufort Sea due to global warming, should the companies behind Mackenzie Gas Project take a hard look at shipping gas by LNG tankers?
9. Peter Drucker believed that the corporation, rather than the state, was the ideal candidate for the leadership of a rapidly changing modern industrial society. Have resource companies or Crown corporations best demonstrated such leadership?
10. Has the Alberta government shot itself in the foot by supporting the Keystone XL pipeline project? After all, this pipeline will have two effects: transporting more and more bitumen to heavy oil refineries in the United States; diminishing the prospects for new upgraders in Alberta.

Notes

1. Following World War II, the rich Mesabi Iron Range in Minnesota was near exhaustion, and US iron and steel plants began to look abroad for alternative sources. Iron ore deposits in northern Quebec and Labrador were close at hand but required enormous investment in infrastructure. The previously worthless ore bodies of the Labrador Trough, known for over a half-century, suddenly became a valued resource. By 1947, plans were laid by six American iron and steel companies for large-scale open-pit mining around Knob Lake in northern Quebec and at other locations in the region. Several mining towns were built, including Gagnon and Schefferville in Quebec and Labrador City and Wabush in Labrador. In 1954, Schefferville was connected by a 571-km railway to the port of Sept-Îles on the north shore of the St Lawrence River. Iron concentrate was shipped by rail to Sept-Îles and then by ship from Sept-Îles to the iron and steel plants in Ohio and Pennsylvania. Boom conditions took hold until the 1980s, when a downturn in the demand for US steel resulted in a sharp drop in iron ore shipments and lower-priced foreign steel from Asian firms began to displace American steel. Several mines closed, and the result was a depressed northern mining landscape of abandoned operations.

2. Under the terms of Confederation, the federal government had no responsibility for highway construction in the provinces. Ottawa did become involved in railway-building in order to unite the country. After World War I, political pressure from western farmers caused Ottawa to build the Hudson Bay Railway to Churchill, Manitoba. Completed in 1929, it was designed to provide an alternative route to European markets for western grain. Prairie farmers believed that the shorter Hudson Bay rail route to their major European customers would reduce their total shipping costs. The new railway never became an important export route for grain because of the high cost of marine insurance in the ice-ridden waters of Hudson Strait and Hudson Bay. Instead, the Hudson Bay Railway was instrumental in unlocking the mineral, forest, hydro, and fishing resources of northern Manitoba.

    By the late 1950s, Ottawa began to play a role in northern transportation, thus abandoning its laissez-faire policy. The reason was straightforward: the federal government saw northern development as a way to strengthen the nation and to ensure Canadian sovereignty. With Prime Minister Diefenbaker's Roads to Resources program (see Chapter 3) for northern areas of the provinces and a similar program in the territories, roads were built to help private companies reach world markets. One example was the road built to connect the asbestos mine at Cassiar (now closed) in northern British Columbia with the port of Stewart. Other highways were constructed to provide major northern centres with a land connection to southern Canada. For example, the Mackenzie Highway from Edmonton to Hay River was extended to Yellowknife. In all, from 1959 to 1970, over 6,000 km of new roads were built under this program at a cost of $145 million (Gilchrist, 1988: 1877).

3. The Klondike gold rush of 1897–8 was the first major mining operation in the North but was the opposite of megaprojects, which are controlled by corporations. The Klondike gold rush, on the other hand, drew thousands of individual prospectors into the North seeking the gold nuggets found in the streambeds of the Klondike River and its tributaries. Placer mining involved the recovery of auriferous deposits found in streambeds by simple and inexpensive technology called panning or surface sluicing. This technology enabled seasoned prospectors and greenhorn amateurs to recover the most accessible nuggets and fine sand-like gold particles from tributaries of the Klondike River. Within a few years, placer gold became harder and harder to find. Gold was buried in frozen ground well below the surface but individuals had neither the capital nor the organization necessary to mine this

gold. This type of mining was much more capital-intensive, requiring expensive dredging and hydraulic equipment along with separate, highly organized water-supply systems and electric power supplies. Large companies gradually took over the gold-mining industry in Yukon. The Yukon Gold Corporation was the largest of these operations. To supply its mining sites on Bonanza and Eldorado creeks with water and electrical power, it built a hydroelectric dam and a 100-km-long water distribution system (Rae, 1968: 99–100). This mining was limited to the summer months when the frozen ground (permafrost) could be thawed and the gravel sorted. During the long summer days, the dredges operated around the clock, although the existence of permafrost made hydraulic operations more difficult and time-consuming.

4. Funded by the Petroleum Incentive Program (PIP), drilling for petroleum in frontier areas like the Beaufort Sea and the Sverdrup Basin by Canadian firms was a by-product of the National Energy Program (NEP). More than 200 wells were drilled in the Mackenzie Delta–Beaufort Sea region. Approximately one-third of these wells were drilled from ships or floating platforms. By the early 1980s, there were nearly 200 billion m³ of recoverable natural gas and around 120 million m³ of recoverable crude oil (Nassichuk, 1987: 279). The potential reserves are thought to exceed the proven reserves by perhaps as much as 20 times (Procter et al., 1984). In 1984, the discovery of the Amauligak 'elephant' field gave some indication of the potential size of the Beaufort Sea hydrocarbon deposit. This single discovery doubled the proven oil reserves of the Beaufort Sea deposit. Two years after the discovery of the Amauligak field, a token shipment was sent by tanker to Japan. The significance of this shipment was twofold: that oil could be produced commercially from the Beaufort Sea deposit, and that an Arctic tanker route could be used to ship its oil to world markets.

5. Other large hydroelectric facilities were also constructed in the 1950s and 1960s, including Churchill Falls in Labrador, the Manicouagan–Outardes complex on the Quebec North Shore, Sir Adam Beck station on the Niagara River in Ontario, W.A.C. Bennett Dam on the Peace River in northern British Columbia, and the Columbia River dams in the southern part of British Columbia.

6. Three oil-drilling rigs specially outfitted for the Mackenzie Delta in 2001 cost a total of $50 million, five times more than rigs used in southern Canada. They were manufactured in Edmonton and the nearby town of Nisku. *National Post*, 15 Feb. 2001, C7.

7. The Norman Wells oil field, discovered in 1920, has proven oil reserves of around 100 million cubic metres of light grade oil (Esso, 1980). A small refinery was erected at the site and oil was produced to meet the needs of the residents of the Mackenzie Valley. In 1925, operations ceased because the demand for oil was too small. In 1932, production recommenced because the mining operation at Port Radium on Great Bear Lake needed fuel oil. In 1936, the gold mine at Yellowknife also required fuel oil. Given the slow growth of a market for oil in the Mackenzie Basin, production increased very slowly until, for a short time during World War II, the Canol Project (see Chapter 3) brought oil from Norman Wells by pipeline to Whitehorse and then on to Fairbanks, Alaska. After the war, the refinery at Whitehorse was closed, the new wells at Norman Wells were capped, and the pipeline was abandoned. By 1947, the pipeline, pumping equipment, and support vehicles were sold as surplus war assets. Imperial Oil purchased the refinery and moved it to Edmonton, where it processed oil from the newly discovered Leduc field in central Alberta. Norman Wells returned to supplying customers in the Mackenzie Valley. From 1946 on, the local market grew as Indians and Métis moved into settlements where their houses were heated by fuel oil. The demand for fuel oil increased as community infrastructures grew and more mines appeared.

References and Selected Reading

Abele, Frances, Thomas J. Courchene, F. Leslie Seidle, and France St-Hilaire. 2009. *Northern Exposure: Peoples, Powers and Prospects in Canada's North*. The Institute for Research on Public Policy, vol. 4. Quebec City: Marquis Book Printing.

Alberta. 2011. 'Oil Sands Projects in the Regional Municipality of Wood Buffalo', *Labour Market Information* (Feb.). At: <www.woodbuffalo.net/linksFACTSOG.html>.

Alberta Energy. 2010. 'Fact and Statistics', 16 Aug. At: <www.energy.alberta.ca/OilSands/791.asp>.

———. 2011. *Alberta's Oil Sands Projects and Upgraders*, Jan. At: <www.energy.alberta.ca/LandAccess/pdfs/OilSands_Projects.pdf>.

Alberta Treasury Board. 2010. *Responsible Actions: A Plan for Alberta's Oil Sands*. At: <www.treasuryboard.gov.ab.ca/ResponsibleActions.cfm>.

Antwerp Facets. 2010. 'Financial Crisis Shuffles Rankings of Diamond Producers', 18 Aug. At: <www.antwerpfacetsonline.be/financial-crisis-shuffles-rankings-diamond-producers>.

Barrett, Tom. 2002. 'Big Salaries, Boredom Lead to Trouble', *Edmonton Journal* (Special Fort McMurray Economic Report), 10 May, EJ15.

Bell, Jim. 2010. 'Mine's Alchemy Turns Nunavut Poverty into Hope', *Nunatsiaqonline*, 20 June. At: <www.nunatsiaqonline.ca/stories/article/98789_agnico-eagles_alchemy_turns_nunavut_poverty_into_hope/>.

Berger, Thomas R. 2006. *The Nunavut Project: Nunavut Land Claims Agreement. Implementation Contract Negotiations for Second Planning Period 2003–2013*. Conciliator's Final Report. Ottawa: Indian and Northern Affairs Canada. At: <www.ainc-inac.gc.ca/al/ldc/ccl/fagr/nuna/lca/nlc-eng.asp>.

Bone, Robert M. 1984. *The DIAND Norman Wells Socio-economic Monitoring Program*. Report 9–84. Ottawa: Department of Indian Affairs and Northern Development.

———. 1998. 'Resource Towns', *Cahiers de Géographie du Québec* 42, 116: 249–59.

Bouw, Brenda. 2010. 'A Eureka Moment for a Stubborn Prospector', *The Globe and Mail*, 16 Aug. At: <www.theglobeandmail.com/report-on-business/industry-news/energy-and-resources/eureka-moment-for-stubborn-prospector/article1675008/>.

Bradshaw, Ben, and Jason Prno. 2007. 'Program Evaluation in a Northern, Aboriginal Setting: Assessing Impact and Benefit Agreements', presentation at the annual meeting of the Canadian Association of Geographers. 30 May.

Byrd, Craig. 2006. *Diamonds: Still Shining Brightly for Canada's North*. Ottawa: Statistics Canada, Catalogue no. 65-507-MIE, No. 007. At: <www.statcan.ca/english/research/65-507-MIE/65-507-MIE2006007.pdf>.

Canada Nuclear Energy Commission. 2009. 'Uranium Mines and Mills in Canada', 1 Aug. At: <www.cnsc-ccsn.gc.ca/eng/about/regulated/minesmills/>.

Canadian Coalition for Nuclear Responsibility. 2005. Map of Uranium Mining Activities in Canada, from 'Uranium Discussion Guide'. At: <www.ccnr.org/uranium_map.html>.

CBC News. 2007. 'The Big Commute', 29 Oct. At: <www.cbc.ca/nl/features/bigcommute/>.

———. 2010a. 'Vale Ramping Up Long Harbour Work', 28 June. At:<www.cbc.ca/canada/newfoundland-labrador/story/2010/06/28/vale-long-harbour.html>.

———. 2010b. 'Oil Sands Poisoning Fish, say scientists, fishermen' 16 Sept. At: <www.cbc.ca/canada/calgary/story/2010/09/16/edmonton-oilsands-deformed-fish.html>.

Centre for Energy. 2007. 'How Are Oilsands and Heavy Oil Formed?' At: <www.centreforenergy.com/generator.asp?xml=/silos/ong/oilsands/oilsandsAndHeavyOilOverview03XML.asp&template=1,1,1>.

Clapp, R.A. 1999. 'The Resource Cycle in Forest and Fishing', Canadian Geographer 42, 2: 129–44.

Consolidated Thompson. 2010. 'News and Media', 9 Sept. At:<www.consolidatedthompson.com/s/NewsReleases.asp>.

De Beers Canada. 2007. 'Snap Lake: Project Development—Phase One'. At: <www.debeerscanada.com/files_2/snap_lake/snaplake-development_photo-enlarge_a.html>.

Diavik Diamond Mines Inc. 2007. 'Diavik Diamond Mine'. At: <www.diavik.ca/loc.htm>.

Drucker, Peter F. 1939. *The End of Economic Man: A Study of the New Totalitarianism*. New York: John Day.

———. 1993. *Post-Capitalist Society*. New York: HarperBusiness.

Duerden, Frank. 1992. 'A Critical Look at Sustainable Development in the Canadian North', *Arctic* 45, 3: 219–25.

Duffy, Patrick. 1981. *Norman Wells Oilfield Development and Pipeline Project: Report of the Environmental Assessment Panel*. Ottawa: Federal Environmental Assessment Review Office.

Esso. 1980. *Norman Wells Oilfield Expansion: Development Plan*. Calgary: Esso Resources Canada Ltd.

Fekete, Jason. 2010. 'China Pumps More Investment into Alberta Oilsands', *Vancouver Sun*, 13 May. At: <www.vancouversun.com/business/China+pumps+more+investment+into+Alberta+oilsands/3025041/story.html>.

Fenge, Terry. 2009. 'Economic Development in Northern Canada: Challenges and Opportunities', in Abele et al. (2009: 375–93).

ForestTalk.com. 2010. 'With New Labour Agreement, Conifex Sawmill Can Reopen in Mackenzie, B.C.', 20 Sept. At: <foresttalk.com/index.php/2010/09/20/with-new-labour-agreement-conifex-sawmill-can-reopen-in-mackenzie-b->.

Gagnon, Jeanne. 2010. 'More Quality Rough Diamonds Needed', *Northern News Service On-Line*. 16 June At: <www.nnsl.com/northern-news-services/stories/papers/jun16_10d.html>.

Gautier, Donald L., et al. 2009. 'Assessment of Undiscovered Oil and Gas in the Arctic', *Science* 324, 5931: 1175–9

Geological Survey of Canada. 2001. *Pipeline–Permafrost Interaction: Norman Wells Pipeline Research*. At: <sts.gsc.nrcan.gc.ca/permafrost/pipeline.htm>.

———. 2007. 'Norman Wells Pipeline Research'. At: <gsc.nrcan.gc.ca/permafrost/pipeline_e.php>.

Gibben, Robert. 2009. 'China Makes Strategic Investment in Quebec Iron Ore', *Financial Post*, 7 Sept. At: <www.financialpost.com/story.html?id=1969515>.

Gilchrist, C.W. 1988. 'Roads and Highways', in James H. Marsh, ed., *The Canadian Encyclopedia*, 2nd edn. Edmonton: Hurtig, 1876–7.

Greenwood, John. 2007. 'Surging Prices Separate Wheat from the Chaff', *National Post*, 15 June, FP1.

Hasselback, Drew. 2002. 'Voisey's Bay Deal Is Done, Next Comes Big Writedown', *National Post*, 12 June, FP7.

Hayter, Roger, and Trevor J. Barnes. 2001. 'Canada's Resource Economy', *Canadian Geographer* 45, 1: 36–41.

Hoefer, Tom. 2009 'Diamond Mining in the Northwest Territories: An Industry Perspective on Making the Most of Northern Resource Development', in Abele et al. (2009: 395–414).

Huston, Robert, and Ronnie Yu. 2005. *The Alaska Gas Pipeline Challenge: Getting It Right*, Report of the Centre for Applied Business Research in Energy and the Environment. At: <www.business.ualberta.ca/cabree/pdf/Completed%20Cabree%20Projects/Huston,Yu-alaskadoc15.pdf>.

Hydro-Québec. 2005. 'La Grande Rivière'. At: <www.hydroquebec.com/generation/hydroelectric/la_grande/index.html>.

———. 2006. 'Eastmain-1: Stage 3: Cofferdams and Dam'. At: <www.hydroquebec.com/eastmain1/en/batir/etapes_photos.html?list=4>.

———. 2007a. 'James Bay Project: Geographic Location'. At: <www.hydroquebec.com/visit/virtual_visit/>.

———. 2007b. 'Construction Projects in Quebec'. At: <www.hydroquebec.com/eastmain1/en/batir/fiche_6.html>.

Innis, Harold A. 1930. *The Fur Trade in Canada*. Toronto: University of Toronto Press.

Joint Review Panel. 2009. *Foundation for a Sustainable Northern Future: Report of the Joint Review Panel for the Mackenzie Gas Project*, Dec. At: <www.ngps.nt.ca/report.html>.

Joseph, Chris. 2010. 'The Tar Sands of Alberta: Exploring the Gigaproject Concept', presentation at Prairie Summit Regina for the Canadian Association of Geographers, 4 June.

Kelly, E.N., J.W. Short, D.W. Schindler, P.V. Hodson, M. Ma, A.K. Kwan, and B.L. Fortin. 2009. 'Oil Sands Development Contributes Polycyclic Aromatic Compounds to the Athabasca River and Its Tributaries', *Proceedings, National Academy of Sciences* 106: 22346–51.

Krugel, Lauren, 2010. 'NEB Decision on Mackenzie Pipeline Likely Coming in November at the Earliest', *Winnipeg Free Press*, 30 Sept. At: <www.winnipegfreepress.com/business/breakingnews/neb-decision-on-mackenzie-pipeline-likely-coming-in-november-at-the-earliest-104088244.html>.

Légaré, André. 2000. 'La Nunavut Tunngavik Inc.: Un examen de ses activités et de sa structure administrative', *Études/Inuit/Studies* 24, 1: 197–224.

Macdonald, Cindy. 2010. 'Mackenzie Pulp Mill Back in Business', *Pulp & Paper Canada*, Sept. At: <www.pulpandpapercanada.com/issues/story.aspx?aid=1000387079&type=Print%20Archives>.

Mackenzie Gas Project. 2005. *Environmental Impact Statement (EIS) in Brief*. At: <www.mackenziegasproject.com/theProject/regulatoryProcess/EISInBrief/EISInBrief.html>.

———. 2007. 'Mackenzie Gas Project Overview'. At: <www.mackenziegasproject.com/theProject/overview/index.html>.

Manitoba Hydro. 2007. 'Jenpeg Generating Station'. At: <www.hydro.mb.ca/corporate/facilities/gs_jenpeg.shtml>.

———. 2010. 'Conawapa Generating Station'. At: <www.hydro.mb.ca/projects/conawapa.shtml?WT.mc_id=2608>.

Manta. 2011. 'Oil Sands Companies in Alberta', Mar. At: <www.manta.com/world/North+America/Canada/Alberta/oil_sand_mining--E313708D/>.

Nassichuk, W.W. 1987. 'Forty Years of Northern Non-Renewable Natural Resource Development', *Arctic* 40, 4: 274–84.

National Energy Board. 2006. *Canada Oil Sands Opportunities and Challenges to 2015*. At: <www.neb.gc.ca/clf-nsi/rnrgynfmtn/nrgyrprt/lsnd/lsnd-eng.html>.

Natural Resources Canada. 2007. 'Canada #1 Supplier of Oil, Natural Gas to US'. At: <www.rncan-nrcan.gc.ca/media/newsreleases/2007/200732a_e.htm>.

———. 2010a. *The State of Canada's Forests: Annual Report 2010*, 21 Sept. At: <canadaforests.nrcan.gc.ca/rpt>.

———. 2010b. *Mineral Production of Canada, by Provinces and Territories*. 1 Nov. At: <mmsd.mms.nrcan.gc.ca/stat-stat/prod-prod/ann-ann-eng.aspx>.

Nexen. 2010. 'Long Lake Phase One'. At: <www.nexeninc.com/en/Operations/OilSands/LongLake/PhaseOne.aspxSAGD>.

Northern Gas Pipelines. 2005. Maps Archive. At: <www.arcticgaspipeline.com/Reference/Maps/APGmap7-02.jpg>.

Northwest Territories (NWT). 2007. *Diamond Facts: 2006 Diamond Industry Report*. At: <www.iti.gov.nt.ca/diamond/pdf/diamondfacts_2006.pdf>.

Northwest Territories (NWT) Bureau of Statistics. 2010a. 'Value of Mineral Shipments, NWT', *Statistics Quarterly*, Mar. At: <www.stats.gov.nt.ca/publications/statistics-quarterly/sqmar2010.pdf>.

———. 2010b. *Statistics Quarterly* 32, 1. Yellowknife: Bureau of Statistics.

Oil Sands InfoMine. 2008. *State of the Industry Review*. At: <oilsands.infomine.com/commodities/soir/oilsands/>.

Poelzer, Greg. 2009. 'Education: A Critical Foundation for a Sustainable North', in Abele et al. (2009: 427–65).

Procter, R.M., G.C. Taylor, and J.A. Wade. 1984. *Oil and Natural Gas Resources of Canada, 1983*. Geological Survey of Canada Papers 83–31. Ottawa: Minister of Supply and Services.

Quebec. 2011. *Plan North*, 6 July. At: <www.plannord.gouv.qc.ca/english/messages/index.asp>.

Rae, K.J. 1968. *The Political Economy of the Canadian North: An Interpretation of the Course of Development in the Northern Territories of Canada to the Early 1960s*. Toronto: University of Toronto Press.

Selleck, Lee. 2009. 'Mackenzie Pipeline Gets Green Light from Panel', CBC News North, 30 Dec. At: <www.cbc.ca/news/canada/north/story/2009/12/30/jrp-mackenzie-pipeline.html>.

Slocombe, D. Scott. 2000. 'Resources, People and Places: Resource and Environmental Geography in Canada', *Canadian Geographer* 44, 1: 56–66.

———. 2008. *Canadian International Merchandise Trade*. Catalogue no. 65-001-XIB. At: <www.statcan.ca/english/freepub/65-001-XIB/65-001-XIB2007012.pdf>.

Sudol, Stan. 2011. 'Rails to the Ring of Fire', 30 May. At: <www.thestar.com/printarticle/998685>.

Suncor Energy. 2008. At: <www.suncor.com/start.aspx>.

Strategy West. 2010. *Existing and Proposed Commercial Oil Sands Projects*, Sept. At: <www.strategywest. com/downloads/StratWest_OSProjects_201009.pdf>.

Thompson, John. 2011. 'Victoria Gold Proposes New Mine', *Yukon News*. 4 Jan. At:<yukon-news. com/business/21162/>.

TransCanada. 2010. 'Keystone Pipeline Starts Deliveries to U.S. Midwest', 30 June. At: <www .marketwire.com/press-release/Keystone-Pipeline-Starts-Deliveries-to-US-Midwest-TSX>.

University of Alberta. 2010. 'Oilsands Mining and Processing Are Polluting the Athabasca River, Research Finds', *Science Daily*, 30 Aug. At: <www.sciencedaily.com/releases/2010/08/ 100830152536.htm#>.

US Energy Information Administration. 2010. 'Canada: Oil'. At: <www.eia.doe.gov/cabs/canada/ Oil.html>.

Vanderklippe, Nathan. 2010. 'The North Scraps the Bottom', *The Globe and Mail*, 9 Apr., B9.

Watkins, Melville H. 1963. 'A Staple Theory of Economic Growth', *Canadian Journal of Economics and Political Science* 29: 160–9.

———. 1977. 'The Staple Theory Revisited', *Journal of Canadian Studies* 12, 5: 83–95.

Weber, Bob. 2010. 'Panel Assails Ottawa's Changes to Mackenzie Pipeline Report', *The Globe and Mail*, 14 Oct., B8.

Welch, Michael J. 2010. 'Raglan, 2010', 18 Mar. At: <www.infomine.com/minesite/minesite. asp?site=raglan>.

Wellstead, Adam. 2007. 'The (Post) Staples Economy and the (Post) Staples State', *Canadian Political Science Review* 1, 1: 8–25.

Wikipedia. 2007. 'Nelson River Hydroelectric Project'. At: <en.wikipedia.org/wiki/ Nelson_River_Hydroelectric_Project>.

Wuskwatim Power Limited Partnership. 2010. 'About the Partnership'. At: <www.wuskwatim.ca/ partnership.html>.

Yukon Bureau of Statistics. 2009. *The Yukon Statistical Review, 2008*. Whitehorse: Yukon Government Executive Council Office.

6

Environmental Impact of Resource Projects

Setting the Stage

Resource projects impact the environment. The North, with its fragile environment, is a particularly vulnerable region. In its role as a provider of energy and raw materials to global markets, the Canadian North has seen harsh impacts on its landscape from industrial projects. As Figure 6.1 demonstrates, various forms of industrial damage and pollution are found in many areas of the North. Within this region of Canada, however, the vast majority of resource projects are found in the Subarctic biome and therefore this biome has borne the burden of the most direct damage to its natural environment (see Figures 2.2 and 5.2). Without a doubt, the most serious single impact is taking place in the oils sands of northern Alberta.

While resource companies today are more environmentally responsible, this new attitude did not occur voluntarily (as perhaps Drucker would have expected in his vision of a socially responsible corporate world), but was the result of public pressure that resulted in government regulations. Such legislation was and is necessary to protect the public and the environment. Today, resource companies must comply with an environmental review known as an environmental impact assessment (EIA). The objective of such reviews is not to stop industrial undertakings, but to ensure that their impacts on the environment and local population are acceptable, that is, the environmental and social impacts are minimized to an arbitrary 'acceptable level'. Only a few times have resource projects been rejected outright; more often, they are required to adjust details in their plans to reduce environmental and/or social impacts. With governments making the final decision of approving (or rejecting) proposed resource projects, the EIA process is a critical element. The **Canadian Environmental Assessment Agency** (2011) states that its 'role is to provide Canadians with high-quality environmental assessments that contribute to informed decision-making, in support of sustainable development.'

The Ugly Past

Before the formation of the Federal Environmental Assessment Review Office (FEARO) and the federal Environmental Assessment Review Process (EARP) in 1973, resource

---

Vignette 6.1 What Is Sustainable Development?

In 1999, the Canadian Environmental Protection Act (CEPA) became the keystone of Canada's federal environmental legislation aimed at preventing pollution and protecting the environment and human health. The goal of CEPA is to contribute to sustainable development—development that meets the needs of the present generation without compromising the ability of future generations to meet their own needs. Another expression of Ottawa's view of sustainable development is:

The concept of sustainable development was born from the realization that social, economic and environmental health are interdependent. Sustainable development is essentially a shift in thinking, whereby individuals, organizations and governments at all levels understand and take responsibility for the full economic, environmental and social costs of their actions. Canadians require effective and efficient tools to inform them of the potential consequences of their choices and support them in making sustainable decisions. (Canadian Environmental Assessment Agency, 2009)

---

companies, both private and public, evaluated their projects from a viability perspective, but these evaluations paid little or no attention to the environment and the Aboriginal peoples. For them, the North was a wilderness that government and society wanted 'developed'. Hidden from view, however, were environmental and social costs, which, when they surfaced, landed on the desk of the appropriate Canadian government—sometimes federal and other times provincial. Territorial governments rely on Ottawa to pay for the expensive remedial work of cleaning up old development sites. Until Canadian governments passed environmental legislation, companies were not breaking the law by discharging their toxic wastes into the air, land, and water. With the public regulatory agencies and non-governmental agencies now in place, most industrial projects built 40 or more years ago would not pass current standards. Companies that fouled the environment in the past have left the Canadian taxpayer to foot the bill for site cleanup. Even now, companies have an escape from paying for their environmental mess—bankruptcy. In 1999, the Giant gold mine at Yellowknife closed and the owner, Royal Oak Mines, went into receivership, thus avoiding the huge cleanup costs associated with disposing of the 237,000 tonnes of highly toxic arsenic trioxide tailings. Such a public hazard required a solution and the federal government announced the Giant Mine Remediation Project, which, when completed, will likely cost more that $500 million (Indian and Northern Affairs Canada, 2010). Another expensive example—this one dealing with North American Arctic Defence—involves the removal of high levels of PCBs and contaminated soil. These toxic materials were dumped on the ground following the abandonment and initial cleanup of 21 Distant Early Warning radar sites in 1993. In 1998, the federal government signed a $230 million cleanup agreement with Inuit leaders with the goal of completing the project by 2018 (CBC News, 2010).

Aboriginal Canadians have suffered from the effects of industrial pollution in various ways, including polluted water sources, the loss of wild animals and fish that had been primary food sources, and the flooding of valuable hunting grounds. Two

Figure 6.1   Environment Change in the North

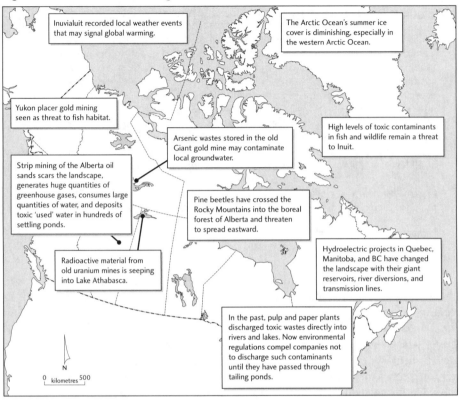

instances of pollution affecting Aboriginal communities stand out. First, those who rely heavily on country food have found their health compromised. Perhaps the most dramatic example involved the Ojibway living along the English–Wabigoon River system near Kenora, Ontario, whose main source of food was fish.[1] The sad story, retold by Shkilnyk (1985), involved the discharge of toxic chemicals with a high content of mercury into the Wabigoon River by the Dryden Paper mill, causing methyl mercury poisoning that incapacitated many people (Roebuck, 1999). Second, Aboriginal workers were involved in dangerous work in the uranium industry. The worst example occurred in the 1950s when Sahtu Dene men were hired to carry sacks of radioactive ore from the mine at Fort Radium on Great Bear Lake to the Mackenzie River.[2] The extremely high incidents of cancer among these workers shocked the community, giving it the name 'Village of Widows' (Markey, 2005; Environment Canada, 2007).

The Emergence of the Environmental Movement

In 1962, Rachel Carson's epoch-making *Silent Spring* appeared. Carson revealed the extent of and the human and other harm caused by pesticides and industrial pollution, shocking Americans and galvanizing many to support and to join environmental organizations. After the publication of Silent Spring, the media reported more examples of industry disposing of toxic wastes in a haphazard manner. While these practices save the companies money, such activities can create ecological disasters. Two examples prove this point. Perhaps the most famous—or infamous—example is the Love Canal chemical waste site in Niagara Falls, New York. This long-abandoned canal provided a handy dumping site for toxic wastes for local industrial firms. In time, the dump was covered with soil and eventually became a housing development. For years, residents complained of ill health and even deaths. By 1977, the cause was determined to be the deadly chemicals seeping to the surface of this residential area. Did the original industrial firms pay for the cost of a cleanup? Of course not. The state had to purchase 250 houses and spent US$250 million on a cleanup operation (McDonald, 2001: 23). In Canada, an abandoned gold mine at Deloro on the southern edge of the Canadian Shield in eastern Ontario had 24 times the allowable level of arsenic in the soil (Damsell, 1999: FP1). In addition, the Deloro site was used as a dump in the 1940s for uranium waste from the Manhattan Project in the US that produced the first atomic bombs and also received radioactive waste from an Eldorado Nuclear uranium refinery in Port Hope, Ontario, a small city on Lake Ontario east of Toronto, which is also severely polluted by harmful radioactivity. With the private operators of the Deloro mine abandoning the site, the Ontario government has spent more than $31 million on the cleanup and these expenses are not likely to end for several years. Still, efforts so far have reduced the amount of arsenic coming off the site by 80 per cent (Ontario, 2010).

As the public became more aware of such environmental disasters, pressure mounted on governments to protect the environment and, ultimately, the quality of human life. Non-governmental organizations goaded governments into action and continue to act as a thorn in the side of resource companies. Environmental legislation in Canada is more complicated than in many countries because Canada is a federation where provinces have control over natural resources and the environment. Under this political arrangement, Ottawa is responsible for the territories and whenever industrial proposals, notably those that involve rivers, cross provincial boundaries. Today, the federal government shares this power with territorial governments and Aboriginal peoples who have concluded comprehensive land claim agreements. Canada's environmental legislation and its environmental agencies are discussed in more detail below.

Environmental Legislation and Resulting Agencies

In 1973, Canada moved towards protecting the environment with the federal **Environmental Assessment and Review Process** (EARP) and the **Federal Environmental Assessment Review Office** (FEARO). However, EARP was not supported by legislation. Over 20 years passed before, in 1995, the **Canadian Environmental Assessment Act** became the law of the land—but it applied only to federal lands. During the same period, provinces passed similar but not identical legislation. Saskatchewan, for example, approved its environmental legislation (Environmental Assessment Act) in 1979–80 while Yukon's Environmental and Socio-economic Assessment Act was enacted in 2003 (Sinclair and Doette, 2010: Table 17.2). Except for the three Maritime provinces, all of the provinces have northern lands that fall under their respective provincial environmental statutes and assessment processes. The territories have a more complicated environmental assessment system that includes direct involvement and partial control by Aboriginal organizations under co-management arrangements with the federal government as a result of the inclusion of environmental matters within comprehensive land claim agreements. Since these agreements establish a legal land base, each Aboriginal land base or settlement area falls under partial Aboriginal control through a co-managed assessment body. Thus, companies or individuals proposing to develop resources within those settlement areas must submit proposals to the appropriate co-managed organization for an assessment. In the Northwest Territories, legislation for environmental assessments takes the form of the Inuvialuit Final Agreement, the jurisdiction of which extends over the northern half of this territory, that is, the land and sea described as the Inuvialuit Settlement Area, and the Mackenzie Valley Resource Management Act, which deals with development proposals in the southern half of the Northwest Territories.

In the case of Yukon and the southern half of the Northwest Territories, their Acts allow for numerous First Nations to conduct co-management environmental assessments when the development proposal impinges on their land claim settlement areas. The Inuvialuit and Nunavut land claim agreements created environmental co-management structures that conduct environmental assessments of development projects that affect their land bases. The four Acts or agreements are:

- The Yukon Territory Environmental and Socio-economic Assessment.
- The Mackenzie Valley Resource Management Act.
- The Inuvialuit Land Claims Agreement, which created two co-management agencies, the Environmental Impact Screening Committee and the Environmental Impact Review Board.
- The Nunavut Land Claims Agreement, which established the Nunavut Impact Review Board.

Under the Canadian Environmental Assessment Act of 1995, amended in 2002, the assessment process consists of three basic stages: screening, comprehensive study, and a review panel. This approach has been adopted by the provinces and territories and is embedded in Aboriginal land claim agreements as co-managed assessment organizations comprised of Aboriginal and federal/territorial officials. Once a project is submitted, it undergoes an environmental screening. The screening procedure has three possible outcomes: (1) the project can proceed without further environmental review other than compliance with existing policies and standards; (2) if the potential adverse environmental effects are not fully known, then a more detailed assessment (comprehensive study) takes place; and (3) if the potential adverse effects are considered significant, then a review panel is established. Most projects, but especially small projects, are screened and then approved quickly. On average, around 5,000 proposals are approved at the screening stage each year. Proposed projects that have significant potential environmental impact must undergo an environmental impact assessment (EIA).

Under this review process, obviously, the legislation and the agencies created by it do not aim to halt industrial projects but rather to ensure that damage to the environment is minimized. Few projects have been rejected or withdrawn. The most famous review, the Mackenzie Valley Pipeline Inquiry led by Thomas Berger, saw the project die on the drawing board (and, in a similar fashion, the Mackenzie Delta Gas Project fell victim to a severe downgrading of its viability due to recently discovered gas reserves, as described in Chapter 5). One reason was that the Berger Report, while presenting Ottawa with a recommendation for development, contained strong reservations, including a 10-year moratorium to settle land claim agreements and to create a conservation area in Yukon. In the early 1980s the Windy Craggy Project, a copper deposit in northwest British Columbia, was actually rejected by Ottawa because its sulphur-rich effluence would have destroyed the pristine Tatshenshini River. With a Cinderella ending, this site turned into a World Heritage Site (Vignette 6.2). Similarly, in November 2010 Jim Prentice, Minister of the Environment, announced that the proposed Prosperity gold and copper mine in the Chilcotin region near Williams Lake, BC, was rejected on the grounds that it would devastate an entire ecosystem around the environmentally and culturally sensitive Fish Lake (Fowlie, 2010). The BC government had already approved the project because the Prosperity mine would stimulate economic growth in the area. This pattern of provincial approval and federal sober second thought also happened with the Oldman River Dam Project in Alberta and the Rafferty–Alameda Dam in Saskatchewan, but in both cases the provinces pushed ahead and built the dams, arguing that economic benefits far outweighed environmental costs.

The field of environmental assessment and management is changing quickly. A broad overview of these changes can be found in Hanna (2009) and Noble (2010).

Vignette 6.2   From a Mining Proposal to a World Heritage Site

The pristine Tatshenshini River nearly became the site of a copper mine, but instead this river basin was named a World Heritage Site. This marks one of the few times that an industrial project has been rejected for environmental reasons. In the late 1980s, Geddes Resources Limited, a Toronto-based mining company, wanted to develop the huge Windy Craggy copper deposit in the mountainous corner of northwest British Columbia near Alaska, only some 80 km downstream from Glacier Bay National Park and Preserve and 32 km from Kluane National Park. This rich copper/gold deposit was valued at $8.5 billion. Geddes applied to the British Columbia environmental agency for permission to proceed with development. The original submission was unacceptable and the company withdrew it. Major flaws in the initial proposal were the danger of highly acidic water draining from the sulphide-rich tailings and the open-pit mine into groundwater and local streams, including the Tatshenshini River. The revised mine plan, released in January 1991, was estimated to increase development costs by $100 million, making the total cost around $600 million. It proposed to accomplish three goals: (1) reduce the amount of waste rock by 50 per cent; (2) separate sulphide-bearing rock from other waste rock (which would still be dumped on the nearby glacier); and (3) place the sulphide-bearing waste rock in a tailing pond. The revised proposal was not accepted, causing the company to abandon its plans. Efforts by environmental groups such as the Western Canada Wilderness Committee, World Wildlife Fund, Sierra Club, National Audubon Society, and Tatshenshini Wild played a critical role in shaping the final outcome. In 1993, the Tatshenshini–Alsek Wilderness Park was created. The following year the park was designated as a World Heritage Site.

Sources: Environment Canada (1990); Geddes Resources Ltd (1990); Robinson (1991: B1); Searle (1991); Draper (2002: 357).

For northern places, co-management has emerged as a progressive move to include Aboriginal residents in this process. As well, co-management may also promote further local involvement in the design, operation, and ownership of resource projects. As discussed later in this chapter, the Nisichawayasihk Cree Nation is involved in the Wuskwatim hydroelectric project in northern Manitoba.

Types of Environmental Impacts

Resource projects tend to have three types of spatial impacts on the landscape. Each varies by duration and magnitude, with megaprojects having a longer construction period and a great capital investment. The direct impacts are expressed on the landscape as linear, areal, and accumulated impacts. Indirect or secondary effects refer to related developments that support the project, such as a paved highway to a new school, but that are not part of the project's construction budget.

Linear effects are associated with projects such as highways, seismic lines, and pipelines. While small in total area, these changes to the landscape have a greater effect than their areal extent would suggest. For example, wildlife is affected by such developments, especially migrating animals such as caribou. Also, access by hunters to wildlife is enhanced by new roads. In the late 1970s, the Berger Inquiry called for a wilderness zone along the Arctic coastal plain stretching from Alaska to Yukon so that the Porcupine caribou herd would not be disturbed by the presence of the proposed Mackenzie Valley Gas Pipeline. While this pipeline was never built, this recommendation set a precedent. Now concern for the Porcupine caribou herd has shifted to the recently approved (January 2011) Mackenzie Delta Gas Pipeline.

Areal effects are associated with industrial developments that affect huge geographic areas. A hydroelectric project, for example, can have an impact on entire river basins by diverting rivers, creating giant reservoirs that flood extensive areas, and reversing

---

Vignette 6.3 Pollution from Abroad

Pollution from the highly industrialized countries in the northern hemisphere reaches the seemingly pristine lands and waters of northern Canada in the form of minute particles and gases. These tiny substances originate from factories, farms, and urban centres in Russia, China, and other industrial nations. As these pollutants enter the global atmospheric and oceanic circulation systems, some eventually are deposited in Canada's North, where they are absorbed into the vegetation cover and water bodies. Gradually but inevitably, these chemical compounds enter the food chain. Most toxic pollutants originate as residue from the spraying of chemicals on crops, exhaust from automobiles, and effluent produced by industrial factories. A substance is declared toxic when it has an inherent potential to cause adverse effects to living organisms, including humans. Arctic haze, a form of smog caused by aerosols and by carbon particles from coal-based industries in Europe and Asia, is an example of global pollution affecting the North. The danger of global pollution in the North is difficult to determine but is now considered a serious threat to the health of Inuit, who consume large quantities of meat from seal and caribou.

the seasonal peak flows of the rivers, thus transforming the natural landscape into an industrial one. Hydroelectric development in northern Quebec has had a massive impact on La Grande Rivière Basin—some describe the project as an ecological disaster (Berkes, 1982, 1998; Peters, 1999; Roebuck, 1999; Rosenberg et al., 1987). Similar impacts have occurred in northern British Columbia, Manitoba, and Labrador at large-scale hydroelectric sites. In the first decade of the twenty-first century, construction of more hydroelectric dams and reservoirs continued in the James Bay region with the diversion of the Eastmain River into La Grande Basin.

Cumulative environmental impacts represent a combination of impacts over time that result in greater damage to the environment than a single impact. Often, cumulative impacts occur in association with a megaproject. In other words, besides the primary impact of a large-scale project, other secondary impacts often are generated by other firms or government agencies, thus adding to the primary impact. These secondary effects take two forms. First, businesses associated with the construction of the megaproject spring up to service the project. Second, after the construction of a megaproject, different business firms and/or public institutions emerge. These various business activities add their own impacts to those of the major project.

Bram Noble (2010: 133) describes the significance of this complex type of impact:

> Environmental effects that are cumulative in nature—that is, they add to or interact with an already existing adverse impact to detract additively or synergistically from environmental quality—are more likely to be deemed significant compared to impacts that do not affect already affected environmental components.

Cumulative environmental damage is more widespread than generally recognized. At first, environmental impact assessment focused on direct impacts. As the concept of EIA progressed, serious attention was paid to the combined effect of several industrial projects within a single geographic area. The first major examination of cumulative effects took place in the 1980s when concern was raised over the discharge of toxic chemicals from pulp mills located along the Athabasca River and its tributaries, thus forcing the environmental assessment review of the Alberta-Pacific pulp mill to expand its geographic area. Much political pressure came from the government of the Northwest Territories, which was concerned about the downstream impact on its Aboriginal population that harvested the fish from this river. The results were astounding: not only did Alberta-Pacific replace one chemical agent, chlorine, which is a major source of dioxins and furans, with another one, hydrogen peroxide, which produces much lower levels of dioxins and furans, but the Alberta government announced that other pulp mills would be required to lower the level of dioxins and furans discharged into streams flowing into the Athabasca River (O'Reilly, 2005).

Why Is the North So Vulnerable to Environmental Impacts?

The Subarctic and, to a lesser degree, the Arctic have been hit hard by past industrial pollution. In many cases, the greatest damage is done by exploration parties who abandon their buildings, equipment, chemicals, and diesel fuel tanks because of the high cost of removing them by air. On the other hand, mining operations today are required by law to restore the site. One measure of the impact is found in the study conducted by Duhaime, Bernard, and Comtois into the startling number of abandoned mining operations in Nunavik in Arctic Quebec. The authors feel confident that a similar pattern of abandonment exists across the North (Duhaime et al., 2005).

An important question is: 'Why is there so much concern about industrial pollution in the North?' While the answer is complicated, three factors stand out—the fragile environment, the high costs of cleanup, and the negative impact on Aboriginal peoples. First, the North's environment takes longer to heal from wounds caused by industrial projects. This issue of fragility is related to three primary factors:

- The North's cold climate means that its resulting biological regime requires a much longer time to repair itself from various forms of industrial damage.
- Permafrost, a unique natural feature of the North, is easily disturbed by industrial construction projects. Removal of the vegetation cover triggers gelifluction and thermokarst topography, which are but two outcomes. Gelifluction occurs when the upper layers of soil thaw during the warmer months so that water-saturated soil moves down slope, resulting in slumping. Melting of ice-rich terrain results in a hummocky landscape termed thermokarst topography (Vignette 2.12).
- The global atmospheric and oceanic circulation systems bring fine particles of industrial pollutants to northern lands and waters from huge industrial complexes in Asia, Russia, and elsewhere. Indeed, recent research conducted by a team of Canadian, Norwegian, and Chinese scientists has shown that the increased melting of the Arctic ice pack and warmer temperatures are releasing into the atmosphere and immediate environment long-banned pesticides such as DDT, PCBs, and other persistent organic pollutants, all of which had been trapped in the ice and in the waters beneath the ice (CBC News, 2011).

Second, the cost of cleanup is much higher in the North than in the more populated areas of Canada. With the Distant Early Warning radar sites, for example, cleanup cost the federal government well over $200 million and a long time—over 20 years—is required for remedial work.

Third, the dependency of Aboriginal peoples on country food remains a cultural constant and thus makes industrial pollution a major health problem. Industrial pollutants from distant countries land in the Arctic through global atmospheric and oceanic systems, enter the food chain of wildlife, and eventually are consumed by Aboriginal peoples. Inuit women who are pregnant have been advised not to eat seal meat and other marine food because it contains high mercury levels.

Co-management

Aboriginal participation in the management of the lands, waters, and wildlife in their settlement areas represents a major step forward. This is not a perfect solution, however, because conflicts do surface in co-management agencies, often between those who hold traditionalist views and/or support Aboriginal **traditional ecological knowledge** (TEK) and development-oriented members who subscribe to Western, science-based assumptions about resource development. Most importantly, co-management boards for environment impact assessment have strengthened the Aboriginal hand in the approval process for industrial proposals and in limiting potential impacts on the environment/wildlife in their land claim settlement areas. As a consequence, the focus of the assessment process has shifted from 'national' issues to 'local' concerns. National issues tended to reflect on larger matters, such as the net benefit of the proposed resource project to the nation; local concerns look at how the impacts will affect the environment, wildlife, and the harvesters of those resources. Co-management boards consist of appointed members from the beneficiaries of land claim agreements and members of the general public selected by the federal, territorial, or provincial governments. These boards, funded by governments, have two functions: to manage the environment and wildlife and to make decisions related to the use of resources.

Two bodies—screening committees and review boards—assess environmental impacts. The screening committee within a particular jurisdiction examines all industrial proposals, to approve those with acceptable impacts and to recommend those with potential significant impacts to a review board. Co-management review boards have exercised considerable power over the resource development process in settlement areas of comprehensive land claim agreements. In 1991, for example, Gulf Canada's proposal to drill four oil wells some 40 to 70 km offshore in the Beaufort Sea was rejected by the review board (Vignette 6.4). In most other cases, review boards have approved industrial proposals, but often with considerable modification of the original project.

Without a doubt, co-management of the environment and wildlife represents an enormous step forward for the beneficiaries of comprehensive land claim agreements. As a result, such Aboriginal groups have gained a measure of control over resource

---

Vignette 6.4   Gulf Canada's Kulluk Drilling Proposal

Gulf Canada had drilled many wells in the Beaufort Sea without encountering problems. In 1990, Gulf proposed to drill four new wells in the offshore waters of the Beaufort. The objective was to add to the known oil reserves of the Amauligak field. The proposed well sites were located in the ice transition zone, i.e., between the normal extent of land-fast ice and the polar ice pack. For the first time, Gulf Canada's proposal was assessed by a co-management board consisting of Inuvialuit and non-Inuvialuit members in Inuvik rather than by the Canadian Environment Assessment Agency based in Ottawa. Local members of the Environmental Impact Review Board (EIRB) were particularly concerned about how Gulf would deal with an oil spill caused by a blowout of a well. Gulf calculated that, in the worst-case scenario, 2.6 million barrels of oil would escape into the Beaufort Sea. Gulf estimated that cleanup costs would reach $400 million. The EIRB learned that the maximum liability under federal regulations for such a blowout was a maximum of $40 million. Gulf's proposal was rejected because of the potential damage of a blowout to the marine environment, wildlife, and the harvesting of marine country food. The board also faulted the federal regulatory agency, the Canada Oil and Gas Lands Administration, for failing to develop adequate oil spill contingency plans, including a much higher liability for oil companies.

Source: Bone (2003: 390–1).

development, land-use management, and wildlife harvesting. Co-management means that governments and Aboriginal organizations manage the environment together. This newly won power is restricted to their settlement areas (areas determined by each comprehensive land claim agreement), but operation of co-management boards goes far beyond the concept of public involvement because they are anchored in the culture of the particular Aboriginal group (Noble, 2000: 106–8; Bone, 2003). The net result has had fundamental implications—the Canadian North has undergone a political realignment in the field of environmental assessment. This process began in 1984 with the transfer of the environmental impact assessment organization from Ottawa to Inuvik under the Inuvialuit Final Agreement. Out of this realignment, a new co-management arrangement for environmental impact assessment in the Inuvialuit Settlement Region was achieved. The co-managed agencies (Environmental Impact Screening Committee and the Environmental Impact Review Board) have equal Inuvialuit and non-Inuvialuit representatives. The Inuvialuit Game Council nominates the Inuvialuit members while the federal and territorial authorities select their representatives. Unlike the Ottawa-based agency, these two co-managed boards place much more emphasis on possible negative impacts of proposed industrial projects on the local environment and wildlife. This new emphasis reflects the Inuvialuit culture and the continuing importance placed on wildlife harvesting.

Within their traditional territories, six Aboriginal groups (Inuvialuit, Gwich'in, Sahtu/ Métis, Inuit, Yukon First Nations, and Tlicho) have reached land claim agreements. In 1992, for example, the Gwich'in concluded their land claim agreement for some 22,000 km² along both sides of the Yukon–NWT border, and that agreement spawned the Gwich'in Renewable Resource Board (GRRB). The GRRB is a co-management board that manages the renewable resources in the Gwich'in settlement area and deals with environmental impact assessments. The Gwich'in Tribal Council and various government departments nominate an equal number of board members. In this way, power is shared between the Gwich'in and governments. Similar arrangements exist with other Aboriginal groups who have negotiated comprehensive land claim agreements.

Other forms of co-management frequently focus on a single resource but they often only have advisory powers. The Beverly-Qamanirjuaq Caribou Management Board, for example, is concerned with resource management of caribou over several jurisdictions in the territories and provinces and among several Aboriginal user groups, and it must deal with the contentious issue of caribou harvesting by Dene and Inuit communities. The board was formed in the early 1980s in response to the apparent decline of caribou. This board consists of Aboriginal members from communities that harvest caribou, and government biologists and resource managers. Their primary function is to exchange information and to ensure the survival of caribou by protecting the herd's habitat and, as needed, limiting the harvest by persuasion. While this board has only an advisory role, the various governments responsible for caribou tend to accept its recommendations. As Kendrick (2000: 2) states:

> Caribou co-management is a process of cross-cultural learning. Ideally, the knowledge of the caribou-hunting communities and government biologists and managers complement each other with the aim of achieving the sustainable use of a culturally and economically important resource.

The Mackenzie Delta Gas Project represents the largest construction project facing the Northwest Territories. The Joint Review Panel (JRP) was formed in 2004 to examine the potential impacts on the environment and the lives of the people in the project area. The nine-person panel included two members nominated by the Inuvialuit Game Council, three by the Mackenzie Valley Environmental Impact Review Board, and four by the Minister of the Environment Canada. With a majority of northern members, the interests of residents of the Northwest Territories were secured somewhat in the manner of co-management boards. Thus, the Joint Review Panel contained a strong element favouring traditional views about the environment and wildlife, and at least one member, Dr Peter Usher, was a member of the Berger research team of the 1970s. Traditionalist views were expressed and documented at the many Native community hearings of the Joint Review Panel (representing another link to the Berger approach to impact assessment). Such an approach takes time and is the primary reason why the JRP report ran well past its original deadline—to the dissatisfaction of the proponents and both federal and territorial governments. The panel submitted its comprehensive but controversial report at the end of 2009. The federal and territorial governments did not accept the report because they felt that certain recommendations went beyond the panel's mandate and would have committed the two governments to a more active role in environmental and social matters associated with the Mackenzie Gas Project rather than leaving the marketplace to sort out loose ends.

Another Step Forward: Cleanup and Decommissioning

Since the late 1990s, governments have required companies to restore the natural habitat found at their closed mine sites. Now, all mining proposals require closure plans that must be approved by the appropriated public regulatory agency. The final closure plan may not be exactly the same as returning the site to the original conditions, but the aim is for a minimal environmental footprint.

The closing of uranium mines, including decommissioning, is a special case and falls under the Nuclear Safety and Control Act, which places the responsibility on Ottawa. Not surprisingly, the closure of uranium mines is much more demanding and involves considerable decommissioning costs. Detailed decommission plans are embedded in their original proposals. Often, these decommission plans take the form of environmental management agreements between the company undertaking a resource project and the appropriate territorial and/or provincial government. Such an agreement between the Saskatchewan government and Areva, a French-owned uranium mining company, resulted in the successful decommissioning of the Cluff Lake mine in northwestern Saskatchewan. After 22 years of production, the Cluff Lake mine ceased operations in 2002 when the ore reserves were depleted. At that time, the company's decommissioning program began. According to Areva (2006):

> When decommissioning is complete, the site is expected to have no significant adverse effects on the environment in the short or long term. The people of the region will be able to use the site for traditional purposes. As the first of its generation of uranium mines to be decommissioned, Cluff Lake aims to set a high standard in the industry for returning a site to its natural state.

The Berger Inquiry Sets the Standard

As discussed in Chapter 5, the Berger Inquiry into the proposed Mackenzie Valley Pipeline Project was the first environmental and social assessment of a major resource project. While other northern EIAs followed the Berger Inquiry,[3] Justice Berger broke new ground and set the standard for future inquiries. Most importantly, Berger moved the hearings from the boardroom to Aboriginal communities and the central discussion from technical aspects of the project to local concerns about the potential impact of the project on wildlife and Aboriginal culture. One community hearing participant summed up the issues at stake quite succinctly:

> I wonder how people in Toronto would react if the people of Old Crow went down to Toronto and said, 'Well, look, we are going to knock down all those skyscrapers and high rises . . . blast a few holes to make for lakes for muskrat trapping, and you people are just going to have to move out and stop driving cars and move into cabins.' (Quoted in Page, 1986: 212)

With such explosive accounts from local figures, the community hearings were closely followed by the national media and in this way the Canadian public became aware of the issues surrounding this massive pipeline project.

The media became so interested in the proposed Mackenzie Valley pipeline because of Berger's strategy of shifting the focus from technical issues to a much broader examination of the consequences of industrial development on the environment and the culture of Aboriginal peoples. In a sense, Berger was educating the Canadian public about northern development, the environment, and Aboriginal peoples. The outcome was a remarkable shift in public attitude towards industrial projects—a shift from acceptance of the inevitable to one of sharp questioning about potential impacts.

Four elements of the Berger Inquiry are now found in environmental impact assessments and reports:

- Industrial proposals are more carefully scrutinized by the federal/provincial/co-managed environmental assessment and review inquiries.
- The role of public participation has been greatly enlarged.
- Hearings are held in the communities affected by proposed projects as well as larger cities.
- Environmental issues affecting Aboriginal peoples have become a critical area of assessment because environmental impacts caused by industrial projects could have a particularly harsh effect on habitat that supports hunting, trapping, and fishing.

Environmental Impact of Pipelines

In the 1980s, little was known about constructing and operating a pipeline buried in frozen ground. In that sense, the Norman Wells Oil Expansion and Pipeline Project of 1982–5 was a gamble that proved successful because, over 25 years, no major rupture of the pipe has occurred.

What was the engineering challenge that threatened the environment? First of all, pipeline engineers know that breaks will occur. The challenge is to manage those breaks. In January 2011, for example, crews scrambled to repair a rupture in the Alyeska pipeline in Alaska (Vanderklippe, 2011a). One challenge facing engineers is how to accommodate a buried 'warm oil' pipeline constructed in discontinuous permafrost, which means the pipeline passes through frozen (freezing temperatures) and unfrozen terrain (above-zero temperatures). On the one hand, the warm pipe could melt the permafrost, which could lead to a pipeline break. On the other hand, lowering the temperature of the oil pipe could reduce its viscosity and thus slow the rate of flow. Other concerns were:

- The hilly topography, combined with the melting snow in the spring, could erode the earth around the buried pipe and even undermine its ground support, thus leading to a break (Figure 6.2).
- The ice in the permafrost below the pipe could melt in the summer as the exposed ground along the pipeline route warms because of the long period of daily sunlight. Again, ruptures could occur.

The Norman Wells pipeline was designed to allow oil to flow at near 0°C, which, it was hoped, would minimize the possibility of the pipe either thawing or freezing the ground around the pipe (Duffy, 1981: 33–4). In fact, oil was initially chilled to −2°C before entering the pipe at Norman Wells to minimize ground disturbance such as subsidence caused by frost heave or the melting of ground ice (MacInnes et al., 1989; Burgess and Riseborough, 1989). After 1993, the company requested a summer/winter temperature regime with the oil chilled to −4°C in winter and raised to 12°C in summer (Burgess et al., 1998: 95). Over the years, the Norman Wells pipeline has maintained its integrity, though the pipe has shifted up to a metre in many places and water erosion has exposed the buried pipe. The company has responded by filling in these depressions and by covering exposed steep slopes with wood chips to reduce surface heating of the ground.

With the measurements of the Geological Survey of Canada, hard evidence reveals two effects of a warm pipeline in a permafrost landscape: (1) the degree of ground subsidence increases, and (2) the extent of the zone of thaw increases.

Canada's first Arctic natural gas pipeline, the Mackenzie Delta Gas Project, may see gas flowing to southern markets by 2018—or it may not, as discussed in Chapter 5. This pipeline is expected to transform the Mackenzie Valley into an energy corridor, but at the cost of environmental and social impacts. Because of the complex legislative arrangement in the Canadian territories, the task of assessing these impacts fell to two agencies—the **Joint Review Panel** (JRP) and the **National Energy Board** (NEB).

Each had a different role. The Joint Review Panel attempted to represent the voice of the people of the impact zone through a series of community hearings. Their mandate was to examine the potential environmental, socio-economic, and cultural effects of the project. Two key questions were asked: (1) What would such a pipeline mean for the environment and wildlife? (2) How would it affect the Aboriginal peoples who still depend on game for country food? On the other hand, the National Energy Board

Figure 6.2   Wood Chip Slope along Norman Wells Pipeline

The Norman Wells Pipeline right-of-way lies in the discontinuous permafrost zone. Permafrost is found in approximately 75 per cent of the terrain near Norman Wells but this reduces to about 35 per cent at the southern end near Zama in northern Alberta. To prevent thawing of the ice in the ground from the heat produced by the warm oil in the pipeline and from solar radiation heating the exposed surface of slopes, the pipeline company (Interprovincial Pipe Line Ltd) insulated the slopes with blankets of wood chips up to 1.8 metres in depth made from trees harvested along the right-of-way. The photograph, taken at Canyon Creek several years after oil began to flow to Zama, records the collapse of the ground above the pipeline and slumping along the steep slope leading to the partially frozen creek.

Source: National Research Council Canada.

Figure 6.3   Geological Cross-Section of the Norman Wells Pipeline near
Fort Simpson, NWT

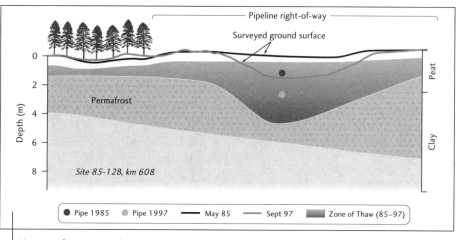

'A series of instrumented sites were established as part of the Geological Survey of Canada's monitoring program. This cross-section, from a site located on a peat plateau south of Fort Simpson, shows the increase in thaw depth, and the associated ground surface settlement and pipe settlement that has occurred over 12 years'

(Natural Resources Canada, 2006). Reproduced with the permission of Natural Resources Canada 2008, courtesy of the Geological Survey of Canada.

focused on hard aspects of the project, such as engineering, feasibility, safety, and cost. Both reports were completed in December 2009. The federal government accepted the findings of the NEB, but Ottawa was clearly uncomfortable with the wide-ranging set of recommendations in the JRP report. Perhaps Ottawa and Yellowknife were concerned about the panel's concept of northern sustainability, which, if accepted, would commit the governments to a much greater role in the environmental and social stewardship of this region (JRP, 2009: 589, Table 19.2). But most irritating recommendation was the JRP demand for an independent agency to monitor how well the governments kept their environmental and social promises. Through a series of consultations in early 2010 between the JRP and Ottawa, this impasse was resolved by the removal of recommendations that Ottawa saw as intruding on its powers. Now the NEB could amend its two-volume report, with the 'watered down' version of the JRP findings presented in one chapter and the NEB taking on the role of the government's version of an independent monitoring agency, per the original JRP recommendation (NEB, 2010).

Environmental Impact of Mining

Mining, by its very nature, has a harsh impact on the local landscape. But the mining of the Alberta oil sands is different. First, the massive size of these operations creates an extensive geographic impact area. Second, industrial pollution from these activities ranges from water contamination to emissions of greenhouse gases and drastic alterations to the landscape through giant **tailings ponds** and open-pit mines. Third, the oil sands, originally called **tar sands** and, by detractors of such development, 'dirty oil', have taken on enormous geopolitical significance as the principal target of environmental groups in North America and as the source of energy security for the United States.

Located in Canada's boreal forest, the Alberta oil sands contain much of the world's oil reserves and account for most of Canadian oil production (see Chapter 5, 'Alberta Oil Sands'). In spite of the wealth generated, the mining and processing of tar sands (Figure 6.4) come at a high cost to the environment. Consequently, oil sands development flies in the face of environmental efforts to create a cleaner world and to slow down global warming. The political rhetoric between opponents and supporters of oil sands development, in which **'dirty oil'** is contrasted with **'ethical oil'**, and the heated media debates in North America have both environmental and geopolitical consequences. For instance, the seemingly neutral mandate of environmental monitoring of Alberta's oils sands by the industry-funded Regional Aquatics Monitoring Program (RAMP) has turned into a battleground between industry and environmentalists. Confidence in RAMP within the scientific community sank so low that the former federal Minister of the Environment, Jim Prentice, appointed a panel of independent scientists to quickly assess the situation and determine by December 2010 if a first-class, state-of-the-art monitoring system is in place and serving the public interest by providing the necessary information to protect the environment. As described below, the panel's answer was clearly and unequivocally negative.

Since each mine has passed an environmental assessment, why have the oil sands become the symbol for environmental degradation and a critical source of greenhouse gases? The answers are found in the vastness of this industrial undertaking and in its inherent nature, which is so hard on the environment. The problem confronting industry is that bitumen is difficult to access and turn into heavy oil without great expenditure of water and energy and huge environmental impacts. Simply stated, tar sands cannot be pumped from the ground in a natural state, but must be mined or extracted by underground heating (steam injection) and then subjected to processing by upgraders to produce a synthetic **heavy oil**. This poses a dilemma. The black gold from the oil sands is driving the Canadian economy, but its extraction and processing ruin the environment by impacting the landscape, the water regime, and the atmosphere.

Key questions are:

- Has industry done all it can to curtail the spewing of gases into the atmosphere, the release of toxic water into tailings ponds, and the scarring of the landscape?
- Have the federal and provincial governments been on top of the situation?
- What are the current state of the environment, the quality of the monitoring system, and the prospects for restoration?

Ultimately, as with all development projects, the basic question remains: *Can the environmental consequences be limited to an acceptable level?* While all projects have been approved by environmental assessments, environmental groups believe the bar is set too low (Vignette 6.5). Both the federal and provincial governments express concern about the level of pollution, especially about the toxic tailings ponds and the fouling of the Athabasca River and its reduced streamflow, but neither government wants

---

Vignette 6.5　Is the Bar Too Low?

The approval in January 2011 of the Joslyn oil sands mine located some 70 km north of Fort McMurray by the federal/provincial review panel made it the ninth approved mine in Alberta's oil sands regions. If construction goes as planned, the extraction of bitumen from this open-pit mine will commence in 2017 (Vanderklippe, 2011b). Environmentalists, as in the past, objected that the level of acceptable environmental impacts is far too low for oil sands projects. While the review panel for the Joslyn project did recommend that the French energy giant Total SA pay particular attention to reclamation of toxic mine waste and the protection of species at risk and valued wildlife, John Bennett, Sierra Club Canada executive director, decried the decision, stating that:

> Alberta and the federal government have not seen a tar sands project that they didn't love. We have no faith in this process whatsoever, because it's a system geared to encouraging the creation of more and more of these things, rather than to monitoring and controlling their environmental impact. (Quoted ibid.)

---

to see production curtailed. In fact, production is increasing mainly because of the strong demand from the United States. The reasons are clear: the United States sees Alberta oil as a long-term secure source and its Gulf coast refineries, which were formerly supplied by tankers from Venezuela's oil sands fields, now depend on oil from Alberta. This dependency will increase with the completion of the second leg of the Keystone XL pipeline (see Vignettes 5.11 and 6.6). However, as of June 2011, the Keystone Pipeline expansion has not yet been approved by the US State Department. The delay is due to opposition from the US Environmental Protection Agency, which is concerned about Alberta's 'dirty oil' having adverse impacts on the US environment as well as flying in the face of the US **clean energy policy**.

The extraction of bitumen, as noted in Chapter 5, involves two methods: open-pit mining and steam injection. The physical nature of the oil sands helps explain why pollution and its impacts are so great. The first impact of open-pit mines involves the removal of the surface material to access the oil sands. Open-pit mines are only possible in a relatively small geographic area where the overburden is less than 75 metres. Beyond that depth, steam injection wells are used. The first step in surface mining is to drain and remove the muskeg and scrape off the overburden. The exposed oil sands are then scooped up, crushed, and transported, via conveyor belt or truck, to the extraction plant.

The second step associated with open-pit mining is to remove the bitumen from the sand, clay, and water. This process takes place in an upgrading plant where the oil in bitumen is separated from sand and water and fine particles are removed to produce a synthetic crude oil. Hot water is used to separate the oil from the sand. The effluent, a mixture of water, clay, sand, and residual bitumen, flows into a tailings pond. Huge tailings ponds are designed to recycle water in the mining operation, but their main purpose is to allow the toxic particles in the water to settle to the bottom of the pond.

Figure 6.4 Alberta's Oil Sands Deposits

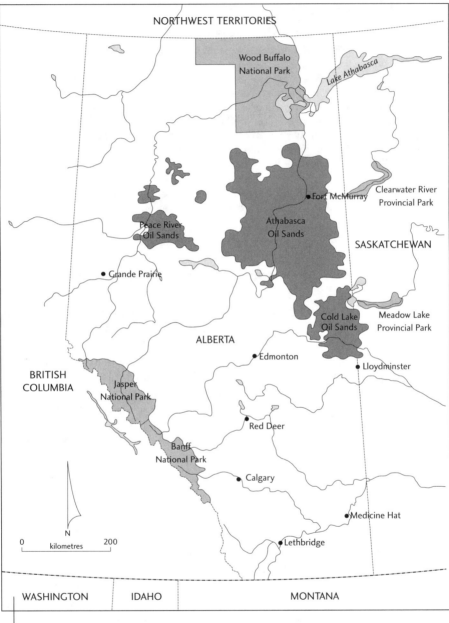

Source: Einstein (2006).

Under the environmental assessment agreement, companies must restore the tailings pond to a 'sustainable landscape'. These ponds, in total, extend over 170 km² and this number is increasing as more mines come into production. By 2011, no pond had yet been restored. The procedure involves draining the pond, burying the toxic waste at the bottom of the pond, reshaping the topography, and then covering with topsoil.[4] Trees, grasses, and shrubs are then planted. When the former tailings pond becomes a sustainable landscape, the Alberta government will release the company from its obligation and the land will be returned to the province.

Suncor Energy (2011) has reached the first stage in this long process to restore a tailings pond into a sustainable mixed-wood forest and a small wetland environment capable of supporting a variety of plants and wildlife.[5] The final stage, however, is decades away because recently planted vegetation needs time to establish itself as a mixed-wood forest. Still, naysayers doubt that such restoration will recreate the boreal forest. The Pembina Institute, an environmental advocacy group, argues in a 2008 report by Jennifer Grant that restoration will not restore a form of the boreal forest; rather, the landscape will consist of non-native trees that can survive in this artificial landscape.

Since about 10 per cent of the Athabasca oil sands area is covered by less than 75 meters of overburden, open-pit mining is possible. In fact, most production occurs at giant open-pit mines operated by Suncor, Syncrude, Albian Sands, and Canadian Natural Resources. Imperial Oil's Kearle mine is expected to begin operations in 2012, but the plan to ship 200 massive Korean-built modules along the Columbia River to the inland port of Lewiston, Idaho, and then by highway through Idaho, Montana, and Alberta to Fort McMurray has hit another snag—a judge in Montana ordered the state to halt issuing travel permits until 'the scope of the project' and its environmental impact are determined (Vanderklippe, 2011c). Thus, this bold and controversial plan to transport needed equipment using scenic US highways has proven troublesome and costly. American environmental groups threw up legal roadblocks that forced Imperial to reorganize their huge truckloads into smaller ones and to extend their environmental review. All of this has delayed construction of the $8 billion Kearle oil sands plant for at least a year and has increased construction and transportation costs (Gilbert, 2011).

All six of the currently operating open-pit mines are located in the Athabasca oil sands area where oil sands are close to the surface. These mines are the principal source of damage to the boreal forest and the landscape. They also are main polluters of air and water. Steam injection mining (in-situ SAGD) takes place where the oil sands are found at much deeper depths. While steam mining is more expensive, environmental impacts on the boreal forest and the landscape are reduced. Nonetheless, steam injection mining demands large quantities of water and this demand is affecting local surface water supplies and groundwater. The environmental impact associated with this huge industrial undertaking cannot easily be described. As the chair of the Oil Sands Advisory Panel wrote: 'And as a final note, our site visits had an indelible impact. It is hard to forget the sheer extent of landscape disruption, the coke piles and the ubiquitous dust' (Dowdeswell et al., 2010).

As noted above, the extraction and refining of oil from the oil sands deposits place pressure on the environment. Upgraders emit sulphur dioxide, nitrogen oxides, particulate matter, and metals to the atmosphere, while the extraction processes require huge quantities of energy and water. Natural gas provides most of the energy, and in the burning of this fuel source greenhouse gases are released into the atmosphere. Water, drawn from the Athabasca River, presents a different environmental problem. Once the extraction is completed, these now contaminated waters must be placed in

Figure 6.5   Oil Sands Extraction

Drawing of oil sands extraction by surface mining and high-pressure steam injection system.
Source: Adapted from Cantox Environmental (2006).

settling ponds. In addition, water levels in the river and downstream in Lake Athabasca have fallen precipitously since exploitation of the oil sands began.

The impact of oil sands projects on the Athabasca River has remained at the centre of controversy for years. While the companies have long claimed that their operations have had no impact on the water quality of the Athabasca River, a recent monitoring study of the Athabasca River directed by Professor Schindler of the University of Alberta suggests otherwise (Kelly et al., 2010). In fact, these scientists identified 13 toxic pollutants in water samples from the Athabasca River that came from the oil sands by means of air and various water pathways. The authors recommended that a 'robust monitoring program to measure exposure and health of fish, wildlife, and humans should be implemented in the region affected by oil sands development.'

Adding both credence and urgency to this recommendation, the residents of Fort Chipewyan, who consume fish from the river on a regular basis, complained about the effect of these mutated fish on their health, pointing out that an unusually large number of fish caught by local residents are deformed. This news caught the attention of Federal Minister Jim Prentice, who on 30 September 2010 ordered an independent investigation by an appointed Oil Sands Advisory Panel. Their report, *A Foundation for the Future: Building an Environmental Monitoring System for the Oil Sands* (Dowdeswell et al., 2010), concluded that the industry-operated monitoring system, RAMP, was unsatisfactory, and their principal recommendation reflected the earlier recommendation of the University of Alberta scientific team (Kelly et al., 2010). With Prentice leaving government for the private sector, the new Environment Minister, John Baird, said Ottawa accepts the recommendations of the report and will provide the leadership and funding to build a world-class monitoring system (McCarthy, 2010). But with cabinet changes in the past year, this task was passed to Peter Kent, the new Environment Minister.

Vignette 6.6   High-Stakes Geopolitics: US Energy Needs versus
the Environment

Alberta's oil sands would not be developed if the US market did not exist. In fact, this market is growing, thus causing more investments into oil mining. As for regulating land use in the region, Alberta has been asleep at the switch. Much-belated land-use regulations proposed for the Athabasca oil sands finally appeared in March 2011, based on the 2009 Alberta Land Stewardship Act (Tait et al., 2011). This controversial proposal, known as the Lower Athabasca Regional Plan, aims at protecting sensitive habitat, wildlife, and boreal forest on lands not yet leased to oil sands companies— and perhaps on lands leased but not yet developed. In other words, operations already up and running will not be constrained by the regulations. Not surprisingly, oil sands companies strongly oppose the plan as it could limit their access to this valuable resource in the next decade. Alberta Minister Mel Knight wants to have the final draft of the proposed regulations placed before cabinet by the late summer of 2011 (ibid.). Yet, Alberta's efforts were overshadowed on 21 July 2011 by the announcement of a 'world-class' monitoring plan for the oil sands region by federal Environment Minister Peter Kent. Of course, with Ottawa and Alberta responsible for such environmental monitoring, the two governments are obliged to work together to implement this very expensive federal plan. Ottawa expects the oil sands companies to pay for the monitoring. Given this complicated situation, the outcome is far from assured.

So, does energy trump the environment? In the turbulent twenty-first century, the most critical geopolitical question facing the United States is energy security, and such security largely comes from the Alberta oil sands. Yet, is this security worth the high environmental costs? On the one hand, the geopolitical importance of Alberta's oil sands for the US cannot be underestimated. Most imported oil into the US comes from these oil sands. The indisputable fact is that, in the coming decade, the US will need more and more 'secure' oil from Alberta. Yet, US landholders and environmentalists are concerned that a rupture in the pipeline would release toxic chemicals placed in the bitumen to facilitate its flow, and that these chemicals would contaminate groundwater, making it unfit for animal and human consumption. As well, some US politicians and environmentalists have hung the label 'dirty oil' on the oil sands and named its release of huge quantities of greenhouse gases as one of the principal sources of global warming (Nikiforuk, 2010; Marslen, 2008). Approval or rejection of the Keystone expansion pipeline proposal to the Gulf states by the US government will signal who has won this epic struggle.

Environmental Impact of Hydroelectric Projects

Each resource project has a distinct environmental footprint. Most are highly localized. Hydroelectric projects, however, have a 'regional' impact on river basins. Dams flood vast areas of land, alter river courses and streamflows, cause shoreline erosion, and increase the mercury content of the water in reservoirs. In the past, the cost of pre-flood clearing of trees was considered too expensive, and anyone who has visited such a reservoir is struck by the extent of distorted shoreline where slumping has taken place and dead trees poke above the water level. Unfortunately, nature does not correct such man-made problems quickly. Some 70 years after the construction of the

Island Falls Dam on the Churchill River in Saskatchewan, standing snags (dead trees) lie just below the surface of the reservoir (Sokatisewin Lake), while along its shoreline trees have a tilted or 'drunk' appearance.

Hydroelectric projects have an enormous impact on the environment. Crown corporations are the leading agencies in the design, construction, and operation of northern hydroelectric projects. As discussed in Chapters 3 and 5, provincial power companies have built large-scale hydroelectric projects across the provincial norths, and these projects have profoundly altered the physical character of river basins and the seasonal rhythm of streamflow in British Columbia, Manitoba, and Quebec. In all cases, hydroelectric development is an important element of the provincial economy. The purpose of energy development is to supply low-cost electrical power to industrial users and to export surplus power to markets in neighbouring provinces and the US. The physical transformation of the landscape and its impact on wildlife and Aboriginal peoples were considered acceptable trade-offs to governments. Hydroelectric developments in Manitoba and Quebec provide two case studies.

In Manitoba, initial plans called for the construction of hydroelectric projects on the Saskatchewan, Churchill, and Nelson rivers. In 1960, the first northern power station, Grand Rapid, was built on the Saskatchewan River as it enters Lake Winnipeg. In the following years, hydroelectric projects took place on the Nelson River. In 1961, the construction of the Kelsey hydroelectric dam was completed and the nickel smelter at Thompson consumed most of its output. Four other dams have been constructed: Jenpeg, Kettle Rapids, Long Spruce, and Limestone. In the early 1990s, plans to build the Conawapa Dam on the Nelson River were shelved because US markets were unable to absorb the power at high enough prices (the Great Whale River Project in northern Quebec faced a similar problem). In November 2002, the Conawapa Dam project gained fresh strength because of a combination of growing energy needs in Ontario and the pressure to reduce greenhouse gases, which eliminates the possibility of new coal-fired thermal electric plants in Ontario. For the Conawapa Dam to proceed to the construction phase, the high cost of building a hydroelectric power line from Manitoba to southern Ontario would have to be resolved. If this line is conceived of as the start of a national power grid, Ottawa might help with its financing (Benzie, 2002: A4).

In the early 1970s, Manitoba decided to increase the water flowing through the Nelson River and thus increase the generation of power by existing power stations. The plan was to divert most of the water from the Churchill River into the Nelson River. A control dam was built at the outlet of Southern Indian Lake to prevent water from continuing to flow to the mouth of the Churchill River, and a channel was dug between the southern edge of the lake and the Rat River, which flows into the Nelson River. The water level on Southern Indian Lake rose by three metres and in 1976 water began to flow into the Nelson Basin. While the amount of electric power generated by existing power stations on the Nelson River increased, the diversion of water from the Churchill River to the Nelson River is seen as an ecological disaster (Rosenberg et al., 1987: 81). Submerged vegetation still chokes the lake, causing oxygen depletion during the winter, and mercury levels in fish have risen, making frequent consumption unwise. Changes in the depth of water in the lake have greatly diminished the white-fish population, resulting in the collapse of the commercial whitefish fishery. Those fishers living in the Indian village of South Indian Lake received special compensation through the Northern Flood Agreement with Manitoba Hydro (ibid., 83). Overall, five First Nations in northern Manitoba were adversely affected by flooding arising from hydroelectric projects on the Nelson and Churchill rivers, and were to receive compensation from Manitoba Hydro through the Northern Flood Agreement. However, this agreement proved difficult to implement. In the 1990s, Tataskweyak Cree Nation, York Factory Cree Nation, Nisichawayasihk Cree Nation, and Norway House Cree Nation signed comprehensive implementation agreements that, with more precise terms, are proving easier to implement. As well, resource co-management boards (Manitoba and these First Nations) have been established as part of the agreements. Most importantly, enlightened partnerships between Manitoba's First Nations and Manitoba Hydro are now a fact of life. For instance, the Nisichawayasihk Cree Nation and Manitoba Hydro formed a partnership in 2006, the Wuskswatim Power Limited Partnership (see Chapter 5). One outcome has been an environmental assessment that stressed Aboriginal traditional knowledge to minimize project impacts, including limiting the flood area (Wuskwatim Power Limited Partnership, 2010).

La Grande Project in northern Quebec had the greatest geographic impact on the northern landscape with its extensive flooding of river valleys to create giant reservoirs and with its diversion of waters from one river to another to maximize the water flow through existing power stations (Vignette 6.7). Environmental impacts caused by damming La Grande Rivière (the first phase of the James Bay Project) are well documented (Gill and Cooke, 1975; Berkes, 1982; Rosenberg et al., 1987; Cloutier, 1987; Gorrie, 1990; Hornig, 1999). Coppinger and Ryan (1999: 69) played down the direct ecological impact of this massive project and suggested, rather, that newly constructed roads in the James Bay area have a much greater impact on the natural vegetation and wildlife because they increase the number of southern visitors and expand the hunting range of local people. In their words:

> As we have pointed out, the La Grande complex is not going to be an ecological disaster. But the roads leading to it and the increasing human population could be. People can have a very real impact on productivity, diversity, and species survival. The increasing number of people, be they miners, foresters, developers, electric workers, vacationers, campers, hunters, anglers, or residents, all change the nature of the region. Habitat loss through construction and development reduces the region's bioproductivity. It is the increasing stresses placed on individual species because of their commercial, ritual, or recreational value that further reduce productivity and diversity.

Fikret Berkes (1998: 106) substantiates the threat of road access. He reports that the Chisasibi Cree in 1983–4 took advantage of access to an area near the LG-4 dam where a herd of caribou had congregated: 'Large numbers were taken (the actual kill was unknown), even though the caribou stayed in the area only for a month or so. Chisasibi hunters used the road, bringing back truckloads of caribou. There was so much meat that, according to one hunter, "people overdosed on caribou".' This hunt involved 'shooting wildly, killing more than they could carry, and not disposing of wastes properly'. Like other caribou hunting peoples, the Quebec Cree strongly believe that caribou must be respected, meaning that hunters must only take what they need.

Vignette 6.7   **The Geographical Extent of the James Bay Project**

The scope of the James Bay hydroelectric project is staggering. It involves the trans-formation of the northern landscape in Quebec and a new lifestyle for the Cree and Inuit living in this area of the Subarctic. The aim of the project is to eventually harness the energy of all the rivers flowing through 350,000 km² of northwestern Quebec, thereby producing up to 28,000 megawatts of electrical power. At least a dozen rivers would have their water diverted, leaving shrunken waterways with pools of water and dried-up riverbeds. The water is to be collected in vast reservoirs and directed towards generating stations where electrical energy is or will be produced. This massive project involves some 20 rivers and about one-fifth of the total area of Quebec. The intent, since its inception 40 years ago, is to develop three major river basins: La Grande, the first phase of which was completed in 1985; Nottaway, which has proceeded with major diversions into the La Grande system since the 2002 sign-ing of the Paix des Braves by the Quebec Cree and the provincial government; and Great Whale, which was indefinitely postponed after a coalition of the Cree and environmental groups put forth a successful public relations campaign against the project (see Chapter 5).

This concept of a mutual relationship between the Cree and the caribou is strongly etched into Cree spirituality. Cree hunting leader Robbie Matthew translated this cul-tural tradition into a simple but foretelling phrase: 'show no respect and the game will retaliate' (ibid.). The following winter, 1984–5, no caribou appeared on the road where thousands were seen the previous year. The Elders recalled that a similar disaster took place at the turn of the century when hunters with repeating rifles lost all self-control and slaughtered the caribou at a crossing point along the Caniapiscau River. Then, too, the caribou disappeared. Elders, who know that change occurs in cycles, said that the caribou would return but that the hunters must take only what they need. When the caribou returned in 1985–6, the Chisasibi hunt was conducted in a controlled and responsible manner, illustrating that the Cree can exercise self-management.

Environmental Impact of the Forest Industry

In 2010, the Canadian Boreal Forest Agreement came into force, helping chart a new path for the forest industry and for the conservation of the boreal forest and its wildlife by requiring sustainable forestry practices. Signed by nine environmental organizations and 21 of Canada's forestry companies, the agreement covers the 72 million hectares licensed to the Forest Products Association of Canada, with 30 mil-lion hectares of woodland caribou habitat now off-limits to road-building and logging (Boychuk, 2011: 44). As a major concession to the forest industry, environmental groups (Canopy, David Suzuki Foundation, ForestEthics, and Greenpeace) have sus-pended their 'do-not-buy' campaigns in foreign markets and are adopting a pro-buy campaign of 'green' forest products. In sum, the heart of the agreement calls for four major developments (Canopy, 2011):

- a commitment to comprehensive land-use planning, which will reinforce the ecological-driven harvesting plan outlined in the agreement;
- more sustainable forest practices, including full utilization of trees into commercial products;
- greening the marketplace by convincing buyers to purchase higher-priced 'green' forest products because of the ecological soundness of their harvesting;
- an increased number of government-legislated protected areas.

Will this truce between formerly bitter rivals hold? The two goals are sound: on the environmental side, to conserve the boreal forest and its wildlife; for industry, to produce forest products that will sell in the global marketplace. The success of the agreement largely depends on the conversion of forest industry practices to become more sustainable and ecologically sound. Forest companies, in turn, believe that with the active support of the environmental groups, global markets will accept higher-priced green forest products.

The path now taken by the forest industry has slowly emerged because of pressure from local residents, Aboriginal groups, and NGOs who objected to air pollution from pulp mills and who witnessed extreme cases of water erosion due to clear-cutting close to rivers or on steep slopes. Stronger federal and provincial regulatory legislation followed. In the late 1980s, the federal government introduced stiffer regulations affecting the discharge of untreated or toxic industrial wastes into the environment, thus forcing the forest industry to meet much higher environmental standards. At that time, the forest industry alone was responsible for half of all the industrial wastes discharged into water bodies and 6 per cent of those sent into the atmosphere (Sinclair, 1990: 39). As discussed earlier, a shift towards better practices was noticeable in Alberta in the 1980s, when public concern and an environmental review board forced Alberta-Pacific to change its milling operation to reduce the volume of toxic chemical residues of dioxins and furans discharged into the Athabasca River. Other pulp mills on the Athabasca River—Weldwood at Hinton, Millar Western and Alberta Newsprint near Whitecourt, and Alberta Energy Company at Slave Lake—also had to meet the new standards. Ironically, the federal water monitoring system on the Athabasca River has reported that toxic substances from the mills have not affected the quality of the river water, but this federal monitoring effort has not tested for toxic wastes associated with the oil sands.

No industry is more vulnerable to public pressures than the forest industry because it depends on Crown-owned timber leases for its raw material. The forest industry realized that it must be more environmentally sensitive or else it could be denied timber leases, particularly in the more fragile northern environments. While clear-cutting has been banned or restricted in many parts of the developed world, the argument for clear-cutting is economic efficiency. Forest industry officials maintain that sustainable, balanced development of forests can best be achieved by clear-cutting because natural and artificial regeneration can then take place. Environmentalists disagree, arguing that regeneration is best after selective cutting. Other environmental problems associated with clear-cutting occur in mountainous and hilly terrain where serious erosion and flooding problems may result. Catastrophic floods and mudslides that take a large human toll are, unfortunately, common occurrences in many developing countries where clear-cutting in the past has affected slope stability and the ability of the land to absorb heavy rainfall.

In Canada today, the size of clear-cut areas is limited, and clear-cutting is banned on steep slopes. Yet, the contentious issue of clear-cutting remains a sore point with environmental groups. Foresters argue that clear-cutting emulates natural disturbances in the boreal forest and that it fosters silviculture (Pitt, 2009). One step in the right direction is found in the historic Canadian Boreal Forest Agreement, whereby 20 major timber companies and nine environmental groups agreed on a plan to protect 170 million acres of Canada's boreal forest, including a logging moratorium to protect endangered caribou. Yet, outside of the protected areas, clear-cutting remains a common logging practice. Equally unclear is whether this new agreement will convince European 'greens', who have urged the European Parliament to pass legislation boycotting Canadian forest products. Forest exports to Europe have declined in recent years from $2.3 billion in 2005 to $1.8 billion in 2008 (Natural Resources Canada, 2007b). Thus, it remains to be seen if exports under the 'green' approach to forestry will increase.

Figure 6.6   Clear-Cutting in the Boreal Forest

Clear-cutting, while the most efficient form of logging, strips away the forest, leaving the exposed land vulnerable to various types of erosion. As shown in the photograph, clear-cut logging is prohibited near water bodies because in times of heavy rainfall sediments and other debris will be carried into the stream or lake, causing both severe gullying and heavy deposition. The previously logged area left of the road indicates the slow nature of forest regeneration.

Source: Greenpeace. Photo by J. Henry Fair.

Vignette 6.8 **The Pine Beetle**

The pine beetle, *Dendroctonus ponderosae*, is a small insect, less than a centimetre long, that lives most of its life under the bark of pine trees, including lodgepole, ponderosa, and western white pine. The lifespan of an individual mountain pine beetle is about one year. Pine beetle larvae spend the winter under bark, feeding on the tree, which is often a mature lodgepole pine. The adult pine beetle emerges from an infested tree and seeks another host. Normally, when hard winter cold snaps maintain the beetle population at an acceptable level, these insects play an important role in the life of a forest. They attack old or weakened trees, speeding the development of a younger forest. Milder winters, however, have seen the mountain pine beetle population expand and increase their geographic range.

Sources: Carroll et al. (2006); Natural Resources Canada (2007a).

Environment Impact on Country Food: An Impending Health Disaster?

Harvesting food from the land and waters runs deep among Canada's Aboriginal peoples (Tait, 2001). This cultural practice remains strong in northern Canada, especially in the Arctic where Inuit still rely heavily on country food for much of their diet. Yet, industrial and urban pollution, whether from local mines, distant coal-burning plants, or contaminated sources of water, has introduced a health threat to those who consume country food on a daily basis. As Stout, Stout, and Harp (2009) point out, anecdotal reports bolstered by scientific evidence clearly indicate a growing health problem. One recommendation—both unpopular and an affront to traditional cultures—is to reduce consumption of country food, yet this is what has sustained Canada's original northern peoples for centuries, and a shift from country food to more processed foods, junk food, and more sugar in their diets has been a contributing factor in the high rates of diabetes among Aboriginal populations.

Contamination of wildlife presents a hidden risk to the consumption of country food. Industrial clusters of activity around the world but especially in Asia produce minute toxic particles belonging to chlorine-containing organic compounds that, by means of the global atmospheric and oceanic circulation systems, find their way into Arctic lands and waters where they become part of the food chain and thus have a negative impact on the health of Aboriginal peoples. Starting with single-celled algae, these toxic substances move up the food chain, becoming more concentrated with each step. This biological process is known as **biomagnification**. The main sources of these pollutants are the pesticide known as DDT and a chemical waste from industrial processes referred to as PCBs (polychlorinated biphenyls). By the 1970s, federal scientists were well aware of the presence of toxic elements in the Arctic food chain and its potential hazard to human life. Efforts by NGOs, especially the Canadian Arctic Resources Committee, succeeded in alerting the general public to this problem, and more recent studies have confirmed the danger to humans (Tesar, 2000; Delormier and Kuhnlein, 2000; Spears, 2002; Furgal et al., 2005). In 1989 a federal government official announced that a study of PCBs found relatively high levels in seals and caribou, the principal game consumed by Inuit. Furthermore, the study revealed that virtually all of the PCBs and related organochlorine contaminants found in southern Canada are also present in the Arctic, though usually at much lower levels. Beluga whales in the St Lawrence River, for example, have PCB levels 25 times higher than those in the eastern Arctic. At that time, Ottawa raised its concerns with other countries, and in 2001 Canada became the first country to sign and ratify the Stockholm Convention on limiting the use of organochlorine chemicals.

For the Inuit, the presence of organochlorine chemicals in wildlife is disturbing and is a potential serious health hazard. Government officials have stressed that they did not believe there was any immediate danger to those eating country food and that these traditional foods have greater nutritional value than many of the prepared foods available in the local stores. However, federal officials did caution pregnant women to limit their intake of country food. In 1990, a study of Arctic beluga whales in the western Arctic revealed high levels of cadmium and mercury. While the federal government has not yet set guidelines for cadmium, the Health Protection Branch in Ottawa recommends that people not eat fish containing more than 0.5 parts mercury per million. Since the whales captured by Inuit hunters had levels of mercury well above the guideline, the consumption of these mammals by the Inuit may be injurious to their health. But the scientific world expressed conflicting views over the threat of mercury to northern peoples. According to Caulfield (2000: 491–2):

> Preliminary research suggests that mercury levels in some regions of the Arctic are already high enough to put children at risk. People in northern Greenland and Inuit and Cree in the eastern Canadian Arctic have the highest exposures. Fully 29 per cent of Canadian Inuit in the eastern Arctic have daily intakes above limits recommended by the World Health Organization (WHO). Analysis of archaeological samples from the fifteenth-century Arctic shows that current mercury levels are three times higher than 500 years before.

Roebuck (1999) recognizes the trade-off between low levels of mercury in fish and the nutritional value of fish to Cree. He worries that without fish in their diet the Cree's nutritional and psychological well-being will suffer, and recommends that those who already have high levels of mercury change their diet, i.e., reduce the number of fish eaten, but that others could continue to consume fish. Roebuck (ibid., 88–9) describes this dilemma:

> All food supplies contain undesirable chemicals such as mercury. Because mercury is naturally present in the environment, it has undoubtedly been present in (i.e., has contaminated) fish throughout the world for eons. It was present in the fish of the James Bay region when the ancestors of the present-day Cree first migrated to the area. What is different now is a general and widespread elevation in concentration of mercury in fish due to global and regional industrial activity, and localized elevations from the formation of reservoirs.

Northern peoples are confused and even suspicious of this scientific information. Scientists are concerned that the message about food contamination is not reaching the Aboriginal population (Myers and Furgal, 2006; Tyrrell, 2006). In any case, how could they stop eating the traditional foods that Mother Earth provided to them? What would they substitute for country foods? In the Inuit community of Salluit in northern Quebec, Poirier and Brooke (2000) reported that local resistance to scientific information on contaminants in fish and wildlife is strong. Hunters trust in their own ability to detect a sick animal and the notion of 'poisoned' animals or fish, as described by scientists, is difficult for them to comprehend. But more than that, Inuit ontology is based on a trust relationship between the Inuit and the animal world. As an Inuk woman explained, 'We do not inflict injustices on the animals and trust that they in turn are good for us, we who do not wrong them in some way' (ibid., 87). Fortunately, the tide may be turning. The 'Monitoring Our Mothers' project looked at contaminant levels in the blood of pregnant Aboriginal women from 14 communities and found that levels of contaminants were lower than in a similar study in 2000 (Weber, 2007).

Conflicting World Views and the Environment

World views are derived from cultural beliefs and values. Western and Aboriginal cultures represent two distinct ways of viewing and understanding the world. The discussion so far has taken place using a Western paradigm. Aboriginal peoples have a different paradigm, steeped in their long-standing relationship with the land that evokes a different spiritual view of the place of humans in the natural world. While Western culture sees humans at the top of a hierarchical arrangement of living creatures, Aboriginal culture views the world from a holistic perspective where all life is seen as a series of relationships among equals. Can this non-Western approach to the land and wildlife help in addressing the issues noted above and in formulating a gentler and more respectful approach to the environment when the prospect of resource development enters the equation? At the Ninth Circumpolar General Assembly in Kuujjuaq, Nunavik, Sheila Watt-Cloutier (2002: 2), then president of Inuit Circumpolar Conference Canada, succinctly described the Inuit view of the relationship between the natural world and health:

> As Inuit, we think in holistic ways. We know that everything is interrelated—the threads of our lives are woven into a garment that is inherently sustainable. Our culture reflects our values, spirit, economy, and health. Our land and natural resources sustain us, and the health of these resources affects our health. If we use and develop these resources with respect, our environment will remain healthy and so will we. The process of the hunt is invaluable, [for] through it we learn what is required to survive and how to gain wisdom—the key to living and acting sustainably.

The original inhabitants of the North have much knowledge about their environment through their long-standing relationship with the land, the waters, the ice, and the wildlife. This local expertise, called traditional ecological knowledge (TEK), is an essential element of the holistic nature of the larger traditional knowledge of a specific Aboriginal culture. Unfortunately, TEK does not fit well within Western scientific approaches to resolving environmental problems. Still, TEK has become part of environmental investigations and offers both an alternative view of how to identify potential threats to the environment associated with a proposed industrial project and help in formulating research hypotheses. One example was provided by Fikret Berkes in a public lecture at the University of Saskatchewan (Berkes, 2003). Berkes revealed that during his work on global warming with the Inuvialuit at Sachs Harbour, NWT, an Elder mentioned that, as a child, thunderstorms were rare events but now were much more frequent. When Berkes relayed this information to a climatologist, the climatologist recognized its meteorological significance, namely that the Arctic storm track may have shifted further north. Climatologists can use such TEK as plausible hypotheses to launch scientific studies.

Vignette 6.9   **Traditional Knowledge and Traditional Ecological
Knowledge**

Traditional knowledge (TK) is a much broader concept than traditional ecological knowledge (TEK). Traditional knowledge is akin to the concept of culture; traditional ecological knowledge is one component of that culture. Graham White (2006: 405) has expressed the difference between TK and TEK:

Traditional Aboriginal knowledge and values about natural environments, including detailed understandings of the land and the behavior of animals in addition to ethical codes governing proper relations of humans to the land and animals—traditional ecological knowledge—are clearly crucial elements of TK. However, TK is a far broader concept than TEK (rendered in this way), encompassing as it does analyses and prescriptions for all manner of social interaction among peoples as well as deeply spiritual and philosophical precepts (often implicit and unspoken).

Peter Usher (2000) argues that TEK must be an essential element in environmental assessment and management. While Usher's view is specific to Nunavut, the concept, if valid in Nunavut, could complement Western scientific approaches in other jurisdictions as well. Progress is slow, but research conducted under the 2007–8 International Polar Year program contained a TEK element and, in a world of increasing environmental assessment, TEK is finding a place. In its assessment of the Mackenzie Gas Project, the Joint Review Panel made extensive use of TEK in their interpretation of potential environmental and social impacts.

How the Western capitalist and TEK paradigms can be meshed remains troublesome because attitudes, customs, and values in the two cultures are so different. Perhaps a bridge between the two paradigms, based on the principle of respect for the land, does exist, but progress along these cultural divides is slow and uncertain (Ellis, 2005; Woo et al., 2007; Christensen and Grant, 2007; Natcher, 1999, 2007). After all, the earlier account of an out-of-control Chisasibi Cree hunt must seem curiously superstitious to most non-Aboriginal readers, yet for northern Native hunters and Aboriginal people generally, this experience conveys an inviolable truth regarding the relationship between humans and the rest of the natural world.

Conclusion

Society—or that part of capitalist society that maintains power—has chosen development over the environment, but within limits. Still, Aboriginal people, who form the majority in many parts of the North, are feeling the negative fallout. In fact, the choice of development over the environment is best exemplified by the Alberta oil sands, and this choice colours Ottawa's international global environmental policy as exemplified by its restrained performance at international conferences on global warming.

The good news is that environmental assessment has gained strength over the years. Through the imposition of more demanding environmental impact assessments, the environment is better shielded from excessive industrial impacts, and, within the northern context, Aboriginal peoples are now more involved in resource projects and their environmental assessment. But without a doubt, economic development continues to affect the environment. The controversial oil sands projects provide such an example where massive operations threaten the environment. Efforts by the federal and Alberta governments to establish an effective monitoring system for the oil sands remain in the formative stage.

The landscape remains scarred by projects that took place prior to environmental assessment, and federal, provincial, and territorial governments have to address these environmental problems. Federal funding for the remediation of past environmental disasters is provided under the Federal Contaminated Sites Action Plan. For example, in 2007, a remediation plan was put in place to clean up the entire Giant mine site, including the long-term containment and underground storage of the 237,000 tonnes of highly toxic arsenic trioxide dust. Current resource projects must return sites to the former environmental condition. Whether this is realistically possible is another matter. The environmental assessment process seeks to determine if projected impacts fall within acceptable limits. Through this process, society has exerted its power to reject, control, and put constraints on resource proposals. Some 50 years ago, the environment was a virtual dumping ground for industrial wastes. No effective controls were in place, but no one worried because society had yet to recognize the pollution problem. The law was no help because it left private individuals and companies affected by pollution problems caused by others to seek a negotiated settlement with the polluter or to turn to the courts—always a costly proposition. Governments did not protect Crown land because they were too interested in promoting industrial development to pay attention to industrial pollution. Aboriginal title had not been recognized and so traditional homelands suffered.

Today, society is more conscious of the hidden costs of industrial projects, and governments have been pressured to pass legislation to protect the environment and to require developers to factor these hidden costs into their project proposals. Company proposals must be submitted to an environmental impact assessment panel for approval, rejection, or modification. Under the comprehensive land claim agreements, environmental impact assessment for land claim settlement areas is shared between the particular Aboriginal group and governments. Co-management of environmental impact assessment has allowed local issues, especially those affecting wildlife, to receive much more attention than under the former system controlled by Ottawa. Regional solutions have also emerged, with the federal Mackenzie Valley Resource

Management Act (1988) leading the way. Six boards, such as the Mackenzie Valley Land and Water Board, 'enable residents of the Mackenzie Valley to participate in the management of its resources for the benefit of the residents and of other Canadians' (Department of Justice Canada, 2007: 9.1).

The assessment and management process, while not perfect, is a vast improvement over the former lack of regulation. Co-management represents one improvement, thus providing local input into management decisions. Innovations in co-management are taking place, indicating a growing acceptance of this form of management in the North where Aboriginal peoples form a significant portion of the population and their reliance on country food places them in jeopardy (Armitage et al., 2007). Yet, one challenge remains: how to include global warming and its potential impact on Inuit and other northern cultures into an assessment and management process. While Canadians are aware of the changing nature of climate, sea ice, and permafrost, the Inuit especially have expressed strong concerns about global warming, particularly its negative effect on shore ice and on the hunting of a major food staple, seals. Perhaps with two Aboriginal cabinet ministers from Canada's North in the Conservative government, Canada will 'punch above its weight' in global environmental affairs.

Challenge Questions

1. Temperate areas of Canada are subjected to more industrial pollution, so why is there so much concern about such pollution in the North and its effect on Aboriginal peoples?
2. Under the Canadian Environmental Assessment Act of 1995, the environmental assessment process consists of three basic stages. What are these three stages?
3. What is the objective of 'decommissioning' a mine site?
4. In July 2011, Ottawa announced a 'world-class' monitoring plan for the Alberta oil sands. Why might this plan fail to materialize—at least in the original form suggested by Ottawa?
5. Explain why a basic tenet of the Ojibway culture is a trust in their environment.
6. What factors in co-management strengthen the place of Aboriginal peoples in the assessment process?
7. Oil sands projects are very expensive. Imperial Oil sought to minimize its construction costs by having much of its Kearle project produced in Korea and then shipped to Fort McMurray via the Columbia River and mountain highways in the US and then to Fort McMurray, Alberta. What went wrong with this plan? Do these unexpected extra transportation costs and opposition from US environmental groups make a case for such manufacturing to take place in Alberta?
8. Why are the federal and territorial governments reluctant to approve all the recommendations of the Joint Review Panel for the Mackenzie Gas Project?
9. Why do the Inuit feel that global warming is affecting their culture so strongly?
10. How have large hydroelectric projects in Manitoba and Quebec created more Aboriginal participation in these projects, which, in turn, has created a cultural tipping point for both the Crown corporations and northern residents?

Notes

1. Between 1962 and 1975 Dryden Chemicals Ltd, a subsidiary of Reid Paper Ltd, produced chlorine and other chemicals used as bleach in the pulp and paper mill of Reid Paper at Dryden, Ontario. The mill flushed its waste products into the Wabigoon River. The mill effluent contained a relatively high level of mercury, which worked its way into the aquatic food chain of the river system. In 1970, the Ontario government discovered that the level of mercury found in fish in a 500-km stretch downstream from the pulp and paper mill was dangerous to health, and advised the Ojibway communities at Grassy Narrows and Whitedog reserves not to eat fish from these rivers. It also banned commercial fishing on all lakes and tributaries of the English and Wabigoon rivers. The impact on the Ojibway was staggering. First, 90 members from the Whitedog and Grassy Narrows reserves exhibited serious neurological symptoms characteristic of methyl mercury poisoning (Roebuck, 1999: 80). Second, the Ojibway lost their economic base with the collapse of commercial fishing and guiding income. Third, they could no longer trust their environment because one of their traditional sources of food, fish, was the source of their illness. This loss of trust in a basic tenet of their culture was psychologically devastating.

2. Until the Port Radium mine was closed in 1960, Sahtu Dene workers were hired to carry sacks of radioactive ore around the rapids on the Bear River, which were reloaded on barges for the long trip south along the Mackenzie River. When cancer cases started showing up among the former workers, the community became alarmed. Deline, NWT, became known as the 'Village of Widows'. The Sahtu Dene brought their concerns to the public and the federal government decided to support a joint investigation into the possible impacts of the Port Radium mine operation on the local people and their environment. According to the 2005 report on possible impacts on the health of Sahtu Dene men who handled the sacks of ore, 'it is not possible to know for certain if the illness or death of any individual Deline ore carrier was directly caused by radiation exposure' (Markey, 2005). The following year, the federal government approved funds for the cleanup of the mine site at Port Radium. The cleanup includes sealing mine openings, safely disposing of equipment and demolishing structures on the site, dealing with exposed tailings, and continued environmental monitoring (Environment Canada, 2007).

3. Arctic pipeline proposals and inquiries certainly did not end with the Berger Report. In 1977, Foothills proposed an alternate route for Alaskan gas—the Alaska Highway Gas Pipeline Project. This project was approved in 1982 but has not yet been built because the cost of delivering natural gas from Prudhoe Bay to the Chicago market is still not commercially attractive (though this project is currently being reviewed and may receive financial assistance from the US government).

4. Next in line was a joint project by Esso Resources Canada Ltd and Interprovincial Pipe Lines (NW) Ltd to construct a pipeline from Norman Wells to Zama, Alberta, which has been discussed at length in Chapter 5. In this case, the proposed project was approved and completed in 1985. The oil industry saw this project as a test case for northern pipelines and, if successful, as a model for much larger pipeline projects to bring Mackenzie Delta gas and Beaufort Sea oil to North American markets. So far, the Norman Wells pipeline has been successful.

5. Gateway Hill, a former Syncrude surface mine, began its restoration in 1983. In 2010, Alberta Environment has certified Gateway Hill restored to the equivalent of its pre-mine landscape. However, this restoration was not a tailings pond but a shallow open-pit mine. For more on this, see Testa (2010).

6. Suncor's first tailings pond, built in the late 1960s, was in operation for 40 years. In 2006, the pond began the long process of reclamation into a sustainable landscape. First, excess

water and fine tailings were removed from the bottom of the pond and transferred to an active tailings pond. Then, the pond was filled with 30 million tonnes of reclaimed tailings sand. At that point, drainage systems were established and low tracts of land were used to manage water runoff. Landscaping then began with 1.2 million cubic metres of topsoil placed over the surface, to a depth of 50 centimetres. Trees, shrubs, and grasses were planted, allowing Suncor to reach the first step in its restoration commitment.

References and Selected Reading

Areva. 2006. 'Cluff Lake'. At: <www.arevaresources.com/operations/clufflake.html>.

Armitage, D.R. 2005. 'Environmental Impact Assessment in Canada's Northwest Territories: Integration, Collaboration and the Mackenzie Valley Resource Management Act', in Hanna (2005: 185–211).

———, Fikret Berkes, and Nancy Doubleday, eds. 2007. *Adaptive Co-Management: Collaboration, Learning and Multi-Level Governance*. Vancouver: University of British Columbia Press.

Benzie, Robert. 2002. 'PM touts Manitoba Hydro Dam to Eves', *National Post*, 27 Nov., A4.

Berger, Thomas R. 1977. *Northern Frontier, Northern Homeland: The Report of the Mackenzie Valley Pipeline Inquiry*. Ottawa: Department of Supply and Services.

Berkes, Fikret. 1982. 'Preliminary Impacts of the James Bay Hydroelectric Project, Quebec, on Estuarine Fish and Fisheries', *Arctic* 35, 4: 524–30.

———. 1998. 'Indigenous Knowledge and Resource Management Systems in the Canadian Subarctic', in Berkes and Folke (1998: 98–129).

———. 2003. 'Inuit Observations of Climate Change: Designing a Collaborative Project', public lecture at the University of Saskatchewan, 9 Jan.

——— and Carl Folke, eds. 1998. *Linking Social and Ecological Systems: Management Practices and Social Mechanisms for Building Resilience*. Cambridge: Cambridge University Press.

Bone, Robert M. 2003. 'Power Shifts in the Canadian North: A Case Study of the Inuvialuit Final Agreement', in Robert B. Anderson and Robert M. Bone, *Natural Resources and Aboriginal People in Canada: Readings, Cases and Commentary*. Concord, Ont.: Captus Press: 382–93.

Boychuk, Rick. 2011. 'The Boreal Handshake', *Canadian Geographic* 131, 1: 30–43.

Bronson, J.E., and B.F. Noble. 2006. 'Health Determinants in Canadian Northern Environmental Impact Assessment', *Polar Record* 42, 4: 1–10.

Burgess, Margaret M., and Daniel W. Riseborough. 1989. *Measurement Frequency Requirements for Permafrost Ground Temperature Monitoring: Analysis of Norman Wells Pipeline Data, Northwest Territories and Alberta*. Geological Survey of Canada, Paper 89–1D. Ottawa: Energy, Mines and Resources Canada.

———, J.F. Nixon, and D.E. Lawrence. 1998. 'Seasonal Pipe Movement in Permafrost Terrain, KP2 Study Site, Norman Wells Pipeline', in A.G. Lewkowicz and M. Allard, eds, *Proceedings, Seventh International Conference on Permafrost*. Ste Foye, Que.: Centre d'étude nordiques, Université Laval, 95–100.

Canada and the Northwest Territories. 2009. 'Governments of Canada & of the Northwest Territories: Final Response to the Joint Review Panel Report for the Proposed Mackenzie Gas Project', 15 Nov. At: <www.ceaa.gc.ca/Content/7/1/B/71B5E4CF-8898-42A3-989B-49ADC0426A4F/MGP_Final_Response.pdf>.

Canadian Environmental Assessment Agency. 2009. Sustainable Development Strategy: 2007–2009: 2.1 Role of Environmental Assessment in SD, 10 Mar. At: <www.ceaa.gc.ca/default.asp?lang=En&n=EB9E8950-1&offset=4&toc=show>.

———. 2011. 'Canadian Environmental Assessment Agency', 19 Jan. At: <www.ceaa.gc.ca/default.asp?lang=En>.

Canopy. 2011. *Canadian Boreal Forest Agreement*. At: <canopyplanet.org/uploads/Agreement101ENG-lr.pdf>.

Cantox Environmental. 2006. 'Oil Sands'. At: <www.cantoxenvironmental.com/sectors/oilgas/oilsands/>.

Carroll, A.L., J. Régnière, J.A. Logan, S.W. Taylor, B. Bentz, and J.A. Powell. 2006. *Impacts of Climate Change on Range Expansion by the Mountain Pine Beetle*. Mountain Pine Beetle Initiative Working Paper 2006–14. Victoria, BC: Natural Resources Canada, Canadian Forest Service, Pacific Forestry Centre.

Carson, Rachel. 1962. *Silent Spring*. Boston: Houghton Mifflin.

Caulfield, Richard A. 2000. 'The Political Economy of Renewable Resource Management in the Arctic', in Mark Nuttall and Terry V. Callaghan, eds, *The Arctic: Environment, People, Policy*. Amsterdam: Harwood Academic Publishers, ch. 17.

CBC News. 2007. 'Governments Spending $25M to Clean Up Uranium Mines', 3 Apr. At: <www.cbc.ca/canada/saskatchewan/story/2007/04/03/uranium-mines.html>.

———. 2010. 'Dewline Clean-up Continues in Nunavut', 29 Nov. At: <www.cbc.ca/canada/north/story/2010/12/29/dew-line.html>.

———. 2011. 'Toxic Chemicals Released by Melting Arctic Ice', 25 July. At: <www.cbc.ca/news/canada/north/story/2011/07/25/science-climate-arctic-ice.html>.

Christensen, J., and M. Grant. 2007. 'How Political Change Paved the Way for Indigenous Knowledge: The Mackenzie Valley Resource Management Act', *Arctic* 60, 2: 115–23.

Cloutier, Luce, 1987. 'Quand le mercures élevé trop haut à la Baie James . . .', *Acta Borealia* 4, 1 and 2: 5–23.

Coppinger, Raymond, and Will Ryan. 1999. 'James Bay: Environmental Considerations for Building Large Hydroelectric Dams and Reservoirs in Quebec', in Hornig (1999: 41–72).

Damsell, Keith. 1999. 'Mining's Toxic Orphans Come of Age', *National Post*, 18 Sept., FP1.

Delormier, Treena, and Harriet V. Kuhnlein. 1999. 'Dietary Characteristics of Eastern James Bay Cree Women', *Arctic* 52, 2: 182–7.

Department of Justice. 2007. Mackenzie Valley Resource Management Act (1998, c. 25). At: <laws.justice.gc.ca/en/frame/cs/M-0.2///en>.

Dowdeswell, Liz (Chairperson), Peter Dillon, Subhasis Ghoshal, Andrew Miall, Joseph Rasmussen, and John Smol. 2010. *A Foundation for the Future: Building an Environmental Monitoring System for the Oil Sands*. Report of the Oil Sands Advisory Panel. Environment Canada, Dec. At: <www.ec.gc.ca/pollution/default.asp?lang=En&n=E9ABC93B-1>.

Draper, Dianne. 2002. *Our Environment: A Canadian Perspective*, 2nd edn. Toronto: Nelson.

Duffy, Patrick. 1981. *Norman Wells Oilfield Development and Pipeline Project*. Report of the Environmental Assessment Panel. Ottawa: Minister of Supply and Services.

Duhaime, Gérard, Nick Bernard, and Robert Comtois. 2005. 'An Inventory of Abandoned Mining Exploration Sites in Nunavik, Canada', *Canadian Geographer* 49, 3: 260–71.

Einstein, Norman. 2006. 'Image: Athabasca Oil Sands', *Wikipedia*. At: <en.wikipedia.org/wiki/Image:Athabasca_Oil_Sands_map.png>.

Ellis, S.C. 2005. 'Meaningful Consideration? A Review of Traditional Knowledge in Environmental Decision Making', *Arctic* 58, 1: 66–7.

Environment Canada. 2007. 'Federal Contaminated Sites Receive Funding'. At: <www.ec.gc.ca/default.asp?lang=En&xml=81941DCD-F8FA-4012-9266-CEC4F186B0F7>.

———. 2009. 'Governments of Canada and Northwest Territories Release the Final Response to the Joint Review Panel's Report for the Mackenzie Gas Project', 15 Nov. At: <ec.gc.ca/default.asp?lang=En&n=714D9AAE-1&news=5FC1AA1E-5DE5-4A50-9C19-E67FA8379A5A>.

———. 2011. 'Canada's Environment Minister Announces an Integrated Plan for Oil Sands Monitoring', 21 July. At: <www.ec.gc.ca/default.asp?lang=En&n=714D9AAE-1&news=DA1E8CBC-D0A6-4304-A1DD-A9206D0818AB>.

Fowlie, Jonathan. 2010. 'Ottawa Halts $815-Million Prosperity Mine', *Vancouver Sun*, 3 Nov. At: <www2.canada.com/vancouversun/news/story.html?id=19531422-0a35-4e37-8ac9-9f3c1c950e11>.

Furgal, C.M., T.D. Garvin, and C.G. Jardine. 2010. 'Trends in the Study of Aboriginal Health Risks in Canada', *International Journal of Circumpolar Health* 69, 4: 322–32.

———, S. Powell, and H. Myers. 2005 'Digesting the Message about Contaminants and Country Foods in the Canadian North: A Review and Recommendations for Future Research and Action', *Arctic* 58: 103–14.

Galbraith, Lindsay, Ben Bradshaw, and Murry B. Rutherford. 2007. 'Towards a New Super-Regulatory Approach to Environmental Assessment in Northern Canada', *Impact Assessment and Project Appraisal* 25, 1: 27–41.

Geddes Resources Limited. 1990. *Windy Craggy Project: Stage 1: Environmental and Socioeconomic Impact Assessment.* Toronto: Geddes Resources.

Gilbert, Richard. 2011. 'Kearle Oilsands Project Faces Protest', *Daily Commercial News and Construction Record*, 11 Apr. At: <dcnonl.com/article/id43847>.

Gill, Don, and Alan D. Cooke. 1975. 'Hydroelectric Developments in Northern Canada: A Comparison with the Churchill River Project in Saskatchewan', *The Musk-Ox* 15: 53–6.

Gorrie, Peter. 1990. 'The James Bay Power Project', *Canadian Geographic* 110, 1: 21–31.

Grant, Jennifer, Simon Dyer, and Dan Woynillowicz. 2008. *Fact or Fiction: Oil Sands Reclamation.* The Pembina Institute. At: <pubs.pembina.org/reports/fact-or-fiction-report-rev-dec08.pdf>.

Hanna, Kevin S., ed. 2009. *Environmental Impact Assessment: Practice and Participation*, 2nd edn. Toronto: Oxford University Press.

Hornig, James F., ed. 1999. *Social and Environmental Impacts of the James Bay Hydroelectric Project.* Montreal and Kingston: McGill-Queen's University Press.

Indian and Northern Affairs Canada. 2010. 'Giant Mine', 12 Aug. At: <www.ainc-inac.gc.ca/ai/scr/nt/cnt/gm/ab/index-eng.asp>.

James Bay Energy Corporation. 1988. *La Grande Rivière: A Development in Accord with Its Environment.* Montreal: James Bay Energy Corporation.

Joint Review Panel. 2009. *Foundation for a Sustainable Northern Future: Report of the Joint Review Panel for the Mackenzie Gas Project*, Dec. At: <www.ngps.nt.ca/report.html>.

Kelly, Erin N., David W. Schindler, Peter V. Hodson, Jeffrey W. Short, Roseanna Radmanovich, and Charlene C. Nielsen. 2010. 'Oil Sands Development Contributes Elements Toxic at Low Concentrations to the Athabasca River and Its Tributaries', *Proceedings, National Academy of Sciences.* At: <www.pnas.org/content/early/2010/08/24/1008754107.full. pdf+html?sid=33a2f9a7-e7ba-42ee-b6f4-d41203d748a7>.

———, Jeffrey W. Short, David W. Schindler, Peter V. Hodson, M. Ma, A.K. Kwan, and B.L. Fortin. 2009. 'Oil Sands Development Contributes Polycyclic Aromatic Compounds to the Athabasca River and Its Tributaries', *Proceedings, National Academy of Sciences* 106, 52: 22346–51.

Kendrick, Anne. 2000. 'Community Perceptions of the Beverly-Qamanirjuaq Caribou Management Board', *Canadian Journal of Native Studies* 20, 1: 1–33.

McCarthy, Shawn. 2010. 'Ottawa, Alberta Blamed for Lax Oil-Sands Oversight', *The Globe and Mail*, 21 Dec., B1.

McDonald, Adrian. 2001. 'Environment at the Crossroads', in Jacqueline West, ed., *The USA and Canada 2002*, 4th edn. Old Woking, Surrey: Unwin Brothers, 22–8.

MacInnes, K.L., M.M. Burgess, D.G. Harry, and T.H.W. Baker. 1989. *Permafrost and Terrain Research and Monitoring: Norman Wells Pipeline.* Environmental Studies No. 64, vols 1 and 2. Ottawa: Minister of Supply and Services.

Markey, Andrea. 2005. 'Deline Uranium Report Released', *Northern News Service.* At: <www.nnsl.com/frames/newspapers/2005-09/sep12_05rad.html>.

Marslen, William. 2008. *Stupid to the Last Drop: How Alberta Is Bringing Environmental Armageddon to Canada (and Doesn't Seem to Care).* Toronto: Vintage Canada.

Myers, H., and C. Furgal. 2006. 'Long-Range Transport of Information: Are Arctic Residents Getting the Message about Contaminants?', *Arctic* 59, 1: 47–60.

National Energy Board. 2010. *Mackenzie Gas Project: Reasons for Decision*, Apr. At: <www.neb.gc.ca/clf-nsi/rthnb/pplctnsbfrthnb/mcknzgsprjct/rfd/rfd-eng.html>.

Natural Resources Canada. 2006. *Norman Wells Pipeline Research.* Geological Survey of Canada. At: <gsc.nrcan.gc.ca/permafrost/pipeline_e.php>.

———. 2007a. *The Mountain Pine Beetle Program.* At: <mpb.cfs.nrcan.gc.ca/biology/index_e.html>.

———. 2007b. 'Statistics and Facts on Forestry'. At: <www.nrcan.gc.ca/statistics/forestry/default.html>.

Nikiforuk, Andrew. 2008. *Tar Sands: Dirty Oil and the Future of a Continent*. Vancouver: Greystone Books.

Noble, Bram F. 2000. 'Strengthening EIA through Adaptive Management: A Systems Perspective', *Environmental Impact Assessment Review* 20: 97–111.

———. 2010. *Introduction to Environmental Impact Assessment: A Guide to Principles and Practice*, 2nd edn. Toronto: Oxford University Press.

North American Council for Environmental Cooperation. 2007. 'Ontario Logging'. At: <www.cec.org/trio/stories/index.cfm?varlan=english&ed=20&ID=216>.

Ontario. 2010. 'Cleaning up the Deloro Mine Site in Marmora', *Newsroom*, 8 Nov. At: <http://news.ontario.ca/ene/en/2010/11/cleaning-up-the-deloro-mine-site-in-marmora.html>.

O'Reilly, Kevin. 2005. 'New Maps Show Impact of NWT Gas Development', CARC news release. At: <72.14.253.104/search?q=cache:BfHT084xNdEJ:www.carc.org/2005/news%2520release%2520backgrounder.doc+Cumulative+environmental+impacts+of+the+Mackenzie+Delta+Project&hl=en&ct=clnk&cd=1&gl=ca&client=firefox-a>.

Peters, Evelyn J. 1999. 'Native People and the Environmental Regime in the James Bay and Northern Quebec Agreement', *Arctic* 52, 4: 395–410.

Pitt, Doug. 2009. 'Harvesting Practices in Canada's Boreal Forest', Natural Resources Canada. At: <canadaforests.nrcan.gc.ca/article/video/boreal/harvesting/trscpt>.

Poirier, Sylvie, and Lorraine Brooke. 2000. 'Inuit Perceptions of Contaminants and Environmental Knowledge in Salluit, Nunavik', *Arctic Anthropology* 37, 2: 78–91.

Robinson, Allan. 1991. 'Exploring the Risks at Windy Craggy', *The Globe and Mail*, 19 Jan., B1.

Roebuck, B.D. 1999. 'Elevated Mercury in Fish as a Result of the James Bay Hydroelectric Development: Perception and Reality', in Hornig (1999: ch. 4).

Rosenberg, D.M., R.A. Bodaly, R.E. Hecky, and R.W. Newbury. 1987. 'The Environmental Assessment of Hydroelectric Impoundments and Diversions in Canada', in M.C. Healey and R.R. Wallace, eds, *Canadian Aquatic Resources*. Ottawa: Department of Fisheries and Oceans, 71–104.

Saskatchewan Environment. 2002. *An Assessment of Abandoned Mines in Northern Saskatchewan*. File R3160. Regina: Clifton Associates Ltd.

Schindler, D. 2010. 'Tar Sands Need Solid Science', *Nature* 468, 7323: 499–501.

Searle, Rick. 1991. 'Journey to the Ice Age: Rafting the Tatshenshini, North America's Wildest and Most Endangered River', *Equinox* 55, 1: 24–35.

Shkilnyk, Anastasia M. 1985. *A Poison Stronger Than Love: The Destruction of an Ojibwa Community*. New Haven: Yale University Press.

Sinclair, William F. 1990. *Controlling Pollution from Canadian Pulp and Paper Manufacturers: A Federal Perspective*. Ottawa: Environment Canada/Minister of Supply and Services.

Sinclair, John, and Meinhard Doette. 2010. 'Environmental Assessment in Canada: Encouraging Decisions for Sustainability', in Bruce Mitchell, ed., *Resource and Environmental Management in Canada: Addressing Conflict and Uncertainty*. Toronto: Oxford University Press, 462–94.

Spears, Tom. 2002. 'Arctic Pollution Causing Natives Serious Health Problems: Study', *National Post*, 19 Aug., A13.

Stout, R., Dionne Stout, and R. Harp. 2009. *Maternal and Infant Health and the Physical Environment of First Nations and Inuit Communities: A Summary Review*. Project #183 of the Prairie Women's Health Centre of Excellence, Winnipeg, Apr. At: <www.pwhce.ca/pdf/AborigMaternal_environment.pdf>.

Suncor Energy. 2011. *Pond 1 Reclamation: 2010 Report on Sustainability*, Jan. At: <sustainability.suncor.com/2010/en/responsible/3508.aspx>.

Tait, Carrie, Nathan VanderKlippe, and Josh Wingrove. 2011. 'Alberta Conservation Plan Stuns Oil Patch', *The Globe and Mail*, 5 Apr., B1.

Tait, Heather. 2007. *Harvesting and Country Food: Fact Sheet*. Statistics Canada 2001 Aboriginal Peoples Survey, Catalogue no. 89–627–XIE, Apr. At: <www.statcan.gc.ca/pub/89-627-x/89-627-x2007001-eng.pdf>.

Tesar, Clive. 2000. 'pops: What They Are; How They Are Used; How They Are Transported', *Northern Perspective* 26, 1: 2–5.

Testa, Bridget Mintz. 2010. 'Reclaiming Alberta's Oil Sands Mines: Forest, Wetlands and a Lake Suggests a Boreal Ecosystem Can Be Rebuilt', *Earth*. At: <www.earthmagazine.org/earth/article/30d-7da-2-16>.

Tyrrell, M. 2006. 'Making Sense of Contaminants: A Case Study of Arviat, Nunavut', *Arctic* 59, 4: 370–80.

Usher, Peter. 2000. 'Traditional Ecological Knowledge in Environmental Assessment and Management', *Arctic* 53, 2: 183–93.

Vanderklippe, Nathan. 2011a. 'Crews Race to Repair Alaska Pipeline', *The Globe and Mail*, 10 Jan., B1.

———. 2011b. 'Total's Joslyn Oil Sands Mine Approved', *The Globe and Mail*, 27 Jan., B1.

———. 2011c. 'Judge Puts Brakes on Oil Sands Equipment', *The Globe and Mail*, 21 July, B4.

Watt-Cloutier, Sheila. 2002. 'Speech Delivered by Sheila Watt-Cloutier, President ICC Canada and Vice-President icc International, at the 9th General Assembly and 25th Anniversary of the Inuit Circumpolar Conference', *Silarjualiriniq: Inuit in Global Issues* 12 and 13: 1–8.

Weber, Bob. 2007. 'Contaminant Levels Dropping among Arctic Mothers, Blood Studies Show', *The Globe and Mail*, 29 Oct., A10.

———. 2009. 'Arctic Pipeline Environmental Study Expected This Week', *The Globe and Mail*, 27 Dec., B1.

———. 2010. 'Mutant Fish Lead to Calls for Ottawa to Monitor Oil Sands', *The Globe and Mail*, 17 Sept., B1.

Wenzel, George W. 1999. 'Traditional Ecological Knowledge and Inuit: Reflections on TEK Research and Ethics', *Arctic* 52, 2: 113–24.

White, Graham. 2006. 'Cultures in Collision: Traditional Knowledge and Euro-Canadian Governance', *Arctic* 59, 4: 401–14.

Woo, M.-K., et al. 2007. 'Science Meets Traditional Knowledge: Water and Climate in the Sahtu (Great Bear Lake) Region, Northwest Territories, Canada', *Arctic* 60, 1: 37–46.

Wuskwatim Power Limited Partnership. 2010. 'About the Wuskwatim Generating Station Project'. At: <www.wuskwatim.ca/project.html>.

7

Aboriginal Economy and Society

Great Strides and Defining Questions

Over the past four decades, **Aboriginal peoples** have taken giant strides towards securing a new place in Canadian society. In that time, they have moved from a restrictive colonial world to a more open post-colonial one where they have gained a measure of control over their lives. And this empowerment keeps growing. Comprehensive land claim agreements have played a critical part in this transformation by allowing a duality—participation in both the market economy and their traditional economy. In the case of the Inuvialuit Agreement in 1984, for example, the Inuvialuit Regional Corporation functions in the market economy while the Inuvialuit Game Council maintains their interest in their hunting economy.

In this chapter, our attention is on the involvement of Aboriginal peoples in the northern resource economy; the challenge to their culture and related social issues; and the impact of global warming on their way of life and on the northern environment. In addition, the emergence of the Inuit **homeland**, Nunavut, and other forms of Aboriginal regional government are examined.

While the northern resource economy forms only one piece of a much larger puzzle, participation in the resource economy, whether as individuals or as corporations, marks a quantum leap into the capitalistic world by Aboriginal peoples. This participation also has implications for the culture, especially their languages, and their social well-being. With the broad parameters for their journey through the twenty-first century already set through land claim agreements, the ramifications for their traditional land-based economy and society are both profound and irreversible. The search for Aboriginal self-government, for instance, loses much of its meaning if this political gain is not accompanied by similar cultural and economic gains. Our discussion begins with an examination of three key events:

- climatic shift and its effect on Inuit hunters and their food supply;
- relocation of Aboriginal peoples from the land to settlements;
- the significance of Aboriginal title for land claims and impact benefit agreements.

Climatic Shift

Without a doubt, the North is warming, with consequences for nature and people (see Chapter 2). The Arctic is the focus of this climatic shift. Snow, ice, and permafrost are in a state of retreat. Not only are summers longer, but later freeze-up and earlier spring breakup translate into challenges for Inuit hunters in their quest for **country food** so necessary for their diet and for sustaining this aspect of their culture. Significant adaptations to the changing Arctic environment may be in the works, perhaps as dramatic as a return to bowhead whale hunting as in the days of the Thule (see Chapter 3).

This climatic shift is a product of our industrial world, and it presents a serious challenge to the Inuit. As the twenty-first century advances, the Arctic climate is anticipated to become much warmer and sea-ice cover is expected to be much less; in other words, the area of the Arctic will significantly decline. Sea ice, so important to Inuit communities and to Inuit hunters, is losing its grip on Arctic waters. Already, land-fast sea ice, so critical for Inuit hunters, lasts for a shorter period of the year and open water exists for a longer period (Aporta, 2010, Laidler, 2010; also see Chapter 8). Inuit marine hunters are faced with a shorter hunting season on sea ice and, when travelling by snowmobile, the danger of falling through usually thin sea ice has become a serious problem (Ford, 2009; Laidler, 2009, 2010). Food security for the Inuit depends heavily on accessing marine life from sea ice. In turn, seal is a critical element in their diet and is essential for their health and maintenance of their traditional culture.

Geographic Shift

The relocation of Indians, Inuit, and Métis from the land to settlements, which took place in the 1950s and 1960s, was a form of social engineering. One unintended but profound consequence was the emergence of a sense of regional consciousness on the part of many Aboriginal groups. The most striking example occurred in the Arctic with **Nunavut**, while the Grand Council of the Crees (Eeyou Istchee) and its Cree Regional Authority have represented another example of regional government. **Nunavik**, in Arctic Quebec, appeared to be on the verge of following the Nunavut example until, in April 2011, the Inuit of the region voted overwhelmingly against ratification of an agreement with the provincial and federal governments (see below).

While this federal relocation policy was well intended, Ottawa failed to recognize the enormous cultural and economic adjustments necessary to make settlement life sustainable. Instead, the movement to settlements marked the beginning of a difficult and often bumpy journey from a nomadic way of life to a sedentary one where a sense of cultural drift and **dependency** on Ottawa took root. People were moved to tiny outposts where fur traders, the RCMP, and missionaries resided. Relocation transformed the Canadian North by creating approximately 200 Native settlements. Some remain small, like Sachs Harbour[1] on Banks Island in the Northwest Territories, which had a population of 122 in 2006; others, like Sandy Bay in northern Saskatchewan and Rankin Inlet in Nunavut, have seen population increases. In 2006, the populations of Sandy Bay and Rankin Inlet were 1,175 and 2,358, respectively.

The immediate purpose of relocation involved food security and access to public services, but the ultimate goal was integration into Canadian society. Once people arrived in settlements, the first crisis was an acute shortage of housing (which remains an issue today). The second crisis was economic—or rather, the absence of a settlement economy. Complicating matters, life in settlements was (and remains) much more expensive than living on the land. Trapping continued, but it did not generate enough cash. Little work was available in the settlements so wage employment was only available to a few. The only alternative was welfare. Clearly, the Achilles heel of the relocation 'strategy' was the absence of a strong settlement economy. Dependency became a way of life. One solution is to subsidize hunters and trappers to remain on the land for part of the year. The rationale is to provide an alternative to settlement living, to minimize welfare payments, and to reduce the dependency mentality, which often leads to social ills such as alcohol and drug abuse. Only Quebec has implemented such an enlightened program, which in 2009–10 cost $19.7 million (Vignette 7.1). The impetus for this social program took place at the time of the James Bay and Northern Quebec Agreement (JBNQA), when the Quebec officials recognized the high economic and human cost of shifting from a land-based culture to a welfare/

Vignette 7.1   The Quebec Income Security Program

Hunting, fishing, and trapping activities today require relatively expensive equipment and transportation. In addition to these expenses, hunters must also continue to pay for maintaining a home in the community. Many hunters combine traditional activities and subsequent benefits paid by Quebec's **Cree Hunters and Trappers Income Security Program** with seasonal employment in order to earn sufficient income. In 2009–10, total program benefits ($19,741,091) represented 70 per cent of the income of 2,432 beneficiaries (1,733 adults and 699 children) enrolled in the program (Quebec, 2010: 37). The remaining $8,402,554 came primarily from employment. This family-oriented program provided an average income of roughly $16,000 per couple, with slightly more income for those with children. Depending on the number of days on the land and the size of the family, the range of income varied from less than $6,000 to more than $24,000 (Quebec, 2010: Figure 3).

Since the implementation in 2002–3 of an insurance fund, revenue losses occurring because of an inability to carry out the harvesting or related activities due to illness or disaster may be partially compensated. In 2009–10, this fund compensated 429 sick-leave days while no event constituting a disaster was claimed during that program year.

Participation in the Income Security Program remains fairly stable from year to year, but older people continue to be major participants. Consultations held in Cree communities in the last year indicate that many older people perceive the program as a viable retirement plan that helps to support their participation in traditional activities following retirement from the workforce. For example, the age structure of the hunters and trappers reveals that those 57 years and older formed 47 per cent of the participants in 2009–10 (ibid., 55). A future challenge therefore will be the recruitment and involvement of younger generations.

settlement lifestyle. Quebec opted for a 'staying on the land' option through its Income Security Program for hunters and extended trapping families.[2]

Settlements like Clyde River on Baffin Island are struggling to survive (Vignette 4.5). Under normal circumstances, urban places that lose their economic function soon die, whether they are single-industry towns or rural communities. Native settlements, however, do not follow this pattern of urban evolution. In fact, most have increased in population since relocation began and, in effect, represent a contemporary institution- alization of the Homeguard Indians who settled around Hudson's Bay trading posts in order to access the security and amenities the HBC posts afforded. The reasons for population increase are simple. People stay because these centres are located within their cultural homelands and offer access to public services. Population is increas- ing because few are leaving, and any out-migration is more than offset by high birth rates. Still, this out-migration sees Canadian cities as the destination of more and more Aboriginal people. While many find their footing, some fall between the cracks and join the urban homeless.

Some 50 years after relocation to settlements took place, has the relocation strat- egy worked? Certainly, these Native settlements are a permanent feature and some are doing well. But the majority of them are not. Faced with high unemployment rates, widespread poverty, poor health, and a host of social problems, many communities are in crisis. Alcohol and drug abuse are widespread. But with little chance to find employment, people become resigned to their lot, lose hope, and become depend- ent on social welfare. The toll has been particularly hard on young people, who are prone to take their lives. Suicide rates for Aboriginal peoples are double those of the rest of the population (Health Canada, 2005; Laliberté and Tousignant, 2009) and in some communities have reached epidemic proportions. In the community of Sandy Bay in northern Saskatchewan, suicides occurred in a cluster in early 2007, with five suicides and 12 attempted suicides. During that time, one day just before noon a 15-year-old Métis girl used the school computer to type a poem, written by an American teen (Warick, 2007). During the lunch break, she hung herself at home. The poem, 'A Simple Smile, a Scream Inside' (Schultz, 2001), was found later, and included the following lines:

I'm screaming on the inside
But a smile is what you see
I'm not content
With this person I seem to be

Copyright © 2001 by Stephanie Schultz. Reprinted with the permission of Health Communications, Inc., www.hcibooks.com.

The geographic shift from the land to northern settlements, while providing easy access to store foods, has not led to a healthy Aboriginal population. The lifestyle of settlement inhabitants is more sedentary and more dependent on processed store foods than that of their grandparents. Store food has been a contributing factor to poor health, including obesity and high rates of diabetes even among teenagers (Vignette 7.3).

Vignette 7.2 From Davis Inlet to Natuashish

If not the worst example of the federal government's relocation program, Davis Inlet certainly is close to the top. Like other northern peoples, the Innu of Labrador were nomadic hunters until the 1960s, when they were relocated into settlements, principally Sheshatshiu and Davis Inlet. What all relocation sites had in common was accessibility by supply ships and thus the goods and services provided by various governments. With the ever-present threat of game shortages, food security was a paramount consideration of governments.

Davis Inlet was a social disaster. This dysfunctional village represented an Aboriginal community in crisis because the old ways were gone and the new way was dependency and welfare. In the case of Davis Inlet, not only did the settlement remove the Innu from their homeland but it left them isolated on an island—making access to traditional hunting grounds impossible for most of the year—without sufficient water and waste disposal systems. In the early 1990s, the Canadian public saw terrible television images of Davis Inlet children sniffing gasoline, yet, despite the efforts of government and community leaders, little changed and the community had 'one of the highest suicide rates in the world' (MacDonald, 2001). The solution was a better site where caribou hunting could take place. At an estimated cost of $150 million, the modern settlement of Natuashish was completed in 2003 on the Labrador mainland, and the nearly 700 Davis Inlet Innu left behind their squalid village for a new town with a sewer and water system (ibid.).

Some people, however, just can't seem to catch a break. In August 2011, the Newfoundland and Labrador government, noting that the George River caribou herd had declined by 80 per cent over the past decade, indefinitely postponed the caribou hunt until further data could be gathered (CBC News, 2011d). And social change takes more than physical relocation. In Natuashish, residents continued to experience the terrible consequences of excessive drinking, gasoline sniffing, and drug abuse. The ugly sight of young children getting high by sniffing gas emphasized a sense of escape from their seemingly hopeless future. Searching for a solution in 2008, the residents voted by a slim majority to ban alcohol. In that year, too, the first two students graduated from the new high school. Although excessive drinking, drug abuse, and gas sniffing did not end, they were limited because alcohol and drugs had to be smuggled into the community and consumed in private. As a result, social conditions improved. Two years later, however, the newly elected chief declared the ban over. The reaction from the community and officials was swift—nurses noted that the number of suicides had dropped by half, and the RCMP had seen a similar drop in crimes. As RCMP Sgt. Wayne Newell noted, 'all types of crimes such as assaults, common assaults, domestic violence, assault causing bodily harm, robberies, sexual assaults, all those type of things, have been down almost half' (CBC News, 2010).

Common sense prevailed. In a plebiscite in 2010, the majority of residents voted to maintain the ban. While some residents are offended by a ban, claiming that their right to purchase and consume alcohol has been infringed and that local bootleggers get rich at their expense, the community is a safer place with a ban.

Vignette 7.3   **Activities, Diet, and Health**

One of the drawbacks of settlement life has been the deterioration of Aboriginal peoples' health. Diabetes, for instance, has affected an unusually high proportion of Aboriginal peoples, including those under 30 years of age. Many are overweight. Health officials point to inactivity and diet as two causes for the poor state of their health. Prior to the 1950s, Aboriginal people were extremely active as hunters, fishers, and food gatherers. Their diet was based on fresh meat and fish (country food) that supplied them with most of their caloric and nutritional needs. As well, country food was an important cultural anchor providing both an economic and spiritual link with the land and wildlife. While tea, flour, sugar, and salt were purchased at the local trading store, it was not until the indigenous peoples moved into settlements that a wide range of foods with high levels of sugar and starch were purchased and consumed. As Collings, Wenzel, and Condon (1998: 303) reported, 'This shift from a high-protein, polyunsaturated fat diet to one high in carbohydrates and saturated fats, especially among the generation(s) raised in the settlement, has not been without adverse health effects.'

The Political Shift

Shortly after World War II a series of world events caught the eye of Aboriginal leaders, political activists, and intellectuals. First, colonial empires were replaced by independent states in Africa and Asia. Second, the American civil rights movement was challenging segregation by taking to the streets. Third, separatist ethno-nationalist groups in some European countries, and in Quebec, were seeking political independence, and their activities often led to violent acts. Relocation of scattered extended hunting families into settlements provided the catalyst for the emergence of a regional tribal consciousness, creating a form of Aboriginal **ethnocentrism**. This newly found unity served as a tipping point for engaging in political change.

The turbulent political atmosphere not only recast world geopolitics, but ideas of independence and self-determination spread to the oppressed Aboriginal minorities in North and South America. In the 1970s, the unthinkable happened in Canada. The Parti Québécois formed the opposition party in the Quebec legislature and then it became the Quebec government in 1976. The question for Aboriginal students and political leaders of the day was simple—what about our political future within the Canadian federal system?

By the 1970s, the Dene tribes in the Mackenzie Basin were moving towards what then seemed like a dangerous and even threatening political position. In 1975, their chiefs announced the Dene Declaration, which for the time was a radical political statement. This declaration called for political independence within Canada (Erasmus, 1977: 3–4). In their land claim negotiations with Ottawa, the Dene demanded their own separate territory and government in the western half of the Northwest Territories. This new state would be called Denendeh.

Ottawa was taken aback by such a bold request for an ethnic form of government, which, it seemed, could lead to a series of semi-autonomous Aboriginal states within Canada and perhaps even tip the balance in favour of Quebec independence. However, by 1990, the Dene land claim had reached the advanced stage of a final agreement, which required approval from the Dene/Métis assemblies. The majority voted against the agreement because it did not deal with self-government. Ottawa discontinued further discussions with the Dene/Métis and offered to begin the process with individual tribes—what might be seen as a divide-and-conquer strategy. The Gwich'in and Sahtu Dene/Métis agreed to separate land claim negotiations.

Thus Denendeh unravelled, putting an end to the political dream of a state within Canada (Vignette 7.4). For the Inuit of the eastern Arctic, however, their political dream was translated into reality with the formation of Nunavut. Though the Inuit had also proposed an ethnic state and government at the same time as the Dene, they were able to adjust to the political realities of the day—and, significantly, in their proposed territory they had a larger majority than did the Dene. In their revised version, the Inuit called for a public government rather than an ethnic one, i.e., a government that in practice would be under Inuit control because of their large numerical majority but where all citizens, regardless of ethnicity, would be treated equally. In 1999, the Inuit wish for a separate territory was fulfilled with the creation of Nunavut.

Vignette 7.4   Denendeh, Self-Government, and Comprehensive Land Claim Negotiations

Prior to 1995, only a few somewhat limited self-government agreements, along the lines of municipal governments, had been concluded. In today's political atmosphere they would be described as 'timid' ventures. Early attempts at self-government included those for the Cree, Naskapi, and Inuit of northern Quebec under the 1975 James Bay and Northern Quebec Agreement and the 1978 Northeastern Quebec Agreement (1978) and for seven Yukon First Nations, pursuant to a 1993 Umbrella Final Agreement.

The concept of Denendeh was different. In 1975, Denendeh represented a bold breakthrough in Aboriginal political thinking with the Dene/Métis negotiators focused on self-government while Ottawa was looking at a land settlement. In this way, Denendeh foreshadowed the acceptance by Canada of a more vigorous version of Aboriginal self-government. Unfortunately for the Dene/Métis, the time was not right because the federal government was strongly opposed to such a radical idea. Even the first comprehensive land claim negotiations, which led to agreements in 1984, 1992, and 1993, did not include the topic of self-government.

A major shift in federal policy towards Aboriginal self-government took place in the Charlottetown Accord. While the Accord was rejected in a national plebiscite in 1992, it called for the entrenchment of the inherent right of Aboriginal peoples to self-government in the Constitution Act, 1982. The rejection of the Charlottetown Accord meant that prospects for constitutional change were no longer in the cards. Using another approach, in 1995 the federal Liberal government recognized the inherent right of self-government as an existing right within section 35 of the Constitution Act, 1982 without entrenching this in the Constitution. From that time on, self-government become part of comprehensive land claim negotiations.

The Emerging Political Map in Arctic Canada

With the addition of self-government to land claim negotiations, this acquired political power is transforming the map of the Canadian North. The dream of Aboriginal
autonomous states within Canada has taken place in the Canadian Arctic. Geography
has played a role in this political evolution because the Arctic has yet to attract many
southern Canadians, thus leaving the Arctic population almost entirely Inuit. At the
same time, its geography has limited commercial activities, so the Arctic with an
extremely weak economic base and very high unemployment rates. The state of the
economy is but one challenge for the Inuit political leaders.

This Inuit political world first took shape in 1999 when Nunavut was formally
carved out of the Northwest Territories. Following on its heels, Nunavik may yet be
destined for a unique form of regional government within a province. While most
forms of self-government take the form of 'ethnic' governments, Nunavut does not.
Nunavut has a 'public' government but in fact represents a de facto ethnic state within
Canada that 'fits comfortably within established traditions of mainstream Canadian
governance' (Hicks and White, 2000: 31). The key factor for the Inuit in accepting a
public form of government was demographic—Nunavut has very few non-Inuit residents. The road to Nunavut is outlined in Table 7.1.

Nunavut

In the late twentieth century, Nunavut was the most innovative political development
to appear on the northern horizon. In 1976, Inuit Tapirisat of Canada (now called
Inuit Tapiriit Kanatami) proposed creating an Arctic territory called Nunavut ('our
land' in Inuktitut) that would represent the political and cultural interests of Canada's
Inuit. It took nearly 25 years, but on 1 April 1999 Nunavut was formally carved out
of the Northwest Territories.

As the largest political unit in Canada, Nunavut extends over 2 million km². At the
same time, Nunavut has the smallest population (estimated at 33,268 inhabitants in
October 2010 (Statistics Canada, 2010), but most (85 per cent) are Inuit. Its population, however, is growing rapidly, at an annual rate of 1.6 per cent, due to its high birth
rate (the highest in Canada). Its people are scattered across the tundra in 26 small,
isolated settlements without a series of highways connecting these places. Given its
extremely weak economy, Nunavut is heavily dependent on transfers from Ottawa.
Transfer payments keep the government operating. Beyond transfer payments, the
mining industry provides some economic growth, but the problem with the mining
industry is the relatively short lifespan of mines, limited employment of Inuit workers,
and the collection of royalties not by Iqaluit but by Ottawa. Most Inuit families rely on
a combination of local employment, social welfare payments, and hunting for country
food. A sign of the dysfunctional nature of the Nunavut economy and limited federal
funding is the acute shortage of public housing, plus the growing number of homeless
Inuit in larger centres and in southern cities.

While the state of Inuit well-being (or the lack thereof) is well documented in
Heather Tait's 2008 report based on Statistics Canada's 2006 Aboriginal Peoples
Survey, solutions are not readily at hand for the economic challenges. In fact, too little

Table 7.1   The Road to Nunavut: A Chronological History

Year	Event
1973	Inuit Tapirisat of Canada (ITC) undertakes a study of Inuit land use and occupancy under the direction of Professor Milton Freeman.
1976	ITC proposes the creation of a separate Arctic territory for the Inuit of the Northwest Territories.
1976	The Inuit of the western Arctic (the Inuvialuit) opt to undertake their own land claim, thus dividing the Inuit land claim and territory into two parts.
1990	Following years of negotiations, the Tungavik Federation of Nunavut, the Northwest Territories, and the federal government sign an agreement-in-principle that includes the formation of a separate territory for the Inuit of the central and eastern Arctic subject to the approval of the division of the Northwest Territories by means of a plebiscite.
1992	The residents of the Northwest Territories approve of its division into two separate territories.
1992	The Tungavik Federation of Nunavut and the federal government sign the Nunavut Political Accord.
1992	The Inuit of Nunavut ratify the Nunavut Land Claims Agreement.
1993	Parliament passes the Nunavut Land Claims Agreement Act and the Nunavut Act.
1993	Nunavut Tunngavik Inc. replaces the Tungavik Federation of Nunavut.
1997	The Office of the Interim Commissioner is established to help prepare for the creation of Nunavut.
1999	The residents of Nunavut elect the members of the Nunavut Legislative Assembly with Paul Okalik as the Premier.
1999	The Territory of Nunavut is formed, and its government begins to function.
1999	The Clyde River Protocol establishes the norms that will govern the relationship between Nunavut Tunngavik Inc.* and the new government of Nunavut.

*Nunavut Tunngavik Inc. receives and administers funds from the Nunavut land claim agreement. It is responsible for ensuring that the Nunavut government respects and follows the land claim, especially those aspects promoting Inuit culture. The Clyde River Protocol requires that the Nunavut government and Nunavut Tunngavik work together to ensure that legislation promotes Inuit culture. In 2002, for example, Nunavut's language commissioner proposed a Quebec-style commercial sign law to ensure that Inuktitut remains the dominant language used on business signs, posters, and commercial advertising. This proposal arose because most of Iqaluit's business signs are in English only.

government revenue and the high rate of natural increase are making problems more acute. Even Nunavut's well-intended 'hire Inuit' program for its civil service, with a target of employing Inuit in 50 per cent of the public jobs, has run into the paucity of qualified Inuit, which takes us back to the Inuit education system. In 2005, 47 per cent of all civil servants were Inuit. However, as Table 7.2 shows, Inuit hiring falls short at the higher-paying jobs—senior management, middle management, and professionals—but exceeds 50 per cent at lower-paying paraprofessional and administrative support levels.

Along with a shortage of jobs in Inuit communities, many people are dependent on welfare, reside in overcrowded housing, and face a bleak future. Colonialism and now neo-colonialism help to explain the roots of this social malaise—the intrusion of Western culture destroyed the Inuit cultural anchors, leaving the people in a state of cultural drift. Some are searching for a safe cultural haven within their political home of Nunavut but note the uphill struggle to keep Inuktitut alive and well in their public schools. Efforts to maintain Inuit culture and hunting practices are under great pressure. Hunting skills and the Inuktitut language remain common among older people but not among the young. This raises an interesting question: Would a federally funded 'on the land' program for Nunavut, similar to Quebec's Income Security Program for Cree hunters and trappers, help to ensure the survival of those aspects of Inuit culture and, at the same time, reduce the high suicide rates? Jack Hicks, who has investigated Inuit suicide rates for several years, made the following observation: 'In northern Quebec, Nunavik has a suicide rate 50 per cent higher than ours in Nunavut, but [Quebec] Cree with whom they [Quebec Inuit] share a land claim do not. They [the Quebec Cree] have essentially national levels of suicide' (CBC News, 2006).

Nunavut Tunngavik Inc., the Inuit land claim organization, has taken on the role to ensure that Nunavut is not only a homeland for the Inuit but also an embodiment of Inuit culture (Légaré, 2000), and is sometimes in conflict with the Nunavut government because Inuit culture and participating within Canadian society do not always mesh well. The challenge for both the government and Nunavut Tunngavik is to ensure that government policies and operations reflect the Inuit way of thinking and living (*Inuit Qaujimajatuqangit*). However, the government is confronted with a limited budget and a trend towards modernization in Inuit life that is moving away from traditional Inuit ways, i.e., engaging in wage employment, living in communities, and attending school and post-secondary institutions. Modernization also has meant greater interaction with the rest of Canadian society, which usually means speaking English.

In 1999, the decision to decentralize government operations to a series of regional centres was an effective but expensive measure in moving towards an Inuit style of government. Another initial step was to have Nunavut Tunngavik form a shadow

Table 7.2 Inuit/Qallunaat Employees in Departments and Boards of the Government of Nunavut, 1999–2005

	June 1999		June 2001		June 2003		June 2005	
	Inuit	Qallunaat	Inuit	Qallunaat	Inuit	Qallunaat	Inuit	Qallunaat
Executive	61	39	48	52	50	50	47	53
Senior management	22	78	20	80	18	82	25	75
Mid-management	24	76	19	81	19	81	23	77
Professional	41	59	23	77	23	77	25	75
Paraprofessional	47	53	54	46	57	43	61	39
Admin. support	64	36	78	22	81	19	86	24
Total depts/boards	44	56	43	57	42	58	47	53
(Number)	(220)	(280)	(925)	(1,243)	(1,014)	(1,395)	(1,233)	(1,387)

Note: Qallunaat refers to non-Inuit employees; more specifically, Qallunaat means 'white people'.

Source: Timpson (2006: 523). Copyright © 2006 International Institute of Administrative Sciences, by permission of Sage Publications Ltd.

government to ensure government legislation, policies, and programs reflect the interests of Inuit. As seen by Nunavut Tunngavik, the cardinal principles are to promote the Inuktitut language and Inuit culture (Vignette 7.5). Even so, three major barriers prevent a smooth transition from a traditional Canadian style of government to an Inuit one. One is the shortage of qualified Inuit for jobs within government. As a result, many in the civil service, especially at the senior levels, are non-Inuit. Second, the use of Inuktitut as a working language within the civil service has not been successful. In fact, English tends to be the working language in the capital of Iqaluit, while Inuktitut is more commonly spoken in the regional centres. Third, political fulfillment of the Nunavut dream is unlikely unless its economy becomes more diversified. At the moment, Nunavut relies most heavily on transfer payments from Ottawa, and such financial dependence—over 90 per cent—may curtail the full flowering of Nunavut's political development. The fear is that these three barriers will cause Nunavut to slide into a version of the two other territorial governments and therefore run counter to *Inuit Qaujimajatuqangit*.

Vignette 7.5 Protecting Inuktitut but Not Willing to Pay the Price

Language is a sensitive issue. Nunavut's government has introduced legislation to protect Inuktitut. The two bills—Bill 6, the Official Languages Act, and Bill 7, the Inuit Languages Protection Act—became law in 2008. Nunavut Tunngavik says the bills are 'a good start' but that the government should do more to protect Inuit language rights, especially in education, where most of the education program should be delivered in Inuktitut as soon as possible (Bell, 2007). The Qikiqtani Inuit Association, which represents the interests of the Inuit of the Baffin Region, the High Arctic, and the Belcher Islands, strongly opposes these two bills because it feels they are 'too weak' and because the Inuit language requires the same status within Nunavut that the French language enjoys in Quebec under that province's famous Bill 101 language law (ibid.). Berger (2006: 1–67) addressed this contentious language issue and recommended that Inuktitut become the language of instruction from kindergarten to Grade 4, with English during these years as a second-language program; commencing in Grade 5, instruction would be in both Inuktitut and English. This proposed bilingual program, according to Légaré (2008a: 365), had three flaws. First, Ottawa was not prepared to fund the $20 million required to support the program. Second, an insufficient number of teachers currently speak Inuktitut. Third, resource materials in Inuktitut are lacking, especially beyond Grade 4. Of course, the last two flaws could be tackled and resolved with sufficient funds—which, as noted, is the central flaw in such a plan. As pointed out in Chapter 6, the paradigm followed by the federal government is more about promoting economic development than about addressing environmental and social issues.

Nunavik: A Regional Government within Quebec?

While most of the world, including Canadians, watched closely as the Arab Spring of 2011 was unfolding, fed and led in many ways by social media, an Inuit Spring was underway in Arctic Quebec. This surprised some observers, or those who were

Figure 7.1   Nunavut: Canada's Newest Territory

Nunavut consists of:
(a) all of Canada north of 60°N and east of the boundary line shown on this map, and which is not within Quebec or Newfoundland and Labrador; and
(b) the islands in Hudson Bay, James Bay and Ungava Bay that are not within Manitoba, Ontario, or Quebec.

LEGEND
○ Territorial capital
● Other populated places
— · — International boundary
········ Provincial boundary
- - - - Dividing line
(Canada and Kalaallit Nunaat)

Source: Adapted from *The Atlas of Canada* (2006). Reproduced with the permission of Natural Resources Canada 2008, courtesy of the *Atlas of Canada*.

paying attention. Just over a decade after the formation of Nunavut, the Nunavik Regional Government for the Inuit of northern Quebec was anticipated to become law in Quebec and Canada (Nunavik Regional Government, 2010). Like Nunavut, the leaders of Nunavik called for a 'public' government, but with 85 per cent of the population Inuit, Nunavik, like Nunavut, in practice would be an Inuit homeland with an Inuit-controlled regional government.

The concept of a semi-autonomous region within a province would break new ground, yet public government requires a cultural leap. Once the Final Agreement was achieved by the three parties—Makivik Corporation on behalf of the Inuit of Nunavik and the federal and provincial governments—the two Makivik negotiators of the agreement, joined by the negotiators for Quebec and Canada, arranged a four-week information tour to all of the 14 Inuit communities in Arctic Quebec (Figure 7.2). This was followed by a 'Referendum Period of one month to allow the people of Nunavik to consider, discuss, and debate the issues' (Makivik, 2011a). In the agreement, a version of regional government using the federal/provincial parliamentary system was embedded within the Quebec political system, giving the new elected assembly of the 11,000 Inuit living in Nunavik responsibility for a wide range of powers, including responsibility for public services such as health and education. While the proposed new government would have the authority to collect taxes, adopt laws, and take out loans, the weak Nunavik economy would dictate that virtually all funding would come from Quebec—a situation similar to that of Nunavut, where

Figure 7.2   Nunavik

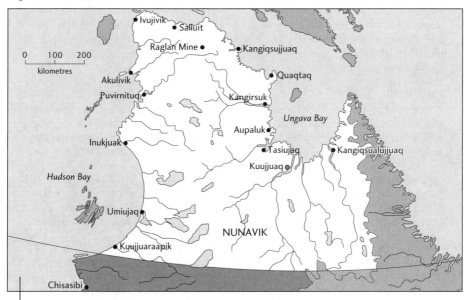

Nunavik is the northernmost region in any of Canada's provinces. Inuit in Nunavik live in 14 villages spread out along the coasts of Hudson Bay, Hudson Strait, and Ungava Bay. Kuujjuaq, the largest of these villages, had a 2006 population of 2,132.

Source: Nunavimmiut Kavamanga (2007). Reprinted by permission of Makivik Corporation.

over 90 per cent of its $1.35 billion revenue for 2011–12 came from Ottawa (Nunavut Department of Finance, 2011).

So what happened during the 'Referendum Period' while the people of Nunavik considered, discussed, and debated the issues? Why was an agreement that seemed to be a fait accompli voted down by 70 per cent of voters in an unexpectedly high

---

Vignette 7.6 From Igloos to Houses

Since 1975, when the Nunavimmiut assumed control over local affairs with the establishment of the Kativik School Board, Kativik Regional Government, the Nunavik Regional Board of Health and Social Services, and Makivik, a development corporation to oversee the funding that came out of the James Bay and Northern Quebec Agreement, the Nunavimmiut have learned how to function in modern institutions and how to shape them to match their culture. One example is shelter. In this new world, igloos are part of the past and now everyone wants a modern house. Housing for the Nunavimmiut faces two challenges: few can afford a house, and house construction in settlements is extremely expensive. How could this very pressing health and social problem be turned into an engine of community growth and job training?

Modern land claim agreements, such as the 1975 JBNQA, provided the Inuit with institutions that had the capacity for regional solutions to housing construction that compares favourably in cost and workmanship with housing construction in southern Canada. Funding for social housing comes from federal and Quebec sources. By channelling these funds to Makivik, social development was also achieved. The Makivik Corporation established its construction division to address the housing and other construction challenges in Nunavik. Klinkig (2005), for example, reports that Makivik's 2004 budget was $12 million for housing and $9 million for construction of docking facilities. Makivik has constructed most of the 2,500 social and staff housing units in Nunavik, while the Kativik Municipal Housing Bureau manages these units (Fournier, 2005). The experience of both organizations has been very positive:

Long-term, comprehensive social housing programs, such as the partnership between the federal and Quebec governments and the Inuit organizations of Makivik and Kativik Municipal Housing Bureau, are the best solution in the Arctic. This approach increases efficiencies and levels of local activity, and yields significant cost savings. The leverage afforded by long-term, stable funding allows for economies of scale through negotiation of bulk purchasing and volume discounts from suppliers. An estimated 15–20 per cent cost savings may be achieved in this fashion. As a non-profit promoter-builder, Makivik has more flexibility in negotiating with contractors and in assuring the maximization of Inuit labour. Oneil Leger, an official with Makivik's Construction Division, estimates that Inuit account for 70–80 per cent of labour on the Makivik-contracted housing sites. Instead of using its own trucks and graders, Makivik enters into contracts with hamlets to use their municipal heavy equipment to prepare pads and move material; this contributes to keeping economic benefits in the communities (ITK, 2004:10).

Makivik Construction is entering its third five-year contract with the governments of Canada and Quebec to build 60 two-bedroom duplexes in the 14 communities. The program calls for approximately 60 two-bedroom duplexes to be built each year over the five-year period ending in the 2015 construction season. As well, this five-year social housing agreement provides funds to the Kativik Municipal Housing Bureau for maintenance and operating costs (Makivik, 2011b).

turnout? In Cairo's Tahrir Square and in other places experiencing democratic uprisings during the Arab Spring, many people flocked to Twitter, Facebook, and other social media to learn what was happening, to share ideas, and to find out what to do and where to go next. In a similar fashion—though certainly in a more peaceful context—the Quebec Inuit (*Nunavimmiut*) used Facebook to share their thoughts about the agreement during the month prior to the referendum. One Facebook site alone had a reported 900 followers, and Facebook was credited by many people as having been instrumental in the high voter turnout (nearly 55 per cent) and in the rejection of the agreement. As one participant said, 'If it was not for Facebook, the majority of the people would have voted in favour [of the agreement], as we would've only heard from people working on the agreement' (CBC News, 2011a).

Is the rejection of the Nunavik Final Agreement simply a bump along the road or a rejection of a 'public' rather than an 'ethnic' regional government? The next step, led by Makivik, is to hold further discussions with the goal of resolving the perceived obstacles to the Nunavik Final Agreement.

The Nunavik model of public government, as set out in the Final Agreement, represents a new type of Aboriginal government that is well suited to the Arctic where the 'permanent' population is almost entirely Inuit. On the other hand, the population of the Subarctic is much more mixed and Aboriginal peoples only hold a large majority in the north of a few provinces, such as Manitoba and Saskatchewan, and parts of Ontario, Alberta, and BC. Then, too, the political overlay of First Nations and Métis governments adds another obstacle to the public government model. Taking Saskatchewan as an example, even though over 80 per cent of the people in the north of the province have declared themselves as Aboriginal, the reality on the ground is quite different because Dene, Cree, and Métis represent distinct cultural groups (Bone, 2001). As well, the Métis and First Nations have their own provincial organizations—the Métis Nation and the Federation of Saskatchewan Indian Nations—as well as local governing bodies.

Comprehensive Land Claim Agreements

The federal government's 1973 Comprehensive Claims Policy was designed to resolve Aboriginal title with respect to ownership, use, and management of lands and resources by negotiating an exchange of land for a clearly defined package of rights, benefits, and a smaller portion of land set out in a settlement agreement. A comprehensive agreement is a legal agreement between a First Nation or other Aboriginal group and the federal government that extinguishes their Aboriginal title in exchange for land and capital. These agreements refer to the total land area claimed as the settlement area. Ottawa, however, only cedes a portion, perhaps around 10 per cent, of the settlement area to the Aboriginal claimants. For instance, the 14 First Nations in Yukon claimed virtually all the territory of Yukon but received ownership of 41,440 km², which represents 8.6 per cent of the land mass of Yukon. The agreement also requires the government to fulfill a number of commitments.

Comprehensive land claim agreements have had a profound impact on Aboriginal peoples, their homelands, and their participation in Canadian society. As discussed in Chapter 3, the recognition of Aboriginal title led to comprehensive land claim agreements (Vignette 3.9). However, the first agreements remained silent on the matter of self-government. The first exception took place in 1993 when the Nunavut Land Claim Agreement was signed, setting in motion the formal establishment of the new territory six years later. While the Inuit had to struggle to have a form of self-government included in the negotiations, the negotiations towards the 2005 Tlicho Agreement for part of the Northwest Territories combined the land claim and self-government from the start (INAC, 2008). Since 2005, comprehensive land claim negotiations automatically consider both land claims and self-government.

For Aboriginal peoples living in the Territorial North, the impact of modern treaties has been revolutionary. Comprehensive agreements have transformed their economic and political lives because they provide the means and structure to participate in Canada's economy and society and yet retain a presence in their traditional economy and society. In conjunction with modern treaties, favourable court decisions have recognized Aboriginal rights to natural resources (Bartlett, 1991; Usher et al., 1992).

Aboriginal peoples in the provincial norths who had signed treaties in the late nineteenth and early twentieth centuries (Figure 7.3) could only turn to specific land claims to seek redress for violations of their treaties, such as the improper loss of reserve land. In the provincial norths, those without treaty in 1970 were the Inuit and Cree of northern Quebec, the Nisga'a and other Indian First Nations in northern British Columbia, and the Labrador Inuit and Innu. Since then, most have negotiated modern land claim agreements. By 2007, 11 such agreements had been concluded, with the Tlicho (Dogrib or North Slavey) Final Agreement, Labrador Inuit Final Agreement, and the Nunavik Inuit offshore islands and waters agreement taking place in 2005, 2005, and 2006, respectively (Table 7.3). The goal to achieve regional government in northern Quebec (Nunavik), as discussed above, is on hold because the April 2011 referendum calling for a regional government was rejected by the Inuit voters. In spite of the support of its business arm, Makivik, the majority feared that Nunavik's power over its lands, culture, and language would be diluted (George, 2011). While the concept of regional government has not been lost, the Nunavimmiut, led by Makivik, will have to regroup and develop a form of regional government acceptable to the majority of Quebec Inuit—and to the provincial and federal governments. The rejected agreement contained four important elements intended to satisfy the residents of the region, but fears about lands, culture, and language won the day. These elements were:

1) an administrative structure and capital resources necessary to function in the Canadian economy;
2) access to natural resources, including subsurface minerals;
3) a co-management role in environmental matters, land-use planning, and wildlife management;
4) self-government (since 2005) of their internal affairs.

Moving from Aboriginal title to a land claim agreement requires a formal process, as defined in the Supreme Court's 1997 *Delgamuukw* decision (for more on this legal ruling, see Hurley, 2000). McNeil (1998: 29) argued, 'Despite its shortcomings, the *Delgamuukw* decision could usher in a new era for Aboriginal rights in Canada. For the first time, the right of Aboriginal peoples to participate as equal partners in resource development on Aboriginal lands has been acknowledged.' Events have proven

Figure 7.3   Historic Treaties

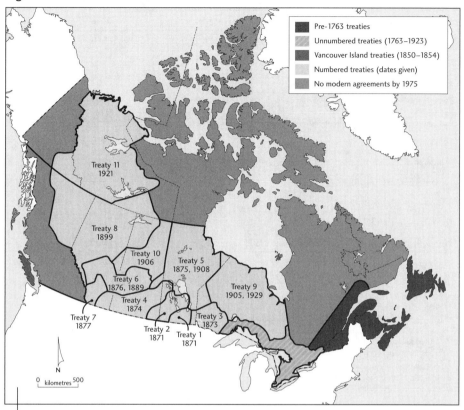

Pre-1763 treaties

Unnumbered treaties (1763–1923)

Vancouver Island treaties (1850–1854)

Numbered treaties (dates given)

No modern agreements by 1975

Prior to Confederation, peace and friendship treaties had been made in New France and British North America, and in the late eighteenth century and first half of the nineteenth century more formal treaties that involved the cession of lands and the creation of reserves were reached on Vancouver Island (the Douglas treaties) and in Upper Canada (the Upper Canada Land Surrenders, notably in the Niagara region, and the Robinson treaties for large tracts east of Lake Huron and north and east of Lake Superior). The so-called numbered treaties, signed in the half-century after Confederation, were designed to secure lands in western Canada for Euro-Canadian settlement. In addition, the Williams treaties involved lands 'missed' in pre-Confederation Ontario land surrenders and were concluded in 1923.

Treaty 11, the last numbered treaty, covered lands in the Mackenzie River Basin north of 60°N. By 1921, roughly half of Canada was not covered by treaties, including lands in Labrador, Quebec, BC, Yukon, and what is now Nunavut. Ottawa hesitated to make treaty with those Aboriginal peoples living in Yukon and the Northwest Territories (which then included Nunavut) because the prospect of white settlers entering these lands was unlikely and because these peoples followed their traditional semi-nomadic lifestyle of hunting, fishing, and trapping. All changed in 1921 when oil was discovered at Norman Wells. Ottawa moved quickly and Treaty 11 was signed the same year.

McNeil correct, though each situation requires proof of Aboriginal title and negotiations between the developer and the Aboriginal group.

Geographers have played a key role in land-use studies that provided the evidence of Aboriginal title and thus for land claims. Freeman (1976) and Usher (1976), for example, helped to develop a system of mapping Inuit traditional land use. Land selection is a critical factor in the claims process, and it involves geographic concerns such as location and quality of land. Frank Duerden (1990: 35–6), a geographer who

Table 7.3 Final Comprehensive Land Claim Agreements*	
Final Agreement	**Date**
James Bay and Northern Quebec Agreement	1975
James Bay Naskapi	1978
Inuvialuit Final Agreement	1984
Gwich'in Final Agreement	1992
Tungavik Federation of Nunavut Final Agreement	1992
Sahtu/Métis Final Agreement	1993
Yukon Indians' Umbrella Final Agreement**	1993
Nisga'a Final Agreement	1999
Tlicho Final Agreement	2005
Nunatsiavut Final Agreement	2005
Nunavik Inuit Agreement (offshore islands)***	2006

*A final agreement is the outcome of successful claim negotiations that have been ratified by the negotiating parties. Canada then passes settlement legislation that gives legal effect to the final agreement. For example, the House of Commons passed the Nisga'a Final Agreement as an Act of Parliament in 2000. The Inuvialuit Final Agreement (1984) was the first comprehensive agreement. The James Bay and Northern Quebec Agreement (1975) took place before Ottawa had established a process for negotiating comprehensive agreements.

**This umbrella agreement involved the Vuntut Gwich'in First Nation (1995); First Nation of Nacho Nyak Dun (1995); Teslin Tlingit Council (1995); Champagne and Aishihik First Nations (1995); Little Salmon/Carmacks First Nation (1997); Selkirk First Nation (1997); Tr'ondëk Hwëch'in First Nation (1998); Ta'an Kwach'an Council (2002); Kluane First Nation (2004); Kwanlin Dun First Nation (2005); and Carcross/Tagish First Nation (2005).

***The Nunavik Inuit will own 80 per cent of the islands in the Nunavik marine region, totalling approximately 5,000 km². These islands provide 85 per cent of the Nunavik Inuit's subsistence harvesting. They will also receive $39.8 million to implement the deal and another $54.8 million for Nunavik Inuit Trust.

continues to work extensively with Yukon First Nations, has outlined four spatial components of the land claims process (see also Duerden and Keller, 1992; Duerden and Kuhn, 1996, 1998; Duerden, 2004):

- mapping areas of use and occupancy in order to develop a case for legitimacy of a claim;
- preparation of maps of land-use potential to identify 'optimum' areas to be retained as an outcome of negotiations;
- evaluation of land selection positions as negotiation proceeds to establish the extent to which they satisfy the goals of the process;
- evaluation of non-ownership agreements regarding land to ascertain the effectiveness of the control they offer and their long-term impact on land-use patterns.

Impact and Benefit Agreements

Impact and benefit agreements (IBAs) are formal, written agreements between development companies and Aboriginal organizations that are designed to compensate Aboriginal peoples for the predicted impacts associated with an industrial development and to permit the development to proceed (Kennett, 1999). Geographers are playing a leading role in the field by studying and assessing IBAs (Galbraith et al.,

Figure 7.4   Modern Treaties

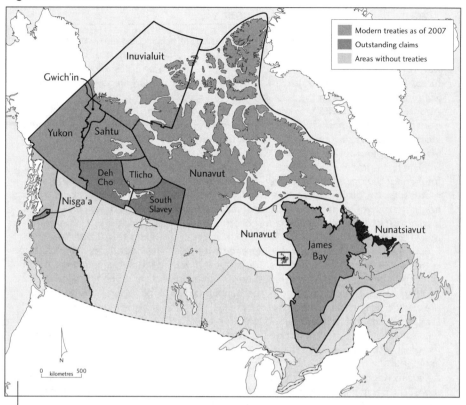

The first modern treaty, the James Bay and Northern Quebec Agreement, was signed in 1975. Since then, modern claims have fallen into two categories: comprehensive and specific. By 2007, the main areas without treaties are found in British Columbia and the Northwest Territories, and portions of Quebec and Labrador.

2007; Prno and Bradshaw, 2007; Noble, 2010). While details of IBAs are confidential, they do provide specific business, employment, and training opportunities for Aboriginal groups with a traditional land claim to the area proposed for industrial development. IBAs vary considerably in their scope and complexity, depending on the scale and nature of the project and the issues identified by the negotiating parties involved. However, most agreements address such topics as:

- environmental protection, including special concerns about wildlife;
- protection of Aboriginal social and cultural values;
- education, training, and employment;
- health and safety;
- business opportunities;
- Aboriginal access to the project site;
- financial compensation;
- dispute resolution mechanism.

Vignette 7.7   A Foot in Both Cultures

From an Aboriginal perspective, comprehensive land claim agreements provide a legal and financial foundation that blends the land-based **subsistence economy** with the commercial economy, thus allowing Aboriginal peoples a foot in two cultures. The underlying premise is that Aboriginal culture, language, and traditions will remain viable within the subsistence economy. But is this premise valid or will Aboriginal youth gravitate to the commercial economy and thus become divorced from their culture and language? Quebec's popular Cree Hunters and Trappers Income Security Program, for example, shows signs of failing to attract younger participants (Vignette 7.1).

The first agreement, known as the Inuvialuit Final Agreement, set the pattern for future negotiations. In the case of this agreement, the Inuvialuit Regional Corporation represents the business structure while the Inuvialuit Game Council is concerned about the environment and wildlife. Co-management of the environment and wildlife, involving the Aboriginal group and government officials, flowed out of the agreement. Unfortunately for the Inuvialuit, Ottawa was not willing to consider self-government as a negotiable item in comprehensive land claims negotiations until 2005.

When resource developments are proposed within the settlement area of an Aboriginal group, the developer must negotiate an impacts and benefits agreement. Such agreements are between the Aboriginal group and the company. As such, they are treated as confidential documents. In 1990, the first IBA agreement, known as the Raglan Agreement, was concluded between Makivik Corporation (representing the Inuit of Nunavik) and Falconbridge Corporation (purchased in 2006 by the Swiss mining giant, Xstrata). Since the mine site and access route to an ocean port are located in Nunavik, lands controlled by the Inuit of northern Quebec, Falconbridge was compelled to negotiate an IBA. Falconbridge gained Inuit approval for an underground nickel/copper mine and an access route to Deception Bay on Hudson Strait. According to O'Reilly and Eacott (1998), the Raglan Agreement contains the following items:

- The Inuit receive $14 million plus 4.5 per cent of mine profits, estimated at $60 million over 15 years.
- Detailed project descriptions trigger renegotiation if the project deviates from original specifications.
- An implementation committee consists of three Falconbridge representatives and three Inuit. Makivik also got agreement to have a representative appointed to the mine's board of directors.
- A joint committee oversees training programs.
- Inuit enterprises, such as Inuit Air, are given preference for contracts and Inuit communities or individuals have entered into joint ventures with companies to seek contracts with the mine.

Impact and benefit agreements vary from one group to the next depending on the political strength and negotiating skills of each group. Under most circumstances, the company has an advantage because its negotiators are more experienced and have the legal and other resources of the company at their disposal. The Aboriginal group has one important advantage—Aboriginal title. The company must reach an agreement in order to proceed with the project. However, not all Aboriginal groups are sufficiently aware of the stakes and some have little experience with such negotiations.

Two cases are considered here: Victor Lake diamond deposit in northern Ontario and the Ekati diamond deposit in the Northwest Territories (the most recent IBA involved Meadowbank gold mine in Nunavut; see Chapter 5). The IBAs for Victor Lake and Ekati were signed in the last decade of the twentieth century and 'broke' the ground for resource developments far from settlements but on traditional lands. In both cases, local residents and Aboriginal officials grasped the significance of an IBA and learned from the experience of the negotiations, but they were, in a sense, learning on the run. At the same time, the resource corporations involved in these projects learned, too, and ultimately recognized that a partnership approach would achieve the best results for both the local residents and their companies.

In the early 1990s, the South African-based diamond company, De Beers, discovered a vast diamond deposit in the Hudson Bay Lowland of northern Ontario, approximately 90 km west of the coastal community of Attawapiskat. The company undertook an environmental assessment and, over a period of several years, negotiated an IBA with the Attawapiskat First Nation since this diamond deposit lies in the traditional land claimed by the Attawapiskat. Both the EIA and the IBA were fraught with problems—largely because First Nations officials were inexperienced and unsure of their grounds (Whitelaw and McCarthy, 2010: 473–4). The Victor Lake Project lies close to two other First Nations—the Moose Cree First Nation and the Taykwa Tagamou Nation, so these groups also were involved. Negotiations began but were complicated. Unlike First Nations that had achieved a regional sense of belonging and therefore had gained a level of sophistication in dealing with the outside world, the First Nations of northern Ontario tended to act independently and, some might say, lacked the capacity to negotiate on an equal footing with De Beers.

The first step saw De Beers and Attawapiskat sign a memorandum of understanding (MOU) in November 1999. After more consideration, in July 2002, the Attawapiskat First Nation declared that it was terminating the MOU, but would like to begin negotiation towards an impact and benefit agreement. During negotiations, they demanded that work halt for at least two months. The reason for the demand was simple: the First Nation leaders were unsure of the consequences of the project and needed time to review technical documents and to consult with band members and advisers. They also believed that their interests would be better served by an IBA, but to achieve such an agreement requires negotiating experience and skills that the First Nation may lack to strike a 'good' deal (Hasselback, 2002: FP3). In 2005, an IBA was concluded between the Attawapiskat First Nation and De Beers. In 2007, the Moose Cree First Nation and De Beers announced that an IBA was signed. Like the Attawapiskat First Nation, the Moose Cree First Nation voted in favour (85 per cent) of accepting the IBA. This agreement is the third one involving the Victor Lake Diamond Project. The Taykwa Tagamou Nation did not sign an IBA but, rather, a working relations agreement (an MOU). These agreements commit De Beers to provide education, training, compensation, and business and other initiatives to maximize opportunities from the Victor diamond mine and to minimize impacts arising from construction and operation. Chief Patricia Faaries-Akiwenzie of the Moose Cree stated that:

> It's important for the Moose Cree citizens to be entering into this IBA with De
> Beers because it shows respect for our Homeland. Our people have been here

since time immemorial and will continue to live in and off this land. Now that De Beers is extracting our resources it is only fair that we have reached an agreement that provides benefits for our First Nation. This Agreement represents a major step forward in our relationship with outside resource users and breaks the trail for other companies who want to utilize our resources in the Moose Cree Homeland. (Beaton, 2007)

Broken Hill Proprietary Company Inc. (BHP), an Australian multinational, has negotiated IBAs with each of the four Aboriginal groups—North Slave Métis Alliance, Akaitcho Treaty 8, Kitikmeot Inuit Association with the Hamlet of Kugluktuk, and Dogrib Treaty 11—whose traditional lands are potentially affected by the Ekati diamond mine. The agreements ensure that these Aboriginal groups will not interfere with the project. In return, the company agrees (1) to ensure that training, employment, and business opportunities are made available to these people, and (2) to minimize any potential adverse environmental and social impacts of the project. Other objectives are to respect local cultures and the land-based economy, to provide a working relationship between the company and the Aboriginal groups, and to determine how the local communities will be involved in employment, training, and business opportunities. Since impact and benefit agreements are private agreements that include a confidentiality clause, exact details, including any cash compensation, are not part of the public record (personal communication with Denise Burlingame, Senior Public Affairs Officer, Ekati Diamond Mine, 2 Aug. 2002; Noble, 2010: 174–5).

The Dynamic Aboriginal Cultures

Since contact, Indian and Inuit cultures have evolved to adjust to new circumstances. In fact, the Métis are a product of early contact between fur traders and Indian women. Culture, being a dynamic phenomenon, constantly changes, though it must retain at least some of its core elements, such as language, belief system, and way of living. This paradox—changing but holding on to core elements—is particularly noticeable in Aboriginal cultures because of the rapid rate of change caused by the intense contact with Western culture since the move to settlements. Cheryl Swidrovich (2001: 99) maintains that 'Aboriginal peoples have, as part of the strength of their traditions, the ability to continually adapt and change as their circumstances demand without losing their sense of cultural integrity.' Swidrovich concludes that 'the adoption of European principles or beliefs should not, in itself, signal a sudden or wholesale change as to who they are as peoples.' David Newhouse (2000: 408), an Onondaga from the Six Nations of the Grand River and professor of Native Studies at Trent University, sees the development of modern Aboriginal societies as a blending of Aboriginal and Canadian cultures:

The idea of one citizen, one vote gives to an individual a role and power that may be at odds with the roles of individuals within traditional societies. These modern notions are being combined with Aboriginal notions of collectivism and the fundamental values of respect, kindness, honesty, sharing, and caring. Yet, despite these combinations, we still recognize the governments as Aboriginal and the people as Aboriginal. What has happened is that our notion of 'Aboriginal' has changed and now incorporates features of both the traditional and modern worlds.

Aboriginal corporations are a business example of responding to new circumstance. Some are products of comprehensive land claim agreements. Designed to manage the funds derived from agreements, these corporations have put the funds into different enterprises and investments. By 2008, they represent a powerful business sector in the northern economy. Other Aboriginal corporations, sometimes with and sometimes without funds from comprehensive settlements, are a product of partnership with private companies. Kitsaki, the business corporation of the Lac La Ronge Indian band in northern Saskatchewan, began its venture in the trucking industry through a partnership with a successful private trucking firm. The Aboriginal Pipeline Group is a partner in the Mackenzie Gas Project. As a business created and owned by Aboriginal groups in the Northwest Territories, the Aboriginal Pipeline Group has secured a right to own one-third of the Mackenzie Valley natural gas pipeline. With the exception of the Deh Cho, who have yet to conclude a land claim agreement, this marks the first time that Aboriginal organizations have joined together to own a share of a proposed megaproject.

Aboriginal Corporations

Now operating within the market economy, Native-owned corporations, like other entrepreneurs, are confronted with many issues, including finding qualified Aboriginal staff, especially senior managers, ensuring that these operations are profitable, which may mean laying off workers in slow times, protecting the environment, and, above all, avoiding overexploitation of renewable resources (Vignette 7.8).

While Aboriginal corporations cannot solve the deep-rooted economic, employment, and social challenges single-handedly, they do represent a start in that direction.

Vignette 7.8   **The Gap between Aboriginal Education and Employment**

Canada faces a shortage of skilled workers. Traditionally, Canada has sought skilled workers from other countries. John Kim Bell, president of the National Aboriginal Achievement Foundation, sees another solution. The demographic explosion among the Aboriginal population has created a large and growing young 'potential' labour force—one that is growing more than twice as fast as the non-Aboriginal workforce. The problem is that most young Aboriginal people are not graduating from high school and, therefore, cannot enter into trade schools or other post-secondary institutions. In the short run, Bell (2002) calls for private industry to draw more Aboriginal workers into their companies by providing on-the-job training. As well, flying in skilled workers from the south is both expensive and fails to recirculate their wages in the local community. Makivik Construction approached this problem by working closely with the Kativik School Board and the Kativik Regional Government to develop a number of on-the-job carpenter training programs for local students who are happy to spend their earnings in their community (Makivik, 2011b).

More importantly, Aboriginal corporations are making progress in breaking the dependency on government, though these corporations often receive financial backing from Ottawa and/or provinces. Such backing, it must be emphasized, is hardly any different from that received by large corporations through their elaborate lobbying efforts.

Just as the Aboriginal political movement of the late 1960s and early 1970s created political leaders, business leaders today represent a new social class within the Aboriginal community. Aboriginal title, however, is an important factor in allowing such businesses to come into being. The Innu of Labrador provide one example. Aboriginal title has favoured the Innu, who have a claim to land around the Muskrat hydroelectric project on the Lower Churchill River, 225 km downstream from Churchill Falls and approximately 300 km from the Labrador–Quebec border. When completed, Newfoundland and Labrador hopes to generate 2,000 megawatts of power, making it one of the largest hydroelectric generating stations in North America with much of the power exported to New England via an underwater transmission line. If Hydro-Québec could strike a deal for transmitting Muskrat power along its existing lines the idea of the more expensive underwater transmission route would be dropped, but this seems unlikely. Given that Newfoundland and Labrador has believed for many years that it received a raw deal in the long-term agreement with Quebec in regard to Churchill Falls power generation, as well as the hard bargaining position of Hydro-Québec, the province might just proceed with the underwater route to break Quebec's hold on its Labrador power.

The province stood to reap major economic rewards, but only if the Innu consent to the project. And they did at the end of June 2011. Under the newly ratified New Dawn Agreement, the economic benefits and royalties from the proposed Muskrat hydroelectric project, plus cash compensation ($2 million annually) for the Churchill Falls dam built nearly 50 years ago (1967–71), were just too attractive for the Innu to reject. On the other hand, Elders were more skeptical about the benefits and more concerned about the impact of the proposed dam on the Lower Churchill River and Innu traditions (CBC News, 2011b). Peter Penashue, an Innu leader and now a minister in the Harper government, said that 'while there are diverging opinions, the deal is in the long-term interests of the Innu communities' (ibid.). For its part, the province was delighted with the New Dawn Agreement: Premier Kathy Dunderdale called it a landmark for the province and its goal of exporting power to the US. While the Muskrat project is only in the proposed stage, the New Dawn Agreement opens the door for the underwater transmission line to Nova Scotia and New England and the prospects of huge exports of power. Tellingly, though, the provincial government was less upbeat about the land claim aspect of the agreement, considering it 'non-binding' and stating that it 'will form the basis for negotiating a final land claims agreement or treaty' that 'will define Innu treaty rights and where those rights will apply in Labrador' (CBC News, 2011c).

Aboriginal title alone provides an opportunity for Native peoples to participate in resource developments, but comprehensive land claim agreements supply the beneficiaries with a business structure as well as a large cash settlement.[3] The cash settlements obtained by Aboriginal groups through comprehensive agreements in the Territorial North exceed $5 billion. In northern Quebec, agreements signed in the 1970s with the Cree, Inuit, and Naskapi totalled $234 million. More recently, in 2002, the Cree and Inuit reached new 'partnership' agreements with the Quebec government (the Paix des

Braves and Sanarrutik agreements). In both agreements, the development of resources will take place within a partnership context. For example, the Inuit will receive 1.25 per cent of the annual value of electricity produced from future hydroelectric developments in Nunavik. Over the next 50 years, the total value of the Cree–Quebec agreement is $3.6 billion. These funds are committed to economic and community development and release Quebec from its responsibilities under the James Bay and Northern Quebec Agreement in these two fields.

The Quebec Inuit reached a $90 million 25-year agreement (the Sanarrutik Agreement) with the Quebec government in April 2002 (George, 2002). Again, the agreement links northern development with community and economic projects in Nunavik. The Quebec government has approval to begin developing the hydroelectric potential of the rivers flowing into Hudson Bay and Hudson Strait (the Nastapoka, Whale, George, and Caniapiscau rivers). In June 2002, the Makivik Corporation received the first payment of $7 million. At the end of 25 years, Makivik Corporation will have received just over $90 million.

Quebec sees these 'partnership' agreements as 'a path to the future', implying integration into the Quebec economy and society. Peter Kulchyski fears that such agreements simply snare the Quebec Cree and Inuit in a process leading to their assimilation into Canadian society (Vignette 7.9). Kulchyski's is not an isolated voice within academe, but within the Aboriginal community the attraction of economic 'deals' with the promise of jobs and business opportunities seems overwhelming. Paul Nadasky (2005) and Graham White (2006) express similar concerns when examining the difficulties of incorporating traditional knowledge into co-management organizations.

Vignette 7.9   Paix des Braves: Shining Model or Trojan Horse?

In November 2001, Quebec Premier Bernard Landry and Cree Grand Chief Ted Moses agreed to the $3.8 billion Nottaway–Broadback–Rupert hydroelectric project in exchange for annual payments over 50 years totalling $3.6 billion. These funds are for economic and community development and will help to address the shortage of housing in the Cree villages and create construction jobs for local workers. As part of the deal the Cree must drop their $8 billion in lawsuits filed against the province of Quebec.

The agreement puts an end to years of disputes over the impact of the hydro-electric projects on the Cree and their claim of unfulfilled promises from the JBNQA. It commits the traditional territory of the Cree to economic development—hydroelectric, forestry, and mining. Some might argue that the Cree have been co-opted. Guy Chevette, Quebec Native Affairs Minister, proclaimed this agreement to be a shining model for the rest of Canada. Chevette stated: 'The path of the future for native people is to give them the opportunity to exploit resources and share in that' (Roslin, 2001). Peter Kulchyski, writing in 1994, warned that 'regardless of the level of power provided to Aboriginal governments, every decision that is made following the dominant logic, in accordance with the hierarchical and bureaucratic structures of the established order, will take Aboriginal peoples further away from their own culture' (Kulchyski, 1994: 121).

Elders, too, have expressed concern. In spite of such concerns, a continuous wave of economic agreements is sweeping across the Aboriginal world, leaving Aboriginal culture in the dust.

But do the Quebec Cree view such agreements more as a means to gain greater control of their destiny and less as a threat to their culture? In July 2007, the Quebec Cree, who for a long time had complained that Ottawa was not living up to its obligations under the JBNQA, signed another agreement with the federal government, which will result in $1.4 billion being paid to the Cree over 20 years. At the same time, the Cree, rather than the federal government, will deliver services in the fields of justice, social development, and economic development (Heinrich, 2007). As well as assuming these federal responsibilities, the Quebec Cree agreed to withdraw their court actions against the federal government. With these issues behind them, Ottawa and the Cree soon began negotiations on self-government (ibid.; see also Rynard, 2001).

A number of years later, regional government appeared to become a reality for the Quebec Cree in 2011 with an agreement-in-principle between the Cree and the Quebec government. Under this arrangement, the Cree are to gain control over 65,000 km² or 20 per cent of the territory covered by the JBNQA, a substantial increase from the 1 per cent of the territory they previously controlled (White, 2011). As well, this political accord resolves the long-standing dispute regarding the province's failure to fulfill its obligations under the JBNQA. The new agreement was the result of Matthew Coon Come, Grand Chief of the Crees of Quebec and former AFN National Chief, supporting the province's **Plan Nord**—an ambitious plan to develop northern Quebec. Whether the Cree follow the example of their Inuit neighbours in northern Quebec and reject ratification, which seems unlikely at this point, will nonetheless be something to follow.

Regional government raises two critical questions for Aboriginal peoples:

1) Will regional governments serve Aboriginal people well by giving them more control over decisions affecting them and their communities?
2) Is this the political direction sought by the June 2011 report of the Assembly of First Nations, *Pursuing First Nation Self-Determination: Realizing Our Rights and Responsibilities*?

Aboriginal enterprises are not immune to business failure. Sometimes such failure is due to global economic conditions. For example, the business arm of the Lac La Ronge Indian band, Kitsaki, was involved in the Wapawekka Lumber sawmill operation with Weyerhaeuser Canada (since merged with Domtar in 2006 in a $3.3 billion deal), the Peter Ballantyne Cree Nation, and Montreal Lake Cree Nation. With the help of the federal government, the partners obtained $22.5 million to build a small-log sawmill just north of Prince Albert. The objective was to make better use of small logs rather than using them as pulpwood. In 1999, the market for lumber in the United States was strong and the Canadian dollar was low in comparison with the US dollar. These economic conditions made the sawmill, which employed 45 workers, profitable. However, when the US imposed duties on Canadian lumber in 2006, the mill was no longer profitable, closed, and went into receivership. The buildings and equipment of Wapawekka Lumber Ltd were sold through an auction as part of the bankruptcy process to Millar Western Forest Products Ltd, which moved them to Alberta.

Figure 7.5   The Wapawekka Sawmill, 2002

The Wapawekka mill, located in the closed boreal forest zone of northern Saskatchewan, produced 2 x 4 and 2 x 6 lumber. This forest operation was a small, marginal player, and in 2006, when the US market collapsed, the mill was forced to close.

Kitsaki and Inuvialuit Regional Corporation

Circumstances have allowed some Aboriginal groups to move boldly into the resource economy. Two Aboriginal corporations, the Inuvialuit Regional Corporation (IRC) and Kitsaki Management Limited Partnership, exemplify this Native entrepreneurial spirit. The IRC is located in Inuvik, NWT, while the headquarters of Kitsaki, the business arm of the 8,000-plus Lac La Ronge band, is at Air Ronge, Saskatchewan.

In 1984, the IRC was formed as a result of the Inuvialuit Final Agreement. Its purpose is to receive and manage the lands and cash settlement of $55 million (in 1979 dollars) flowing out of this agreement. As a company, the IRC sought to invest these funds in profitable enterprises and, by doing so, to provide both employment and dividends for its members. Its strategy has been to form a number of subsidiaries to undertake specific business activities. For example, the Inuvialuit Development Corporation (IDC) is responsible for investing in a wide range of business ventures. By 2007, IDC was a major shareholder in over 20 enterprises and joint ventures in seven sectors—transportation, petroleum services, construction/manufacturing, environmental services, northern services, real estate, and tourism/hospitality. From an initial investment of $10 million in 1987, IDC's asset base has grown to over $135 million, due in large part to its petroleum investments. IDC has expanded each year until 2009 (IRC, 2006: 7). In 2008, the global economic collapse led to a difficult 2009 for IDC. According to the *Annual Report* for 2009 (2010: 7), most companies had positive earnings but three—NorTerra, Aklak Air, and Arctic Oil & Gas Services—suffered heavy

Figure 7.6   Barges of the Northern Transportation Company

Based in Hay River, NWT, Northern Transportation Company has a fleet of 12 mainline tugs, more than 90 barges, and two Arctic Class II charter supply vessels that cover nearly 6,000 km of northern shipping routes, travelling north along the Mackenzie River nearly 1,200 km to reach Tuktoyaktuk on the shores of the Beaufort Sea. From there, the route stretches from Point Barrow, Alaska, to the west all the way to the Boothia Peninsula in Nunavut. In 2006, the company reopened its shipping route north from the port of Churchill, Manitoba, to serve the Kivalliq region in Nunavut. In the same year, the company reopened its shipping route to Fort McMurray, Alberta, from where it first started shipping to Canada's Arctic in 1934. NTCL opened an office in 2005 in Halifax to service the Atlantic coast market and the resource development market in Nunavut. Northern Transportation Company is jointly owned by Inuvialuit Development Corporation and Nunasi Corporation, which represents the Inuit of Nunavut. In the photograph, barges are carrying mobile drilling rig modules along the Mackenzie River, destined for the North Slope of Alaska.

Source: Norterra (2006). Reprinted by permission of Northern Transportation Company Limited.

losses due to the virtual halt to oil and gas activities in the Inuvialuit Settlement Area. As a result, IDC recorded a loss of $14,995,000.

Kitsaki operates at a much smaller scale and without the advantage of a modern land claim settlement. Kitsaki came about when the provincial government insisted that northern uranium mining companies provide employment and business opportunities for northern residents. Unlike the IRC, Kitsaki has little capital and has had to move slowly and cautiously into business opportunities in the mining and forestry industries. Often, the provincial government, as a means of ensuring 'northern benefits', has encouraged forestry and mining companies to offer Kitsaki business opportunities. In 1986, the Lac La Ronge band entered into a joint venture with a private trucking firm to supply goods to the uranium mines in northern Saskatchewan. Without experienced truck drivers, few band members were employed at first, but the trucking firm was willing to train and then employ people from Lac La Ronge. The chief at that time, Myles Venn, saw these new opportunities as a means to achieve 'Aboriginal ownership that will enable us to possess the control we need to secure jobs for our people' (Bone, 2002: 70). By 2010, Kitsaki employed over 800 workers and up to 1,000 seasonal employees. Approximately 25 per cent of their full-time employees are band members, with wages totalling over $6 million in 2009; and almost all of the seasonal employees are young band members (Kitsaki, 2010). Kitsaki's gross revenue for the year ending 31 March 2010 exceeded $100 million from its nine

Figure 7.7   Trucking Enterprise of the Lac La Ronge Band

Northern Resource Trucking is a partnership between the Lac La Ronge band and Trimac Transportation. The company was founded in 1986 to expand Aboriginal participation in northern Saskatchewan's mining industry. Winter driving conditions along roads in the boreal forest to reach remote uranium mines are a challenge.

Source: Kitsaki (2006). Photo by Dale Peacock. Reprinted by permission of the photographer and Northern Resource Trucking.

enterprises: Asiniy Gravel Crushing Limited Partnership; Athabasca Catering Limited Partnership; Canada North Environmental Services Limited Partnership; Dakota Dunes Golf Links Limited Partnership; First Nations Insurance Services Ltd; Kitsaki/ Procon Joint Venture; La Ronge Hotel and Suites Limited Partnership; Northern Lights Foods Limited Partnership; and Northern Resource Trucking Limited Partnership. While most businesses are profitable, a few, like Wapawekka sawmill, failed.

Kitsaki, like the Inuvialuit Regional Corporation, has made only a small dent in addressing employment and social problems among its members. First, the magnitude of the unemployment challenge is beyond the capacity of these two organizations. Second, the skill set and experience found in their labour force fall short of the needs of these two business entities. This mismatch between the Aboriginal labour force and the job needs of the Aboriginal corporations extends well beyond these two groups. The government of Nunavut faces a similar problem (Table 7.2). Consequently, the immediate impact on their respective communities has been limited. The hope is for the long term, with youth of the Lac La Ronge band recognizing that there is a future in wage employment and businesses. At the moment, however, the relatively small number of people benefiting may, in the eyes of the majority, have created a form of Aboriginal elitism. Local support for these business initiatives wavers, partly because of the limited impact on their communities, high expectations, and the investment into enterprises outside of their traditional area. The Inuvialuit Regional Corporation has the resources to make modest annual payments to its beneficiaries (Vignette 7.10),

---

Vignette 7.10   IRC **Distribution Policy to Its Members**

The Inuvialuit Regional Corporation (IRC) reported a net loss of $15 million by the Inuvialuit Corporate Group in 2009. Yet, IRC made the minimum distribution payments to all enrolled Inuvialuit beneficiaries over the age of 18 by drawing on its reserves. In accordance with the IRC distribution policy, each of the 3,939 enrolled Inuvialuit beneficiaries received $400 in 2009, a total distribution of $1,595,600. In comparison, in 2006 the net income was $36.5 million with a distribution payment of $770.12 per beneficiary, for a total distribution of $2.9 million.

Sources: Inuvialuit Regional Corporation (2007, 2010).

---

but Kitsaki, with far smaller resources, is now donating money—$1 million in 2009 and $1.1 million in 2010—to the band for allocation to its five reserves for community projects.

The Future: Finding a Place

Clearly, the economic and political landscape of the North is shifting, with Aboriginal peoples controlling their own governments and creating community and regional businesses. A growing self-confidence is also emerging. As Calvin Helin (2006: 238), a controversial Indian author and spokesperson, puts it: 'What the [Aboriginal business] models examined tell us is that where the greatest successes have been achieved, strong, ethical leadership has been critical in gaining community consensus to move forward toward economic self-reliance.'

This process of economic, political, and social development within the diverse Aboriginal community takes many forms, but, as Helin emphasizes, the key element is moving from a state of dependency to one of self-reliance. Indeed, he has strongly criticized national Aboriginal organizations for maintaining and encouraging a relationship of dependency between First Nations and Canadian governments. How do Indian leaders feel about changing the relationship between First Nations and Ottawa? In July 2011, National Chief Shawn Atleo of the Assembly of First Nations opened the Aboriginal version of Pandora's box by reiterating the key recommendations of a just-released AFN report: phasing out the Indian Act within five years; scrapping the Ministry of Aboriginal Affairs and Northern Development; and establishing two new cabinet ministries, one a senior-level portfolio, that would better serve Aboriginal peoples (AFN, 2011). So far, the federal government has not responded.

With the help of comprehensive land claims, co-management, and impact and benefit agreements, Aboriginal peoples are finding a place in the northern economy. With their land claims behind them, Inuit businesses such as Inuvialuit Regional Corporation and the Makivik Corporation have become powerful economic forces in the continuing process of northern development. Makivik's construction division illustrates a model by generating work and jobs in Nunavik communities through its housing program. As long as Ottawa and Quebec City support public housing in Nunavik with five-year agreements, then Makivik can take advantage of medium-range planning that results in lower costs and greater efficiencies. The shift towards self-government has reached a high level with the creation of Nunavut. But the economic landscape has seen the most profound changes—driven by the need to participate in the market economy or else remain on the margins. While not abandoning their land-based activities, increasingly, Aboriginal groups have ventured into the always risky market economy through a variety of business organizations. Some involve the commercial use of their natural resources; some are assisted by northern affirmative action; and some represent individuals and groups starting a business. Joint ventures, however, have proven to be a quick and usually successful way for Aboriginal business people to gain the expertise necessary for a profitable business.

While great strides were achieved over the last four decades, will the next four decades see similar advances? Aboriginal leadership is faced with defining questions that will shape the future world for northern Aboriginal peoples:

- Can remote Native settlements with growing populations gain an economic foothold?
- Do regional governments provide the political framework for Aboriginal peoples to retain and perhaps redefine their Aboriginal identity, language, and culture while participating in the market economy?
- How can leaders and communities resolve the inherent tension between the demand for self-government and Aboriginal sovereignty and the carrot of huge compensation packages dangled by governments and corporations for agreement to deals involving less than full sovereignty and control?
- With growing Aboriginal populations, how can wildlife—already in flux and under pressure from climate change—continue to serve as country food, which until now has been so necessary for the well-being of northern peoples?
- Will AFN National Chief Atleo's call for a restructuring of relationships with the federal government result in a national debate and perhaps reforms?

The Aboriginal world is gaining more and more control over its affairs, and in the years to come Canada may see an Aboriginal Spring, which will, it is hoped, provide answers to these and other questions.

Challenge Questions

1. In Vignette 7.9, the Paix des Braves is presented as either a shining model for the future (as argued by the Quebec government) or a Trojan Horse (as feared by Peter Kuchyski). Which view do you support and why?
2. Do you agree (or disagree) with David Newhouse that the meaning of 'Aboriginal' has changed over the years and now incorporates features of both the traditional and modern worlds?
3. What two structures in the 1984 Inuvialuit Final Agreement protect their traditional way of life and, at the same time, open the door to participation in the market economy?
4. In what ways could global warming threaten the Inuit way of life?

5. If hunters and trappers in the 1950 and 1960s had received a subsidy (perhaps along the lines of Quebec's Income Security Program), would this have eased the cultural and economic adjustment caused by the federal relocation program for northern Aboriginal peoples? Would such a program, if instituted then, continue to have a positive impact today? Finally, would the federal government balk at adopting such a program because of the cost (Quebec's program cost nearly $20 million in 2009–10)? Explain.
6. Why is diabetes affecting so many Aboriginal peoples living in northern settlements?
7. Resource companies must negotiate impact and benefit agreements with Aboriginal groups on whose ancestral lands proposed projects would occur. Why have some IBAs produce 'good' results for local peoples and others 'less good'?
8. Can population geography explain why the Inuit political jurisdiction of Nunavut enshrined a public government model rather than an ethnic one? Explain.
9. If most Aboriginal languages are disappearing, what are the cultural consequences for Aboriginal peoples and for Canada?
10. Homelessness among Aboriginal populations is a growing problem in the Canadian North. Why? How can it be addressed?

Notes

1. Sachs Harbour was named after the ship *Mary Sachs* of the Canadian Arctic Expedition of 1913. In 1929 a permanent settlement was established when three Inuit families settled there to trap. In 1953 the RCMP set up a detachment. People continue to live a traditional lifestyle hunting muskox, caribou, and polar bear.
2. Salisbury (1986: 76–84) describes changes in hunting among the Cree of northern Quebec. Under the subsidized Income Security Program for hunters and trappers, the increased availability of cash meant that they could make more use of modern technology and transportation. Prior to 1975, hunters took little equipment into the bush and often travelled only short distances. By 1981, the Cree hunters could use chartered aircraft to fly their families into remote bush camps, which now were as comfortable as village housing. As well, hunting has become more mechanized and less physically exhausting because these Cree hunters can now afford to fly a snowmobile to their winter camps. The snowmobile is the workhorse of the hunter. It allows a quick inspection of his trapline and the easy hauling of big game and firewood from the bush to camp. From 1974 to 1979, Salisbury reports that there was a 20 per cent increase in the overall amount of game killed by these Cree hunters and their families (whose number increased by more than 50 per cent over the same time period).
3. Land is a fundamental element in the well-being of Aboriginal peoples. In exchange for surrendering their rights to their traditional hunting lands, treaty Indians received reserve lands, a right to hunt and fish, and other benefits. In some cases, treaty Indians did not receive their correct allotment of land. In such cases, they can file a specific land claim. Some status Indians have not taken treaty. They and other Aboriginal peoples have not surrendered their Aboriginal right to traditional lands. Both can make a comprehensive land claim, a lengthy process that leads to a final agreement. Métis residing in the Northwest Territories are eligible to make a claim but, as of 2001, Métis living in provinces could not.
4. Many treaty Indians claimed that they did not receive their full allotment of land at the time of signing the treaty or that they lost land. To answer these claims, the federal government established the Office of Native Claims in 1974. Since the 1980s, this role has been assigned

to the federal Indian Affairs ministry (in 2011 renamed Aboriginal Affairs and Northern Development). After documenting their case, a band is eligible to make a specific land claim. Specific land claims are often based on a grievance about the federal government's administration of Indian land. Land shortages have occurred in three ways. (1) The original land grant was correct but government officials took some land from them without proper authorization. (2) Lands that should have been granted never were granted. (3) Reserve populations have increased. Most claims are based on 'lost' lands. In Saskatchewan, for example, almost a third of the land placed in reserve under treaty was surrendered to the government and then sold to settlers (Brizinski, 1993: 231). In many cases, since Crown land is either not available or is not suitable, land must be purchased from private land-owners. Given the price of agricultural and urban land, the cost of settling specific claims may run into billions of dollars. The Saskatchewan settlement reached $500 million divided among 27 First Nations.

5. Comprehensive land claims are considered modern treaties. Like earlier treaties, they are based on the concept of Aboriginal title to land. Comprehensive land claim settlements are designed to extinguish Native claim to their traditional land occupancy in exchange for ownership of a block of land, hunting and trapping rights (but not subsurface rights) on selected lands, and monetary compensation.

References and Selected Reading

Aatami, Pita. 2010. *From Igloos to the Internet: Inuit in the 21st Century*. Occasional Paper Series on Canadian Studies, Canadian Studies Center, Henry M. Jackson School of International Studies, University of Washington, 10 Feb. Seattle: University of Washington Press.

Aporta, Claudio. 2010. *Inuit Sea Ice Use and Occupancy Project (ISIUOP)*. International Polar Year Project. At: <gcrc.carleton.ca/isiuop>.

Assembly of First Nations. 2011. *Pursuing First Nation Self-Determination: Realizing Our Rights and Responsibilities*, June. At: <www.afn.ca/uploads/files/aga/pursuing_self-determination_aga_2011_eng%5B1%5D.pdf>.

Atlas of Canada. 2006. 'Discover Canada through National Maps and Facts: Nunavut'. At: <atlas.nrcan.gc.ca/site/english/maps/reference/provincesterritories/nunavut/referencemap_image_view>.

Bartlett, Richard. 1991. *Resource Development and Aboriginal Land Rights*. Calgary: University of Calgary, Canadian Institute of Resources Law.

Beaton, Brian. 2007. 'Moose Cree First Nation Signs Impact Benefit Agreement for Developing Diamond Mine', *Knet Media*. At: <media.knet.ca/node/2996>.

Bell, Jim. 2007. 'qia Wants Language Laws Dumped, Re-written', *Nunatsiaq News*, 29 June. At: <www.nunatsiaq.com/archives/2007/706/70629/news/nunavut/70629_253.html>.

Berger, T.R. 2006. *Nunavut Land Claims Agreement Implementation Contract Negotiations for the Second Planning Period 2003–2013*. Conciliator's Final Report: 'The Nunavut Project'. Iqaluit: Government of Nunavut.

Bone, Robert M. 1988. 'Cultural Persistence and Country Food: The Case of the Norman Wells Project', *Western Canadian Anthropologist* 5: 61–79.

———. 2001. *Nunavik Political Model: Is It an Appropriate Model for Other Provinces—the Case of Northern Saskatchewan*. Report prepared for the Department of Indian and Northern Affairs. Saskatoon: Signe Research Associates.

———. 2002. 'Colonialism to Post-Colonialism in Canada's Western Interior: The Case of the Lac La Ronge Indian Band', *Historical Geography* 30: 59–73.

——— and Milford B. Green. 1984. 'The Northern Aboriginal Labor Force: A Disadvantaged Work Force', *Operational Geographer* 3: 12–14.

Cardinal, Harold. 1969. *The Unjust Society*. Edmonton: Hurtig.

CBC News. 2006. 'Researcher [Jack Hicks] Studies Inuit Suicide Rates', 13 Oct. At: <www.cbc.ca/news/canada/north/story/2006/10/13/suicide-inuit.html>.

———. 2010. 'Natuashish Booze Ban Cancelled: New Chief', 8 Mar. At: <www.cbc.ca/news/canada/newfoundland-labrador/story/2010/08/24/natuashish-court-challenge-innu-824.html?ref=rss>.

———. 2011a. 'Quebec Inuit Vote against Self-Government Plan', 29 Apr. At: <www.cbc.ca/canada/north/story/2011/04/29/nunavik-government-referendum.html>.

———. 2011b. 'Labrador Innu Vote on Contentious Land Claim Deal', 30 June. At: <www.cbc.ca/news/canada/newfoundland-labrador/story/2011/06/30/nl-innu-labrador-claims-churchill-vote-630.html>.

———. 2011c. 'Premier Celebrates Innu Backing of Hydro Deal', 1 July. At: <www.cbc.ca/news/canada/newfoundland-labrador/story/2011/07/01/nl-dunderdale-rxn-hydro-deal.html>.

———. 2011d. 'Labrador Caribou Hunt Start Delayed', 3 Aug. At: <www.cbc.ca/news/canada/newfoundland-labrador/story/2011/08/03/nl-caribou-herd-declining-802.html>.

Collings, Peter, George Wenzel, and Richard G. Condon. 1998. 'Modern Food Sharing Networks and Community Integration in the Central Canadian Arctic', *Arctic* 51, 4: 301–14.

Condon, Richard G., Peter Collings, and George Wenzel. 1995. 'The Best Part of Life: Subsistence Hunting, Ethnicity, and Economic Adaptation among Young Adult Inuit Males', *Arctic* 48, 1: 31–46.

Dahl, Jens, Jack Hicks, and Peter Jull. 2000. *Inuit Regain Control of Their Lands and Their Lives*. Skive, Denmark: Centraltrykkeriet Skive A/S.

Duerden, Frank. 1990. 'The Geographer and Land Claims: A Critical Appraisal', *Operational Geographer* 8, 2: 35–7.

———. 1996. 'An Evaluation of the Effectiveness of First Nations Participation in the Development of Land-Use Plans in the Yukon', *Canadian Journal of Native Studies* 16: 105–24.

———. 2004. 'Translating Climate Change Impacts at Ethnic Community Level', *Arctic* 57, 2: 204–13.

——— and C.P. Keller. 1992. 'GIS and Land Selection for Native Claims', *Operational Geographer* 10, 4: 11–14.

——— and R.G. Kuhn. 1996. 'Applications of GIS by Government and First Nations in the Canadian North', *Cartographica* 33, 2: 49–62.

——— and ———. 1998. 'Scale, Content, and Application of Traditional Knowledge of the Canadian North', *Polar Record* 34, 188: 31–8.

Ellis, S.C. 2005. 'Meaningful Consideration? A Review of Traditional Knowledge in Environmental Decision-making', *Arctic* 58, 1: 66–7.

Erasmus, George. 1977. 'We, the Dene', in Mel Watkins, ed., *Dene Nation—The Colony Within*. Toronto: University of Toronto Press, 3–4.

Ford, J., and L. Berrang-Ford. 2009. 'Food Insecurity in Igloolik, Nunavut: A Baseline Study', *Polar Record* 45, 234: 225–36

Fournier, Watson. 2005. 'Presentation by Makivik Construction Division and Kativik Municipal Housing Bureau to Indian and Northern Affairs Canada', 23 Aug. Ottawa: Indian and Northern Affairs Canada.

Freeman, M.R., ed. 1976. *Inuit Land Use and Occupancy Project*, 3 vols. Ottawa: Indian and Northern Affairs.

——— and G.W. Wenzel. 2006. 'The Nature and Significance of Polar Bear Conservation Hunting in the Canadian Arctic', *Arctic* 59, 1: 21–30.

Galbraith, Lindsay, Ben Bradshaw, and Murry B. Rutherford. 2007. 'Towards a New Super-Regulatory Approach to Environmental Assessment in Northern Canada', *Impact Assessment and Project Appraisal* 25, 1: 27–41.

George, Jane. 2002. 'Nunavik Leaders Celebrate First Quebec Payout', *Nunatsiaq News*, 21 June. At: <www.nunatsiaq.com/archives/nunavut020621/news/nunavik/20621_1.html>.

———. 2011. 'Makivik Corp. Wants to Convene Special Meeting on Self-Government', *Nunatsiaq Online*, 21 June. At: <www.nunatsiaqonline.ca/stories/article/210677_makivik_corp._wants_to_convene_a_special_meeting>.

Harding, Katherine. 2007. 'Visiting Premiers Will Learn of Challenges Facing Nunavut', *The Globe and Mail*, 4 July. At: <www.apathyisboring.com/en/the_facts/news/208>.

Hasselback, Drew. 2002. 'Natives Halt De Beers Diamond Project', *National Post*, 1 Aug., FP3.

Health Canada. 2005. *A Statistical Profile on the Health of First Nations in Canada*. Ottawa: Health Canada.

Heinrich, Jeff. 2007. '$14B Deal Gives Cree Promise of Nationhood', *National Post*, 17 July, A4.

Helin, Calvin. 2006. *Dances with Dependency: Indigenous Success through Self-Reliance*. Vancouver: Orca Spirit Publishing.

Hicks, Jack, and Graham White. 2000. 'Nunavut: Inuit Self-Determination through a Land Claim and Public Government?', in Jens Dahl, Jack Hicks, and Peter Jull, eds, *Inuit Regain Control of Their Lands and Their Lives*. Skive, Denmark: Centraltrykkeriet Skive A/S, 30–117.

Hurley, Mary C. 2000. *Aboriginal Title: The Supreme Court of Canada Decision in Delgamuukw v. British Columbia*. Ottawa: Library of Parliament. At: <www.parl.gc.ca/information/library/PRBpubs/bp459-e.htm#DELGAMUUKW%20v.%20BRITISH%20COLUMBIA(txt)>.

Indian and Northern Affairs Canada (INAC). 2000. 'Fact Sheet—The Nisga'a Treaty', Northern Affairs Program. At: <www.ainc-inac.gc.ca/pr/info/nit_e.html>.

———. 2006. 'Gains Made by Inuit in Formal Education and School Attendance, 1981–2001', *Inuit Social Trends Series*, No. 3. Strategic Research and Analysis Directorate. At: <www.ainc-inac.gc.ca/pr/ra/gmif/gmif1_e.html>.

———. 2008. 'Frequently Asked Questions about the Tlicho Agreement', 28 Oct. At: <www.ainc-inac.gc.ca/ai/mr/nr/j-a2005/02586bk-eng.asp>.

Inuit Tapiriit Kanatami (itk). 2004. 'Backgrounder on Inuit and Housing: For Discussion at Housing Sectorial Meeting, November 24 and 25th in Ottawa'. At: <www.aboriginalroundtable.ca/sect/hsng/bckpr/ITK_BgPaper_e.pdf>.

Inuvialuit Regional Corporation (irc). 2006. *Annual Report 2005*. At: <www.irc.inuvialuit.com/publications/pdf/IRC%20Annual%20Report%202005.pdf>.

———. 2007. '2007 Distribution Payments'. At: <www.irc.inuvialuit.com/beneficiaries/2006payments.html>.

———. 2010. *Annual Report 2009*. At: <www.irc.inuvialuit.com/publications/pdf/2009%20IRC%20Annual%20Report%20-%20Combined.pdf>.

Kennett, S.A. 1999. *A Guide to Impact Benefits Agreements*. Calgary: Canadian Institute of Resources Law, University of Calgary.

Kitsaki. 2006. 'Northern Resource Trucking Limited Partnership'. At: <www.kitsaki.com/nrt.html>.

———. 2010. *Newsletter* (Summer). At: <www.kitsaki.com/Kitsaki%20Updates%20&%20Brochure/Kitsaki%20Update%20Summer%202010.pdf>.

Klinkig, Eileen. 2005. *Presentation by Makivik Construction Division and Kativik Municipal Housing Bureau to Indian and Northern Affairs Canada*, 23 Aug. Ottawa: Indian and Northern Affairs Canada.

Kulchyski, Peter. 1994. *Unjust Relations: Aboriginal Rights in Canadian Courts*. Toronto: Oxford University Press.

Laidler, G.J., J. Ford, W.A. Gough, T. Ikummaq, A. Gagnon, A. Kowal, K. Qrunnut, and C. Irngaut. 2009. 'Travelling and Hunting in a Changing Arctic: Assessing Inuit Vulnerability to Sea Ice Change in Igloolik, Nunavut', *Climatic Change* 94, 3–4:363–97.

Laidler, Gita. 2010. *Inuit Siku (Sea Ice) Atlas*. At: <app.fluidsurveys.com/surveys/gjlaidler/siku-atlas-consultation-1/?code=X45wR&l=en>.

Laliberté, Arlene, and Michel Tousignant. 2009. 'Alcohol and Other Contextual Factors of Suicide in Four Aboriginal Communities of Quebec, Canada', *Journal of Crisis Intervention and Suicide Prevention* 30, 4: 215–21.

Légaré, André. 2000. 'La Nunavut Tunngavik Inc.: Un examen de ses activités et de sa structure administrative', *Études/Inuit/Studies* 24, 1: 97–124.

———. 2001. 'The Spatial and Symbolic Construction of Nunavut: Towards the Emergence of a Regional Collective Identity', *Études/Inuit/Studies* 25, 1 and 2: 143–68.

———. 2008a. 'Canada's Experiment with Aboriginal Self-Determination in Nunavut: From Vision to Illusion', *International Journal on Minority and Group Rights* 15: 335–67.

———. 2008b. 'Inuit Identity and Regionalization in the Canadian Central and Eastern Arctic: A Survey of Writings about Nunavut' *Polar Geography* 31, 3 and 4: 99–118.

——— and Markku Suksi. 2008. 'Rethinking the Forms of Autonomy at the dawn of the 21st Century', *International Journal on Minority and Group Rights* 15: 143–55.

MacDonald, Michael. 2001. 'Innu of Davis Inlet Prepare for Year of Big Changes', Canadian Press, 21 Dec. At: <www.canoe.ca/CNEWS2001Review/1223_innu-cp.html>.

McNeil, Kent. 1998. 'Defining Aboriginal Title in the 90s: Has the Supreme Court Finally Got It Right?' The John P. Robarts Professor of Canadian Studies Twelfth Annual Robarts Lecture, York University, 25 Mar. At: <www.yorku.ca/robarts/projects/lectures/pdf/rl_mcneil.pdf>.

Makivik Corporation. 2011a. 'The Information Tour Preceding the Referendum on the Final Agreement Starts', 14 Feb. At: <www.makivik.org/the-information-tour-preceding-the-referendum-on-the-final-agreement-starts/>.

———. 2011b. 'Housing Development'. At: <www.makivik.org/building-nunavik/housing-development/>.

Natcher, Paul. 2007. 'The Gift in the Animal: The Ontology of Hunting and Human–Animal Sociality', *American Ethnologist* 34, 1: 25–43.

Newhouse, David. 2000. 'From the Tribal to the Modern: The Development of Modern Aboriginal Societies', in Ron F. Laliberte, Priscilla Settee, James B. Waldram, Rob Innes, Brenda Macdougall, Lesley McBain, and F. Laurie Barron, eds, *Expressions in Canadian Native Studies*. Saskatoon: University of Saskatchewan Extension Press, 395–409.

———. 2001. 'Modern Aboriginal Economies: Capitalism with a Red Face', *Journal of Aboriginal Economic Development* 1, 2: 55–61.

Noble, Bram F. 2010. *Introduction to Environmental Impact Assessment*, 2nd edn. Toronto: Oxford University Press.

Norterra. 2006. 'ntcl'. At: <www.norterra.com/oc_1.html>.

Nunavik Regional Government. 2010. *Timeline with Target Dates of Negotiation Activity Leading to the Final Agreement (FA), 2010–11*, 2 Nov. At: <www.nunavikgovernment.ca/en/documents/NRG_Detailed_Timeline_2010-11.pdf>.

Nunavimmiut Kavamanga. 2007. Map of Nunavik. At: <www.nunavikgovernment.ca/en/archives/photos/map_of_nunavik.html> and <nunavikgovernment.ca/en/photos/map_carte1.html>.

Nunavut Department of Finance. 2011. '2011–12 Budget Highlights'. At: <www.finance.gov.nu.ca/apps/authoring/dspPage.aspx?page=budgets>.

O'Reilly, Kevin, and Erin Eacott. 1998. 'Aboriginal Peoples and Impact and Benefit Agreement: Summary of the Report of a National Workshop', Canadian Arctic Resources Committee. At: <www.carc.org/pubs/v25no4/2.htm>.

Prno, Jason, and Ben Bradshaw. 2007. 'Assessing the Effectiveness of Impact and Benefit Agreements in the Canadian North', presentation at the annual meeting of the Canadian Association of Geographers, 31 May. At: <www.usask.ca/geography/cag2007/final_program.pdf>.

Quebec. 2002. 'Complementary Agreement to the James Bay and Northern Québec Agreement—the Quebec Government Allocates an Additional $4.4 Million to the Income Security Program for Cree Hunters', news release, 22 May. At: <communiques.gouv.qc.ca/gouvqc/communiques/GPQE/Mai2002/23/c8749.html>.

———. 2010. *Annual Report of the Cree Hunting and Trappers Income Security Board, 2009-2010*. Departement de l'Emploi et de la Solidarité, 17 Dec. At: <www.osrcpc.ca/images/osrcpc/rapportannuel/2009-2010.pdf>.

Roslin, Alex. 2001. 'Cree Deal a Model or Betrayal?', *National Post*, 10 Nov., FP7.

Rynard, Paul. 2001. 'Ally or Colonizer? The Federal State, the Cree Nation and the James Bay Agreement'. At: <findarticles.com/p/articles/mi_qa3683/is_200107/ai_n8960434/pg_1>.

Salisbury, Richard F. 1986. *A Homeland for the Cree: Regional Development in James Bay, 1971–1981*. Montreal and Kingston: McGill-Queen's University Press.

Schellenberger, Stan, and John A. MacDougall. 1986. *The Fur Issue: Cultural Continuity, Economic Opportunity*. Report of the House of Commons Standing Committee on Aboriginal Affairs and Northern Development. Ottawa: Queen's Printer.

Schultz, Stephanie. 2001. 'A Simple Smile, a Scream Inside', in Jack Caufield, Mark Victor Hansen, and Kimberly Kirberger, eds, *Chicken Soup for the Teenage Soul: Letters of Life, Love & Leaving*. Deerfield Beach, Fla: Health Communications, 56–8.

Slocombe, D. Scott. 2000. 'Resources, People and Places: Resource and Environmental Geography in Canada', *Canadian Geographer* 44, 1: 56–66.

Stackhouse, John. 2001. 'Increasing Raglan's Inuit Workforce', *The Globe and Mail*, 14 Dec. At: <www.globeandmail.com/series/apartheid/stories/20011214-5.htm>.

Statistics Canada. 1998. '1996 Census: Education, Mobility and Migration', *The Daily*, 14 Apr. At: <www.statcan.ca/Daily/English/980414/d980414.htm>.

———. 2003. 'Overview: Canadians Better Education Than Ever', *Education in Canada: Raising the Standard*. At: <www12.statcan.ca/english/census01/Products/Analytic/companion/educ/canada.cfm>.

———. 2010. 'Canada's Population Estimates' *The Daily*, 22 Dec. At: <www.statcan.gc.ca/daily-quotidien/101222/t101222a2-eng.htm>.

Swidrovich, Cheryl. 2001. 'Stanley Mission: Becoming Anglican but Remaining Cree', *Native Studies Review* 14, 2: 71–108.

Tait, Heather. 2008. *Inuit Health and Social Conditions*. Statistics Canada, Catalogue no. 89–637–XWE2008001. At: <www.statcan.gc.ca/bsolc/olc-cel/olc-cel?catno=89-637-XWE2008001&lang=eng>.

Timpson, Annis May. 2006. 'Stretching the Concept of Representative Bureaucracy: The Case of Nunavut', *International Review of Administrative Science* 72, 4: 517–30.

Usher, Peter J. 1976. 'Inuit Land Use in the Western Canadian Arctic', in M.M.R. Freeman, ed., *Inuit Land Use and Occupancy Project*. Ottawa: Department of Indian Affairs and Northern Development, vol. 1, 21–31; vol. 3, 1–20.

———. 1982. 'Unfinished Business on the Frontier', *Canadian Geographer* 26, 3: 187–90.

———. 2002. 'Inuvialuit Use of the Beaufort Sea and Its Resources, 1960–2000', *Arctic* 55 (supplement 1): 18–28.

———. 2003. 'Environment, Race, and Nation Reconsidered: Reflections on Aboriginal Land Claims in Canada', *Canadian Geographer* 47, 4: 365–82.

———, F.J. Tough, and R.M. Galois. 1992. 'Reclaiming the Land: Aboriginal Title, Treaty Rights, and Land Claims in Canada', *Applied Geography* 12, 2: 109–32.

——— and George Wenzel. 1987. 'Aboriginal Harvest Surveys and Statistics: A Critique of Their Construction and Use', *Arctic* 40, 2: 145–60.

——— and ———. 1989. 'Socio-Economic Aspects of Harvesting', in Randy Ames, Don Axford, Peter Usher, Ed Weick, and George Wenzel, eds, *Keeping On the Land: A Study of the Feasibility of a Comprehensive Wildlife Harvest Support Program in the Northwest Territories*. Ottawa: Canadian Arctic Resources Committee, ch. 1.

Warick, Jason. 2007. 'Suicide Epidemic: Northern Community of Sandy Bay Rocked by Five Deaths, 12 Attempts in Recent Months', *Saskatoon Star-Phoenix*, 22 Feb.

Wenzel, George W. 1986. 'Canadian Inuit in a Mixed Economy: Thoughts on Seals, Snowmobiles, and Animal Rights', *Aboriginal Studies Review* 2, 1: 69–82.

———. 1995. '*Ningiqtuq*: Resource Sharing and Generalized Reciprocity in Clyde River, Nunavut', *Arctic Anthropology* 24, 2: 56–81.

———. 2006. 'Cultures in Collision: Traditional Knowledge and Euro-Canadian Governance Processes in Northern Land-Claim Boards', *Arctic* 59, 4: 401–14.

White, Marianne. 2011. 'Crees to Ink Regional Government Deal with Quebec', *National Post*, 26 May, A3.

Whitelaw, Graham, and Daniel McCarthy. 2010. 'Learning from the Victor Diamond Mine Comprehensive Environmental Assessment, Ontario', in Bruce Mitchell, ed., *Resource and Environmental Management in Canada: Addressing Conflict and Uncertainty*. Toronto: Oxford University Press, 473–4.

8

Geopolitics, Climate Warming, and the Arctic Ocean

Setting the Stage

The **geopolitics** of the Arctic Ocean has taken on new significance because of the melting of the Arctic ice cap. While this retreat has only begun, if it continues at the pace of the last two decades, the Northwest Passage could be ice-free perhaps for as long as three months in the summer sometime in the twenty-first century, thus encouraging sea-going ships to take this much shorter route between Europe and Asia. Unlike Antarctica with its 1959 Antarctic Treaty, which established this continent as international territory set aside for scientific research and without military activities, no such arrangement exists for the Arctic Ocean.

Climate warming has changed the political ball game, and more nations are expressing an interest in the High Arctic because of this fundamental 'physical change' in the polar climate and the Arctic ice cap. Two potential consequences are (1) the prospects for a commercial sea route between Asia and Europe through the Northwest Passage increases significantly; and (2) access to the huge offshore oil and gas reserves below the seabed of the Arctic Ocean improves considerably. While the timing of an ice-free Northwest Passage is uncertain, if climate warming continues unabated, open water from Lancaster Sound through M'Clure Strait for much of the summer months could become a reality. Of course, these Arctic waters would be ice-bound for the winter months but this new ice would be similar to the ice that covers Hudson Bay each winter. The geopolitical implications for Canada are enormous—regular Arctic passage with sea access to both Asian and European markets could signal a round of resource development and create additional responsibilities for Canada's managing its Arctic waters, regulating ocean-going vessels in these waters, and charging ships that pollute these waters.

Geopolitics has another dimension, with other nations casting their eyes into these waters and the ungoverned seabed of the Arctic Ocean (see Figure 8.4 for a diagram of the various levels of government jurisdiction over oceans and seabeds). The **Arctic Five**—Canada, Denmark (Greenland), Norway, Russia, and the United States (Alaska)—have the inside track in this race. At a 29 March 2010 meeting in Chelsea, Quebec, the Arctic Five countries confirmed their commitment to following a co-operative approach to resolving outstanding issues in the Arctic, making this race one of co-operation rather than of confrontation (Canada, 2010).

The quest for a short sea route through the Arctic to the East has haunted Europeans for centuries. Yet, the Arctic Ocean had—and still has—a permanent ice pack that prohibits normal navigation by ships. Even so, since the mid-twentieth century, the Soviet Union and now the Russian government have made extensive summer use of its northern sea route to supply its many Arctic communities and mines. Canada, on the other hand, has a much smaller Arctic population and more severe ice conditions. For these two reasons, various Canadian governments have been reluctant to invest heavily in icebreakers and harbours, and this long-standing policy has kept the Northwest Passage in limbo. Yet, as the barrier of ice to shipping slowly disappears, Ottawa is now forced to act. In fact, the increasing number of foreign ships, particularly cruise ships that have entered Arctic waters in the last decade, and the lure of vast 'as yet unclaimed' petroleum deposits in the seabed of the Arctic Ocean represent two forms of international pressure placed on Canada to act, and act decisively.

Figure 8.1   Hans Island—Part of Canada or Greenland?

In 1973, Canada and Denmark formally set the boundary as the midpoint in the waters separating Nunavut and Greenland. However, during those negotiations, no agreement was reach on Hans Island, which straddles the marine boundary. Located in Kennedy Channel of Nares Strait at a latitude of nearly 81°N, this tiny island is claimed by Canada and Denmark. The legal case for each country and possible outcomes are presented by Michael Byers (2009: ch. 2).

While the question of sovereignty over the **Arctic Archipelago** was settled in the nineteenth century when, in 1880, the British government transferred those islands to Canada,[1] questions remain about Arctic waters and the seabed. Inuit use of ice to hunt remains a wild card in Canada's claim for sovereignty over the disputed waters of the Northwest Passage. In sum, except for the tiny Hans Island between Nunavut and Greenland (see Gray, 1997: 69; Byers, 2009: 23), territorial matters have been resolved, leaving questions about the marine boundaries, the Northwest Passage, and the international Arctic seabed as unresolved questions.

The Northwest Passage,[2] for example, represents a top concern. Canada considers the Northwest Passage internal Canadian waters while foreign countries, especially the United States, argue that the Northwest Passage lies in international waters. These issues were dormant until open water became more of a regular feature of the Northwest Passage. Under earlier circumstances, Ottawa was reluctant to invest heavily to secure and manage navigation through the Northwest Passage. More recently, the federal government has announced plans to fund a number of projects that would allow Canada to secure and manage its Arctic waters. One such project is a new icebreaker, named for former Prime Minister John Diefenbaker, but construction of this state-of-the-art ship, announced in the February 2008 budget, has yet to begin. A similar announcement was made in 1986 by Prime Minister Mulroney, but that project was cancelled due to 'budgetary difficulties'.

Besides the Northwest Passage, two boundaries remain contested, one by the United States and the other by Denmark. Confirming that political stalemate, Byers (2009: 6) points out that 'Canada has no disputed land **borders**, but it does have a number of maritime **boundary** disputes, including two in the Arctic.' While negotiations are taking place to resolve these two outstanding boundary issues, there is no guarantee that Canadian negotiators will meet with success. The areas of contention are:

- With Denmark: a 200 km² section of the Lincoln Sea located north of Ellesmere Island and Greenland at latitude 84°N.
- With the United States: the position of the maritime boundary in the Beaufort Sea. Here the contested space is much larger, at 21,436 km².

The international waters and seabed of the Arctic Ocean represent another area of concern for Canada. In this case, Canada is preparing its claim to part of these waters under the United Nations Convention on the Law of the Sea (UNCLOS). In the coming pages, each geopolitical question is explored, but first they are placed within a geographic and political context.

Changing Geography of the Arctic Ocean

The Arctic Ocean is by far the smallest of the world's five oceans, with an area of just over 10 million km².[3] It is distinguished by several unique features, including the Arctic environment, an ice cap, and an encirclement by the land masses of North America, Eurasia, and Greenland. Within the Arctic Ocean, the principal seas are the Barents Sea, Beaufort Sea, Chukchi Sea, East Siberian Sea, Greenland Sea, Kara Sea, and the Laptev Sea, plus the many straits and channels separating the islands forming the Canadian Arctic Archipelago, including the Northwest Passage.

The physiography of the Arctic Ocean consists of three main landforms: shallow continental shelves, submarine mountains, and deep basins. For the most part, Arctic waters are relatively shallow with extensive continental shelves, especially off the coast of Russia. The topography of its ocean floor is dominated by two deep basins, the Eurasia and Amerasia basins, situated on each side of the Lomonosov Ridge. This ridge divides the floor of the Arctic Ocean almost in half, extending as a submarine mountain range some 3,000 metres above the ocean floor and stretching across the floor of the Arctic Ocean from the northern tip of Ellesmere Island and Greenland to the New Siberian Islands, a distance of 1,770 km. The Canada and Makarov sub-basins are located on the North American side of the Lomonosov Ridge while the Nansen and Fram basins are found on the Eurasian side. The deepest waters, in the Fram Basin, reach nearly 4,700 metres below the surface.

The surface waters of the Arctic Ocean mix with those of the North Atlantic Ocean but much less mixing occurs with the waters of the Pacific Ocean because the Bering Strait is both narrow and shallow, having a depth of only 55 metres. For the most part, then, the waters of the Arctic Ocean have limited interchange with those of the Pacific and slightly more with those of the Atlantic Ocean. Cold, dense water flows into the Atlantic Ocean via the Labrador and Greenland currents. The impact of the Labrador Current on mid-latitude coastal temperature is considerable as these Arctic waters chill the coast of Labrador, thus bringing Arctic conditions to relatively low latitudes (52°N). On the other hand, Lancaster Sound, at the high latitude of 74°N, sustains a complex variety of marine life because of its deep waters, regular upwellings, and polynyas. On the other side of North America, warm surface waters from the Pacific Ocean often penetrate into the Beaufort Sea through Bering Strait in the summer, resulting in a relatively long period of open water. For that reason, bowhead whales frequent the Beaufort Sea.

Composed of contiguous sea ice, the Arctic ice cap extends over most of the Arctic Ocean. Its geographic extent varies from winter to summer. In the winter, the ice cap reaches an area of nearly 10 million km²—in other words, it covers practically the entire ocean; in late summer it is reduced to less than 5 million km². Formed from the freezing of sea water into **pack ice** and subjected to the centrifugal forces associated with drifting around the Arctic Ocean, its topography is formed by a combination of broken and refrozen ice creating a rough and irregular surface and large, flat, and relatively undisturbed ice sheets, which have served as ice islands for scientific research on the Arctic Ocean (Vignette 8.1). The major topographic features are pressure ridges caused by the

---

Vignette 8.1 The Nature of Pack Ice

Pack ice is the basic building material of the Arctic ice cap. Pack ice is formed in the following manner: as a crust of sea ice is frozen on the surface of the Arctic Ocean, this sea ice attaches itself to the existing pack ice. In turn, newly formed pack ice adds to the mass of the Arctic ice cap. Without a summer melt, pack ice gains in hardness and thickness, thus providing a challenge to icebreakers. Pack ice, in the form of small and large ice sheets, circulates around the Arctic Ocean.

collision of huge sheets of ice. While pack ice reaches three or more metres in thickness, pressure ridges reach heights two to three times above the level of the pack ice. In the summer, the Arctic ice cap is reduced in size and lies above 75°N; in the winter, the ice cap reaches its maximum extent with seasonal or young ice that extends from land to the permanent **multi-year ice**, thus extending well below 75°N. Seasonal ice, with an average thickness of 1 metre, is much thinner than multi-year ice.

The ice cap, driven by Arctic currents and winds, slowly moves around the Arctic Ocean in a clockwise direction with some ice exiting in the Fram Strait. This clockwise circulation, centred around the Beaufort Sea, is known as the Beaufort Gyre. Another current, known as the **Transpolar Drift**, carries ice from the Russian Arctic across the North Pole into the Greenland Current. Given these dynamic motions, this ice cap is not one solid sheet of ice but consists of huge sheets of pack ice separated by narrow patches of open water. The slow drifting of the ice cap exerts centrifugal forces on it, causing the ice cap to bend, crack, and split. The cracking of the pack ice causes **leads** or open water. In fact, as much as 10 per cent of the ice-covered Arctic Ocean consists of leads and polynyas.

Ice thickness, its spatial extent, and the fraction of open water within the Arctic ice cap can vary rapidly and profoundly in response to changes in wind and temperature. Multi-year ice ranges from two to four metres in thickness while **new ice** is usually less than a metre thick. The spatial extent of the ice cap varies seasonally. In the winter, new ice covers Hudson and James Bays, making the geographic extent of sea ice around 15 million km², but by July the new ice in the bays has melted. By September,

Figure 8.2 Varying Spatial Extent of Arctic Ice Cap in September, 2002 to 2010 (million km²)

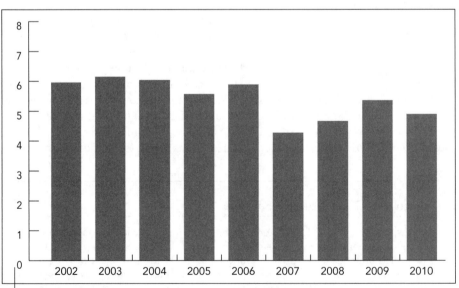

Note: Figures were calculated by Walt Meier, National Snow and Ice Data Center. All values estimated based on the NSIDC Sea Ice Index.

Source: National Snow and Ice Data Center (2010). 'Sea Ice', State of the Cryosphere.

Figure 8.3   Arctic Sea Ice Extent

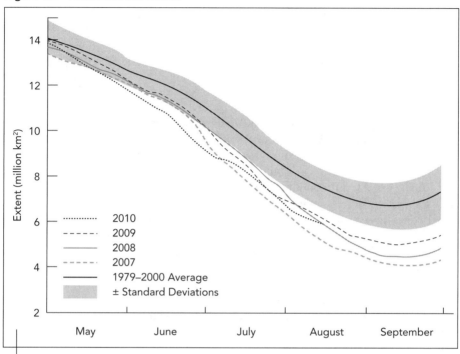

Source: National Snow and Ice Data Centre.

open water reaches its summer maximum and the ice cover has decreased by over half (Figures 8.2 and 8.3). Rothrock, Yu, and Maykut reported that multi-year ice thickness declined by about 1.3 metres between the 1950s and the 1990s (National Snow and Ice Data, 2010).

The central question—still unresolved—is: When will the Arctic ice cap disappear in this century? At the Copenhagen Conference on Climate Change, leading scientists in the field of global warming predicted that the Arctic Ocean will be virtually ice-free by 2050 (Stocker and Plattner, 2009: 4). David Barber, the Canada Research Chair in Arctic System Science at the University of Manitoba (Campbell, 2010), calls for an earlier date—sometime between 2013 and 2030. From his fieldwork, Barber has witnessed the shrinking ice cover—both in its geographic extent and its thickness. With more open water in the summer, a great amount of solar energy is absorbed by the Arctic waters, thus accelerating the rate of melting of the ice cap. From an analysis of their field data, Howell, Duguay, and Markus (2009) concur that the rate of the melting in the last two decades (1979–2008) has increased.

Other evidence supports this hypothesis. Glaciers, for instance, provide a valuable indicator of long-term temperature change. Thus, the retreat of glaciers in the northern hemisphere that began in earnest in the late twentieth century, followed by the shrinking of the Arctic ice cap, provides empirical evidence supporting the hypothesis of climate warming. In the case of the ice cap, the albedo effect is also at play. As discussed in Chapter 2, albedo refers to the amount of sunlight reflected by an object. As the Arctic ice cap melts, two events take place. First, the amount of sunlight reflected is reduced, so more warming of the ice cap surface takes place. Second, with more 'dark' water exposed, the Arctic Ocean absorbs more solar energy, thereby warming its waters, which in turn increases ice melt.

With glaciers and ice caps melting, what will be the effect on ocean levels? The media often refers to the threat of massive flooding resulting from the melting of ice around the world. This threat could become a reality if both the Greenland and Antarctic ice sheets melted, thus causing sea level to rise by more than 70 metres (National Snow and Ice Data Center, 2007). However, the disappearance of the Arctic ice cap alone would add little additional water to the oceans of the world because it is only about three metres thick on average and is already displacing ocean water. The loss of the ice cap alone, then, would only result in a negligible increase in the global sea water level, affecting only low-lying islands and coastlines where elevations are near sea level. In fact, recent increases in sea level are recorded at an annual rate of less than two millimetres—part of that increase is due to the melting of glaciers, the rest a result of ocean thermal expansion (National Snow and Ice Data Center, 2007).

In spite of the solid science and evidence of global warming, a note of caution about predictions is appropriate. First, the complex interaction of the various natural forces that affect the Arctic ice cap, the Arctic Ocean, and its climate are not fully understood and therefore the rate of warming over the past 20 years could slow or even decrease. Second, an unexpected natural event could interrupt this recent warming trend, such as a modern version of the Mount Tambora volcanic eruption of 1815, which chilled the northern hemisphere for several years.[4] Nonetheless, the scientific evidence strongly suggests the Arctic ice cap will disappear in the summer months sometime in the twenty-first century.

Arctic Sovereignty in the Twenty-First Century

The Arctic has long held centre stage in Canadian politics and in the minds of Canadians. In the 1958 federal election, John Diefenbaker's 'Northern Vision' captured the attention of the Canadian public and rallied voters to support his Progressive Conservative Party. But Arctic sovereignty goes beyond a vision that espouses economic development under programs like Diefenbaker's 'Roads to Resources'. Federal governments over the past 30 years have waffled on committing serious dollars to the North. One reason is the absence of a serious threat to Canada's Arctic sovereignty. As Coates, Lackenbauer, Morrison, and Poelzer (2008: 1) put it:

> Arctic sovereignty seems to be the zombie—the dead issue that refuses to stay dead—of Canadian public affairs. You think it's settled, killed and buried, and then every decade or so it rises from the grave and totters into view again.

Today, climate warming has changed the game. With the prospect of ocean shipping passing through the Northwest Passage and the urgency to submit a claim by 2013 for a piece of the unclaimed seabed of the Arctic Ocean, Ottawa is on the hot seat to provide both a vision and funding. In 2007, Prime Minister Harper declared that 'Canada has a choice when it comes to defending our sovereignty over the Arctic.

We either use it or lose it. And make no mistake, this Government intends to use it' (*Victoria Times Colonist*, 2007). One of Canada's leading Arctic experts, Michael Byers, puts it this way in his definitive book, *Who Owns the Arctic?* (2009: 6): 'Sovereignty, like property, can usefully be thought of as a bundle of rights.' Unfortunately, each right calls for substantial public investments, and past federal governments, both Conservative and Liberal, either set higher spending priorities for projects in southern Canada or were confronted with serious deficits that precluded large investments in Arctic sovereignty. Time will tell if the Harper government can deliver on its promises.

Maritime sovereignty, unlike territorial sovereignty, often does not contain a full bundle of rights. For example, under the UNCLOS regime, full sovereignty only extends to 12 nautical miles (22 km) from shore. Described as territorial seas, coastal states have full sovereign rights over these waters, including airspace, seabed, and subsoil. Beyond the 12 nautical miles, the bundle of rights diminishes. The diagram below indicates the complexity of determining the various maritime boundaries (Figure 8.4). The most important points are:

- *Territorial seas*. Full sovereignty exists within 12 nautical miles from shore.
- *Continental shelf*. Limited sovereignty occurs between 12 nautical miles and 200 nautical miles from shore.
- *Extended continental shelf*. Seabed sovereignty goes beyond 200 nautical miles to a limit of 350 nautical miles from shore.
- *High seas*. The so-called 'high seas' lie beyond national sovereignty and are considered international waters.

Figure 8.4   Zones of Maritime Sovereignty

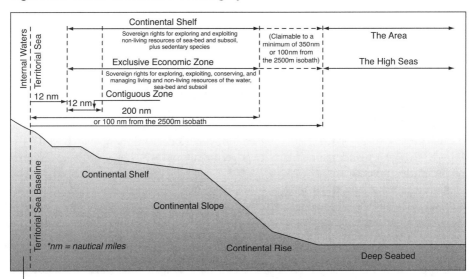

Note: 'The area' is the sea floor beneath the 'high seas'.

Source: Fisheries and Oceans Canada (2010).

Vignette 8.2 Evolving Marine Sovereignty

With water covering about two thirds of the Earth's surface, a basic question confronts the nations of the world: How far does the sovereignty of coastal nations extend into the ocean? While the answer has evolved over time, especially since 1945, the basic trend has seen coastal states gain more control over these waters. The importance for Canada, with one of the longest coastlines in the world, is critical. Canada's Arctic Archipelago and its Arctic continental shelf are playing a key role in determining the geographic range of its ownership of the waters of Arctic Ocean and its seabed. Based on the 1982 **United Nations Convention on the Law of the Sea (UNCLOS)**, which describes the nature of sovereignty and the associated maritime boundaries that coastal states can set, Canada passed its Oceans Act (1991), which delineates its maritime zones (Canada, Department of Justice, 2010). Equally important, Article 234 of UNCLOS—the **Arctic Clause**—allows Canada to enforce its pollution regulations in its **Exclusive Economic Zone**. Another regulation that asserts Canada's sovereignty over its waters is **NORDREG**, which requires that domestic and foreign vessels formally register before entering Canadian ice-covered areas.

- *The area.* Seabed sovereignty belongs to the international community under the control of the **International Seabed Authority**.

With the prospect of ships navigating the Northwest Passage and extraction of resources from the seabed no longer relegated to the back burner by the Arctic ice cap, Ottawa is confronted with a new Arctic reality. On 10 August 2010, the federal government announced its policy in a *Statement on Canada's Arctic Foreign Policy*. Canada's policy is based on three points:

1) resolve boundary issues;
2) secure international recognition for the full extent of our extended continental shelf;
3) address Arctic governance and related emerging issues, such as public safety.

Such statements have a strong appeal to the Canadian public, but when it comes to investing large sums outside of the Canadian voting ecumene, governments are always cautious. While the government has made some announcements and committed some funds to meet these goals, their long gestation period clearly stretches beyond the four- to five-year mandate of a majority government and well beyond the uncertainty of a minority government.

Given the cost and time horizon, how has Canada matched its stated goals with concrete action? The answer is 'slowly but surely'. Over the past seven years, Ottawa has made a series of announcements and provided start-up funding for several Arctic projects. Given a relatively long time frame, adjustments have been made and funds earmarked in several federal budgets for preliminary work. Total costs are difficult to estimate but $20 billion may come close; after that, annual operating expenses will demand more federal funds. During the 2005 and 2007 election campaigns, the

Conservatives promised to build three huge, armed icebreakers (but later reduced this to one, the CCGS *John G. Diefenbaker*, scheduled for completion in 2017 at the Vancouver Shipyards); to create a deepwater port at Nanisivik (construction of Nanisivik Naval Facility was to begin in 2011, but in June 2011 the Defense Department announced that construction will not begin until 2012, with the naval port fully operational in

---

Vignette 8.3   **Space Technology, Sovereignty, and Hard Political Decisions**

One of the most controversial and expensive investments in Arctic sovereignty involves Radarsat satellite surveillance. This space technology provides very detailed daily surveillance of Canada's Arctic waters with the ability to monitor not just shipping but the state of the ice pack and other environmental variables. In 2008, the Canadian government was confronted with a dilemma: to approve or reject the sale of MacDonald Dettweiler Associates (MDA) to the US firm of Alliant Techsystems. Ottawa rejected the sale on the ground of national security, which translated into Arctic sovereignty. Ottawa had funnelled considerable research funds to MDA for its Radarsat project, and this political decision ensures ongoing support through a unique collaboration between government (the Canadian Space Agency) and industry (MDA). Canadian government agencies receive the satellite data and their payments for this service support MDA's next generation of Radarsat projects. For Canada's management of its sensitive marine waterways but especially the Northwest Passage, as well as for ensuring and affirming Arctic sovereignty, the value of this satellite surveillance cannot be underestimated (Vachon, 2010). Not surprisingly, Ottawa will spend almost half a billion dollars on developing the next generation of Radarsat satellites, the Radarsat Constellation, which are to be deployed by 2015.

To maximize the data flow from the three orbiting Radarsat satellites, Canada's first Arctic Satellite Station Facility (ISSF) was opened at Inuvik in 2010. While Inuvik benefits from economic spinoffs from the operation of the satellite station, by far the bulk of the economic benefits will flow to southern Canada, especially the Vancouver area where the MDA plant is located. Southern taxpayers should recognize that large public investments in the Arctic usually benefit industries in southern Canada. The explanation for the spatial pattern of economic benefits associated with northern projects lies in the concept of economic leakage.

2016 (Canadian Press, 2011); to establish a small military training centre at Resolute Bay (2010 funds allocated); to modernize and expand the Canadian Rangers (begun 2007); and to establish a High Arctic Scientific Centre (Cambridge Bay was selected in 2010 and is planned to open in 2017). In addition, Ottawa is heavily committed to Radarsat satellite surveillance of its Arctic waters (Vignette 8.3).

On a smaller scale, the 2009 federal budget designated $2 million for a feasibility study into the site of a High Arctic Research Station. In August 2010, Cambridge Bay was selected over Resolute Bay and Pond Inlet (Harper, 2010). The goal is to open the yet-to-be-constructed station by 2017. The 2010 budget contained $18 million for pre-construction design. This project will also have two other important results:

exercising sovereignty and promoting economic and social development. Similar to the Nunavik Research Centre located at Kuujjuaq in Arctic Quebec, the High Arctic Research Station will create jobs, promote development, and investigate and protect the Arctic environment. In these ways, it is hoped that it will contribute to the overall quality of life for Arctic residents.

Drawing Lines in Water

As noted earlier, Canada has several outstanding international jurisdictional issues related to Arctic waters and the need to draw or redraw lines in the water—no easy task!—to indicate which nations own what. The principal issues here involve the Beaufort Sea, the Northwest Passage, and the Arctic seabed (see Figure 8.5).

Figure 8.5   The Changing Political Landscape: Disputed Waters and Seabed in the Arctic Ocean

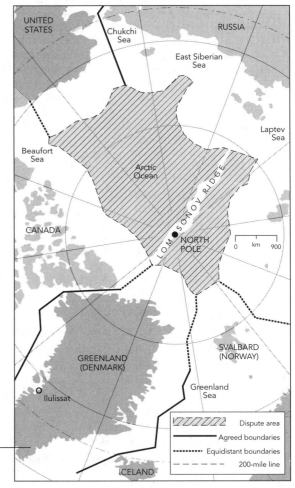

Source: *National Post*. Reprinted by permission of the publisher.

The Beaufort Sea Dispute

The Beaufort Sea dispute between Canada and the United States revolves around a relatively small area of 21,436 km². Attempts at resolving the Beaufort Sea boundary are underway, but success may be elusive. At stake within this contested triangular-shaped area, besides management of the water and fish stocks, is ownership of potentially vast seabed petroleum deposits. The disputed area is tangled in history, with Canada arguing that the 1815 treaty between Great Britain and Russia set the boundary along the 141st meridian west not only for the border between Alaska and Yukon but also for the maritime boundary in the Beaufort Sea. The US, however, argues for a line of equidistance from the coast of each country which would shift the boundary east of the 141st longitude and thus provide the United States with sovereignty rights over the contested area of the Beaufort Sea (Figure 8.6).

Figure 8.6   Contested Area of the Beaufort Sea

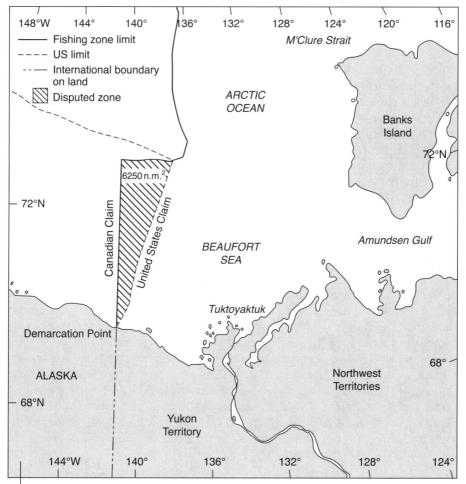

Canada, as with its other boundary disputes, is seeking a negotiated solution. One step in this direction is found in the joint Canada/US seabed mapping program for the Beaufort Sea in the contested area. This political decision for a co-operative approach to hydrographic investigation may lead to a solution. The joint seabed mapping program involves the Canadian Coast Guard icebreaker *Louis S. St-Laurent* and the US Coast Guard cutter *Healy* working together. This effort to map the seabed has three objectives:

- to help resolve the maritime boundary between the two countries;
- to identify potential petroleum and other seabed resources;
- to assist the two countries in preparing their respective claims for a section of the international seabed of the Arctic Ocean.

The Northwest Passage: Internal Waters or International Strait?

For years, Canada's claim to the Northwest Passage as internal waters was solidified by the massive ice pack that occupied much of this waterway. With the Northwest Passage virtually impassable for commercial vessels, ice was Canada's friend. Added to the 'ice as a barrier to commercial navigation' argument was the 'ice as a platform for Inuit hunters to kill seals' argument. From both perspectives, ice reinforced Canada's position.

But with climate warming and a retreating ice pack, the Northwest Passage now is ice-free, albeit for a very short time. Canada continues to argue that the waters between the islands of the Archipelago are internal seas. Countering that position, Washington and other foreign governments claim that the Northwest Passage is an international waterway because it connects two segments of the high seas—the Atlantic and Pacific oceans. Support for this position comes from UNCLOS: the right of transit passage exists 'between one part of the high seas or an exclusive economic zone and another part of the high seas or an exclusive economic zone'. While both perspectives accept that these are Canadian waters, the key difference is that if the Northwest Passage is designated as an international waterway, then foreign ships could pass freely. On the other hand, if the Northwest Passage was declared internal waters of Canada, then 'foreign vessels must have Canada's permission [to pass through the Northwest Passage] and are subject to the full force of Canadian domestic law' (Byers, 2009: 43). In December 2010, Canada gained more credence for its 'internal waters' position by announcing that Lancaster Sound will become a national marine conservation area and, in the future, may become a World Heritage Site. Lancaster Sound serves as the eastern entrance to the Northwest Passage and therefore is a critical piece of the ownership puzzle of the Northwest Passage.

An ice-free Northwest Passage could alter world shipping patterns because it offers a much shorter sea route between Asia and Europe. In fact, an Arctic shipping route through the Northwest Passage would shave roughly 4,000 km off current maritime routes between Europe and Asia. As we saw in Chapter 3, this dream of a shortcut between Europe and Asia was stymied for centuries by the much colder climate of the Little Ice Age and an even more formidable Arctic ice cap (see Vignette 3.5, Figure 3.8, and Table 3.4).

The southern route through the Northwest Passage begins in Lancaster Sound but quickly veers south along the Peel Channel. The attraction of this route is the

Figure 8.7   The *ss Manhattan* Challenges the Northwest Passage

The reinforced Arctic class US supertanker ss *Manhattan* entered the Northwest Passage at Lancaster Sound in the summer of 1969. The question facing the oil producers at Prudhoe Bay, Alaska, was how to transport the oil to market. One possibility was to use the Northwest Passage. The *Manhattan* took the northern route through M'Clure Strait but ran into impassible ice. With the help of the Canadian ice-breaker, *John A. Macdonald*, the *Manhattan* freed itself and then changed course to the Prince of Wales Strait, Amundsen Gulf, and on to Prudhoe Bay.

Source: © Dan Guravich/CORBIS.

presence of open water and less chance of encountering hard, thick ice. This route is named after Roald Amundsen, the Norwegian polar explorer who, from 1903 to 1906, meandered his way across the waters separating the islands of Canada's Arctic Archipelago. His small but agile ship, the *Gjoa*, was able to follow leads in the ice until the ship became lodged in pack ice. After three summers, Amundsen became the first person to sail across the Northwest Passage. The *Gjoa* entered Lancaster Sound and then Barrow Strait. Facing a wall of ice, Amundsen veered his ship almost due south along Peel Sound to reach Queen Maud Gulf and the open coastal waters along Canada's mainland and then proceeded along these coastal waters to the Beaufort Sea.

The second crossing of the southern route of the Northwest Passage began in 1940, when the *St Roch*, under the command of Inspector Henry Larsen of the RCMP, set sail from Vancouver for Halifax across the Arctic Ocean (Table 3.4). Its orders were to assist the Canadian armed forces that were to occupy Greenland.[5] Taking two summers, the *St Roch* was the first ship to cross the Northwest Passage from west to east. Larsen followed Amundsen's route in reverse, with one exception. Instead of proceeding north along Peel Sound to Barrow Strait, Larsen took his small craft through the very narrow Bellot Strait that separates Southhampton Island from Boothia Peninsula and then sailed northward from Prince Regent Inlet to Lancaster Sound.

Over the last decade, a remarkable number of vessels of all shapes and sizes have crossed the Northwest Passage (Table 8.1). In sharp contrast, no ship made the passage in the nineteenth century and only Amundsen and Larsen were successful in the first half of the twentieth century. By the end of the first decade of the twenty-first century, at least one crossing took place each year. Other ships, including research vessels and cruise ships, have entered Arctic waters but have not travelled the entire length.

As Arctic waters provide one of the world's last wilderness areas, cruise ships have found a niche market by scheduling Arctic voyages. Billed as 'Arctic Adventures', these expensive voyages have grown in number. In 2009, Robert McCalla (2010b: Table 2) reported that 15 voyages by 10 cruise ships entered Canada's Arctic Ocean, with two ships crossing the entire Northwest Passage following the southern route. Since

Figure 8.8   The Arctic Ocean and the Northwest Passage

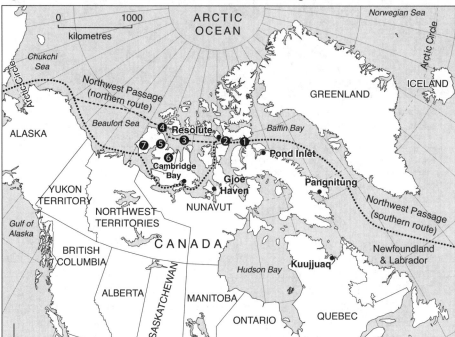

The Northwest Passage is a rarely used sea route across the Arctic Ocean connecting Europe with Asia. The most challenging section lies in the ice-clogged waters between islands in the Arctic Archipelago. Ships enter the Northwest Passage at Lancaster Sound (1 on map) and pass through Barrow Strait (2) to Resolute. At that point, ships following the southern route (dotted line) turn south where more open water is normally found, while those taking the northern route (dashed line) continue in a westerly direction through Viscount Melville Sound (3) and M'Clure Strait (4), which lie north of Victoria Island (6) and Banks Island (7). These two islands are separated by Prince of Wales Strait (5). These sea channels are commonly filled with thick Arctic pack ice. Until recently, ships found it impossible to navigate M'Clure Strait. This strait is named after Commander Robert M'Clure, whose ship, the HMS *Investigator*, was trapped in pack ice in the strait for three winters (1851–3) and finally abandoned during the search for Sir John Franklin. In 1969, the American supertanker SS *Manhattan* had to be freed from M'Clure Strait by a Canadian ice-breaker, was able to turn around, and headed south to more open waters through Prince of Wales Strait.

Source: Adapted from Dawson et al. (2008).

Table 8.1 Key Events: Crossing the Northwest Passage, 1969 to 2011

Date	Event
1957	United States Coast Guard Cutter, *Storis*, crossed the Northwest Passage in 64 days.
1969–70	The Exxon supertanker, the ss *Manhattan*, travels through the Northwest Passage twice to prove it is a viable commercial route for shipping oil from Prudhoe Bay, Alaska, to refineries along the US east coast. The hull of the supertanker is covered with a thick protective steel belt and its reinforced bow is designed to ride up on the pack ice, allowing the weight of the massive ship to break the sea apart. While Exxon had not asked for Ottawa's permission, Canada grants 'unsolicited' permission and provides an icebreaker escort. While the two crossings (one from the east to Prudhoe Bay and the other from the west to the US east coast) are successful, damage to the *Manhattan* is considerable and plans are abandoned for using the Northwest Passage as a shipping route for Alaskan oil.
1977	Willy de Roos sailed through the Northwest Passage in his steel yacht, the *Williwaw*.
1984	The first cruise ship to navigate the Northwest Passage was the MS *Explorer*.
1985	The US Coast Guard icebreaker, *Polar Sea*, traverses the Northwest Passage without asking permission. Ottawa is deeply troubled and approaches Washington for clarification.
1986	Jeff MacInnis and Wade Rowland took their catamaran, the *Perception*, through the Northwest Passage over several summers. In the same year, David Cowper left England in a lifeboat, the *Mabel El Holland*, and, after three winters, reached Bering Strait.
2000	The RCMP patrol vessel, *St. Roch II*, escorted by the Canadian Coast Guard icebreaker, *Simon Fraser*, departed Vancouver and completed the voyage in one summer
2001	Captain Jarlath Cunnane left Ireland in a sailboat, the *Northabout*, and sailed through the Northwest Passage.
2003	Richard and Andrew Wood with Zoe Birchenough took their yacht, *Norwegian Blue*, from the Bering Strait to Davis Strait in one season.
2006	The cruise liner, MS *Bremen*, crossed the passage from east to west.
2007	French sailor, Sébastien Roubinet, maneuvered his ice catamaran, *Babouche*, from west to east in a single season.
2008	Two Canadian resupply vessels, MV *Carmilla Desgagnes* and *Arctic Cooperative*, reached both eastern and western Arctic communities.
2009	Nine small vessels and two cruise ships (MS *Bremen* and MS *Hanseatic*) made the crossing.
2010	Cruise ship MS *Hanseatic* is scheduled to cross the Northwest Passage.
2011	Cruise ship *Kapitan Khlebnikov** plans to follow Amundsen's Route, August 22 to September 13 while on its historic circumpolar voyage. The MS *Bremen* is set to take the Northern Route.

*Prior to 1991, this Russian ice-breaker escorted Russian merchant ships along the Northeast Passage.

these waters are neither fully charted nor necessarily totally free of ice, navigating the Northwest Passage is risky—as the captain of the cruise ship, *Clipper Adventurer*, discovered when his ship ran aground in 2010 on its way to Kugluktuk (formerly

Coppermine), Nunavut. Fortunately, calm weather prevailed and the passengers were transferred to the Canadian icebreaker, *Amundsen*, taken to Kugluktuk, and then flown to Edmonton and on to other destinations (Cohen, 2010). Given the short navigation season, a salvage operation began a few days later. The ship was refloated and towed to Cambridge Bay and then to Nuuk, Greenland, where the ship was repaired (Marinelog, 2010). This misadventure was a relatively minor one, but it underscores the perilous nature of such voyages as well as the need for Canada to expand its presence in the Arctic Ocean.

Canada's Arctic Seabed Submission

The United Nations Convention on the Law of the Sea (UNCLOS) trumps the previous doctrine known as 'freedom of the seas', which limited rights and jurisdiction over the oceans by coastal nations to a narrow strip of three nautical miles. The freedom of the seas has its roots in the seventeenth century when European countries, especially Great Britain, 'ruled the waves'. By the twentieth century, this doctrine was challenged by countries seeking to control wider strips of the continental shelves. In 1945, the United States took the first step to extend its jurisdiction over all natural resources over its continental shelf. Other nations soon followed suit, but with varying distances from their shores. In 1967, the United Nations sought to bring order to this matter. In 1982, the United Nations adopted the Law of the Seas Convention, which regulates oceans by dividing the sea floor into zones of national and international jurisdiction. As Figure 8.4 demonstrates, nations have managerial control to a distance of 200 nautical miles offshore and, under special circumstances, beyond 350 nautical miles. On the other hand, international jurisdiction covers those waters and ocean floor beyond national control. These waters are called 'high seas' and the term for the ocean floor is 'the area'. The United Nations created the International Seabed Authority to manage the use of the area.

    The legal umbrella for Canada to claim more Arctic waters and seabed is UNCLOS. Other circumpolar countries—the US, Denmark, Norway, and Russia—also seek a share of the international part of the Arctic Ocean and, with the exception of the US,[6] these countries have ratified the Convention. Each country's submission relies on the geology and geomorphology of the Arctic Ocean. Under the aegis of UNCLOS, this struggle for a share of the international seabed is heading for a peaceful and scientific-driven solution. Each country has 10 years from the time that it signs the UNCLOS agreement to present its claim to a portion of the extended continental shelf (see Figures 8.4 and 8.5). Since Canada ratified the Convention in 2003, it has until 2013 to present this information to the Commission.

    For several years, Canada has been collecting and analyzing scientific, technical, and legal information in preparation of making its submission. The core of the scientific information is based on the geomorphology and geology of its continental shelf but especially the continental slope and rise leading to the ocean's bottom. Fieldwork is expensive but necessary to map the character of the continental shelf and to determine its outer limits. As shown in Figure 8.9, a provisional analysis made in 2007 estimated that the greatest extended area in the Arctic Ocean was just west of the Queen

Elizabeth Islands. Since 2007, considerable scientific work has taken place, sometimes in conjunction with the United States in the western Arctic and with Denmark in the High Arctic north of Ellesmere Island. Extensive mapping of the ocean's floor gives credence to a much larger Canadian extended continental shelf.

What can Canada expect to gain? Beyond Canada's 200 nautical mile boundary, the areas of extended continental shelf shown in Figure 8.9 consist of approximately 1.5 million km^2 (Fisheries and Oceans Canada, 2010). In comparison to the possible additional territory to be gained from marine boundary disputes with the US and Denmark, this claim is far greater in geographic extent and has the potential to greatly add to Canada's mineral and petroleum reserves. Indeed, the motivating factor behind the race for the Arctic Ocean's seabed is its potential wealth of resources, especially petroleum. An estimate of oil and gas resources within Canada's 200 nautical miles is found in Table 8.2, but much more likely lies in the international seabed—hence the scramble to submit claims to the United Nations.

The success of Canada's submission depends on the quality of its scientific evidence derived largely from fieldwork in the Arctic Ocean. Given the stakes, Ottawa

Figure 8.9   Canada's Extended Continental Shelves, 2007

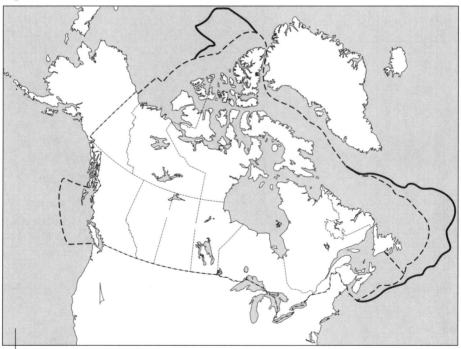

Canada has extensive continental shelves on its Arctic and Atlantic coasts. Within 200 nautical miles, Canada has exclusive sovereign rights to explore and exploit the seabeds of these continental shelves. The waters and seabeds are also known as Exclusive Economic Zones (dashed lines). Under UNCLOS regulations (Article 76), Canada may have control over its extended continental shelves, as indicated by the solid black lines.

Source: Foreign Affairs and International Trade (2007).

Table 8.2 Oil and Gas Resources Within Canada's Continental Shelf

OIL RESOURCES

Region	Discovered Resources		Undiscovered Resources		Ultimate Potential	
	10^6 m³	MMbbls	10^6 m³	MMbbls	10^6 m³	MMbbls
Northwest Territories and Arctic offshore	187.9	1182.5	799.7	5032.6	987.6	6215.0
Nunavut and Arctic offshore	51.3	322.9	371.8	2339.4	423.1	2662.3.0
Arctic Offshore Yukon	62.5	393.8	412.7	2596.8	475.2	2990.6
Total	301.7	**1899.1**	1584.1	**9968.8**	1885.9	**11,867.9**

GAS RESOURCES

Region	Discovered Resources		Undiscovered Resources		Ultimate Potential	
	10^6 m³	trillion cubic feet	10^6 m³	trillion cubic feet	10^6 m³	trillion cubic feet
Northwest Territories and Arctic offshore	457.6	16.2	1542.2	54.8	1999.8	71.0
Nunavut and Arctic offshore	449.7	16.0	1191.9	42.3	1641.6	58.3
Arctic offshore Yukon	4.5	0.2	486.6	17.3	491.1	17.4
Total	911.8	**32.7**	3220.7	**114.3**	4132.6	**146.7**

Note: 1. MMbbls - million barrels (of oil); Tcf - trillion cubic feet (of natural gas).

2. The Arctic offshore includes marine areas offshore Yukon and the Northwest Territories in the Beaufort Sea, and offshore Nunavut in the High and Eastern Arctic. Resources within Yukon are not included.

Source: Indian and Northern Affairs Canada (2010).

has invested heavily—but none too soon. Pure scientific work such as mapping the floor of the Arctic Ocean received little attention until such scientific data were essential to document the claims of Canada (and other countries) to the extended continental shelf. While Canada's scientists are working closely with their American and Danish counterparts, each nation will jealously defend its 'right' to the Arctic seabed. In fact, Canada's best defence is accurate and complete scientific data and mapping. Russia, too, recognizes the importance of the scientific component of its submission, which most likely will overlap with the submission of Canada. Canada and Russia will present conflicting conclusions to the United Nations Commission on the Limits of the Continental Shelf in December 2013.

The state of Canada's submission is revealed in its December 2009 report (Fisheries and Oceans Canada, 2009):

> The Arctic is subdivided into two areas: the Eastern and Western Arctic. The Eastern Arctic required seismic surveys to show the natural prolongation of the Lomonosov and Alpha-Mendeleev Ridges, then bathymetric surveys to map the foot of slope and 2,500 metre depth contour. The Western Arctic required bathymetric surveys to map the foot of slope and seismic surveys to determine sediment thickness. There are few existing data sets in these Arctic areas relevant to the continental shelf submission so the Program must conduct all the work itself.

Conclusion

Geopolitics has again turned the world's attention to the Canadian Arctic. The retreat of the Arctic ice cap is transforming the Arctic—both as a natural region and as a geographic region of Canada. Driven by climate warming, the demise of the ice cap has wide-ranging implications for Canada, its Arctic sovereignty, and the peoples of the Arctic—and those implications will require a considerable adjustment to the realities of a 'warmer' Arctic and the expected increased development and shipping. Inuit are already facing the reality of an Arctic, if not free of ice, with less ice. What happens to their traditional hunting of seals and other mammals on land-fast ice? A shorter period of land-fast ice has implications for food security and hunter safety. But how do Inuit and other Arctic peoples play a role in this process? The answers lie in territorial and federal governments as well as the **Arctic Council**, where indigenous organizations, such as the **Inuit Circumpolar Council**, have permanent participation status.

Without a doubt, the economic and social implications of this new northern world are enormous—and to some degree frightening. But with climate warming relentlessly increasing Arctic temperatures, turning back is not an option.

Moving forward into this warmer northern world, Arctic sovereignty demands a more active Canadian coast guard and military presence. With much greater access to world markets, resource development should flourish and, if crafted carefully through impact and benefit agreements, benefits will accrue locally. For example, besides the commercial traffic passing through the Northwest Passage, mining operations will greatly benefit from longer periods of ocean shipping. The change in ice conditions has been remarkably swift. Ten years ago the Nanisivik and Polaris lead/zinc mines closed. For the 20 years of their operation, these two mines, the former on the northern tip of Baffin Island and the latter on Little Cornwallis Island near Resolute, were truly extreme outposts of the world economy, surrounded by thick ice for all but a short span during which the year's production was exported by the ice-strengthened MV *Arctic* to European markets. In sharp contrast, the proposed Mary River mine project, located on the northern tip of Baffin Island, will provide Europe in the years to come with much needed iron ore throughout the year (Baffin Iron Mines Corporation, 2007).[7]

With such developments close at hand, Canada is under pressure to protect its interest as custodian not only of the Northwest Passage, but also of its other Arctic waters and seabed. Environmental issues loom large in a more active shipping season. Fears of oil leaks and other forms of pollution are not unfounded—the issue of accidents and spills is not a question of 'if' but of 'when'—and Canadian resources must be put in place to manage such pollution and to charge the ship's captain and shipping company each time an incident occurs. That means having feet on the ground and men and women on the water in the Far North. Imagine an Arctic where armed Canadian coast guard vessels patrol these waters and, when necessary, force foreign ships to port to ensure that every ship entering the Northwest Passage has obtained a Canadian permit. In addition, in our uncertain world, terrorists might seek to use the Northwest Passage to strike at targets in North America, forcing Canada to address US concerns about a 'secure' Arctic. The solution is close at hand. Canada's 2013 submission to UNCLOS, which doubtless will meet some challenges, especially from Russia, should settle ownership of waters and seabed. At the same time, Ottawa must invest considerable resources to ensure a secure Arctic and provide enough military muscle to enforce its regulations and to patrol this vast and changing region.

Challenge Questions

1. Why have the Arctic Ocean and the Northwest Passage moved from the outer margins of global geopolitics to become central concerns?
2. What is the physical difference between young ice and multi-year ice and what is the economic significance of this difference?
3. Why is M'Clure Strait, the western end of the northern route of the Northwest Passage, rarely free of ice while Lancaster Sound, at the eastern entrance of the passage, commonly free of ice?
4. Assuming that ocean shipping begins to use the Northwest Passage, why would the northern route be more attractive to super-size vessels than the southern route?
5. Why are predictions of when the Arctic ice cap will disappear subject to such a wide range of answers?
6. From an economic development perspective, why do the large investments designed to defend Canada's Arctic sovereignty largely benefit industries in southern Canada?
7. Climate warming has its downside but could there be upsides? For instance, while the Inuit have already found hunting on ice for seals more difficult, could new hunting opportunities emerge in a much warmer climate such as the Medieval Optimum (700 to 1,200 years ago), when a larger and more sedentary population existed in the Canadian Arctic?

8. Ottawa is faced with many demands for money. If you were Prime Minister, would you stick to your Arctic commitments, such as the very costly construction of the CCGS *John G. Diefenbaker* and the building of deepwater ports, or, like previous prime ministers, would you put such projects on hold in order to balance the budget?
9. Can you make a case that Arctic Canada and its Inuit residents could gain from the impact of climate warming? Explain.
10. Geopolitics is a rough game. The Northwest Passage debate over internal waters and international straits might end with a compromise. Such negotiations may take place in the current high-level discussions between Ottawa and Washington dealing with a continental approach to security. What do you think Canada would gain and lose in such a compromise?

Notes

1. Canada's claim to the Arctic Archipelago was enhanced by an agreement in 1930 with Norway regarding the Sverdrup Islands located just west of Ellesmere Island. This agreement released any claim on these islands by Norway. Such a claim would have been based on the discovery and mapping of these islands by the famous Norwegian explorer, Otto Sverdrup. His expedition into the Arctic Archipelago took place from 1898 to 1902.
2. Ownership of the ice-covered waters of the Arctic Ocean between the islands of the Arctic Archipelago is not entirely resolved, especially for the waters of the Northwest Passage, where the distance between islands sometimes exceeds 12 nautical miles.
3. The definition of the Arctic Ocean may include Hudson Bay. By this definition, the geographic extent of the Arctic Ocean reaches 14 million km².
4. The eruption of Mount Tambora changed the world's climate for several years. In 1815, this volcanic activity filled the atmosphere with volcanic ash and dust, reducing the amount of sunlight reaching the Earth. In 1815, weather stations did not exist so temperature readings were not recorded, but the following summer of 1816 was so cool that it was dubbed 'the year without summer'. By 1991, weather stations existed around the world. Based on their recordings, the impact of the 1999 eruptions of Mount Pinatubo in the Philippines and Mount Hudson in Chile resulted in a mean world temperature decrease of about 0.5°C for the next two years (Rosenberg, 2010).
5. The *St Roch* undertook the difficult task of crossing the Arctic Ocean because of British and Canadian concerns for Greenland. Nazi Germany had already conquered Denmark and had its eyes on Denmark's Arctic colony. At the same time, the US was concerned about British/Canadian influence and even occupation of Greenland. According to Shelagh Grant (2010: 249–53), Ottawa sent the *St Roch* to Greenland from Vancouver through the Arctic Ocean to disguise its true mission to support Canadian occupation forces in Greenland.
6. The United States has not yet ratified the Law of the Sea Convention because it fears US national interests might be curtailed and its power in the international arena weakened. (This same approach has been reflected in the American refusal to adhere to other international conventions and agreements, such as the Ottawa Convention to Ban Landmines, the Kyoto Protocol, and the International Criminal Court.) UNCLOS would require that the US submit to the International Seabed Authority (ISA). While the US Senate continues to oppose ratification, the Obama administration supports becoming a member of this Convention.

7. Plans for a port on Baffin Island associated with the proposed iron mine locate it facing Foxe Basin. It would seem, however, that the federal government should consider financial support to site this port facing Baffin Bay and thus provide, as well, a future base for Canadian coast guard vessels expected to be responsible for monitoring shipping along the Northwest Passage.

References and Selected Reading

Abele, Frances, Thomas J. Courchene, F. Leslie Seidle, and France St-Hilaire, eds. 2009. *Northern Exposure: Peoples, Powers and Prospects in Canada's North, vol. 4, The Art of the State*. Montreal: Institute for Research on Public Policy.

——— and Thierry Rodon, 2007. 'Inuit Diplomacy in the Global Era: The Strengths of Multilateral Internationalism', *Canadian Foreign Policy* 13: 3.

Arnold, Samantha. 2008. 'Nelvana of the North: Traditional Knowledge and the Mythical Function of Canadian Foreign Policy', *Canadian Foreign Policy* 14 (Spring): 95–108.

Baffin Iron Mines Corporation. 2007. 'Mary River Project'. At: <www.baffinland.com/MaryRiverProject/default.aspx>.

Beauchamp, Benoît, and Rob Huebert. 2008. 'Canadian Sovereignty Linked to Energy Development in the Arctic', *Arctic* 61, 3: 341–3.

Berton, Pierre. 1988. *The Arctic Grail: The Quest for the Northwest Passage and the North Pole, 1818–1909*. Toronto: McClelland & Stewart.

Birchall, S. Jeff. 2006. 'Canadian Sovereignty: Climate Change and Politics in the Arctic', *Arctic* 59, 2: iii–iv.

Borgerson, Scott. 2008. 'Arctic Meltdown: The Economic and Security Implications of Climate Change', *Foreign Affairs* 87, 2: 63–76.

Byers, Michael. 2009. *Who Owns the Arctic? Understanding Sovereignty Disputes in the North*. Vancouver: Douglas & McIntyre.

Campbell, Dean. 2010. 'Walking to the North Pole? Hurry Up', *The Globe and Mail*, 4 June. At: <www.theglobeandmail.com/life/travel/walking-to-the-north-pole-hurry-up/article1592323/singlepage/>.

Canadian Coast Guard. 2010. 'The CCGS *John G. Diefenbaker* National Icebreaker Project', 28 Apr. At: <www.ccg-gcc.gc.ca/e0010762>.

Canadian Press. 2011. 'Nanisivik, Nunavut Naval Facility Project Delayed for at Least Two Years', *Daily Commercial News and Construction Record*, 7 Aug. At: <dcnonl.com/article/id44797>.

Charron, Andrea. 2005. 'The Northwest Passage: Is Canada's Sovereignty Floating Away?', *International Journal* 60 (Summer): 831–48.

Coates, Kenneth S., P. Whitney Lackenbauer, Bill Morrison, and Greg Poelzer. 2009. *Arctic Front: Defending Canada in the Far North*. Toronto: Thomas Allen.

——— and William R. Morrison. 2008. 'The New North in Canadian History and Historiography', *History Compass* 6, 2: 639–58.

Cohen, Tobi. 2010. 'Cruise Ship Runs Aground in Nunavut', *National Post*, 29 Aug. At: <www.nationalpost.com/Cruise+ship+runs+aground+Canadian+Arctic/3457290/story.html>.

Dawson, J., P.T. Maher, and D.S. Slocombe. 2007. 'Climate Change, Marine Tourism and Sustainability in the Canadian Arctic: Contributions from Systems and Complexity Approaches', *Tourism in Marine Environments* 4, 2 and 3: 69–83.

Department of Justice. 2010. Oceans Act, 5 Nov. At: <laws.justice.gc.ca/eng/O-2.4/page-2.html#anchorbo-ga:l_I>.

Dodds, Klaus. 2010. 'A Polar Mediterranean? Accessibility, Resources and Sovereignty in the Arctic Ocean', *Global Policy* 1, 3: 303–11.

Elliot-Meisel, Elizabeth. 1999. 'Still Unresolved after Fifty Years: The Northwest Passage in Canadian–American Relations, 1946–1998', *American Review of Canadian Studies* 29 (Fall): 407–22.

Fisheries and Oceans Canada. 2009. 'Canada's Submission to the Commission on the Limits of the Continental Shelf under United Nations Convention on the Law of the Sea (The "Continental Shelf Project")', 23 Dec. At: <www.dfo-mpo.gc.ca/ae-ve/evaluations/09-10/6b060-eng.htm>.

———. 2010. *Canada's Ocean Estate: A Description of Canada's Maritime Zones*, 9 July. At: <www.dfo-mpo.gc.ca/oceans/canadasoceans-oceansducanada/marinezones-zonesmarines-eng.htm>.

Foreign Affairs and International Trade. 2007. 'Defining Canada's Extended Continental Shelf', 31 July. At: <www.international.gc.ca/continental/limits-continental-limites.aspx?lang=eng>.

———. 2010a. 'Arctic Ocean Foreign Ministers' Meeting', 29 Mar. At: <www.international.gc.ca/polar-polaire/arctic-meeting_reunion-arctique-2010_index.aspx>.

———. 2010b. *Statement on Canada's Arctic Foreign Policy*, 10 Aug. At: <www.international.gc.ca/POLAR-POLAIRE/ASSETS/PDFS/CAFP_BROCHURE_PECA-ENG.PDF>.

Gerhardt, Hannes, Philip E. Steinberg, Jeremy Tasch, Sandra J. Fabiano, and Rob Shields. 2010. 'Contested Sovereignty in a Changing Arctic', *Annals, Association of American Geographers* 100, 4: 992–1002.

Globe and Mail, The. 2010. 'Theories Differ, but Climatologists Concur on Need for Action', 14 Dec., CC1 (Climate Change, a special information feature).

Gray, David H. 1997. 'Canada's Unresolved Maritime Boundaries', IBRU Boundary and Security Bulletin 5, 3 (Autumn): 61–70: At: <www.dur.ac.uk/resources/ibru/publications/full/bsb5-3_gray.pdf>.

Griffiths, Franklyn, ed. 1987. *Politics of the Northwest Passage*. Montreal and Kingston: McGill-Queen's University Press.

———. 2004. 'Pathetic Fallacy: That Canada's Arctic Sovereignty Is on Thinning Ice', *Canadian Foreign Policy* 11 (Spring): 1–16.

Hall, C.M., and J. Saarinen, eds. 2010. *Tourism and Change in Polar Regions: Climate, Environment and Experiences*. Milton Park, Abingdon, Oxon, UK: Routledge.

Harper, Steven. 2007. 'Prime Minister Steven Harper Announces New Offshore Arctic Patrol Ships', News Release: Prime Minister of Canada, 9 July. At: <pm.gc.ca/eng/media.asp?id=1742>.

———. 2010. 'PM Announces High Arctic Research Station Coming to Cambridge Bay', News Release: Prime Minister of Canada, 24 Aug. At: <pm.gc.ca/eng/media.asp?category=1&id=3599&featureId=6&pageId=26>.

Heininen, Lassi, and Chris Southcott. 2010. *Globalization and the Circumpolar North*. Fairbanks: University of Alaska Press.

Howell, S.E.L., C.R. Duguay, and T. Markus. 2009. 'Sea Ice Conditions and Melt Season Duration Variability within the Canadian Arctic Archipelago: 1979–2008', *Geophysical Research Letters* 36, L10502, doi:10.1029/2009GL037681.

Huebert, Rob. 2006. 'Canada–United States Environmental Arctic Policies: Sharing a Northern Continent', in Philippe Le Prestre and Peter Stoett, eds, *Bilateral Ecopolitics: Continuity and Change in Canadian–American Environmental Relations*. Aldershot: Ashgate, 115–32.

———. 2007. 'Renaissance in Canadian Arctic Security?', *Canadian Military Journal*. At: <www.journal.forces.gc.ca/vo6/no4/north-nord-eng.asp>.

———. 2008. 'Walking and Talking Independence in the Canadian North', in Brian Bow and Patrick Lennox, eds, *An Independent Foreign Policy for Canada? Challenges and Choices for the Future*. Toronto: University of Toronto Press, 118–36.

———, ed. 2009. *Thawing Ice, Cold War: Canada's Security, Sovereignty, and Environmental Concerns in the Arctic*. Winnipeg: University of Manitoba Press.

———. 2010. *The Newly Emerging Arctic Security Environment*. Prepared for the Canadian Defense and Foreign Affairs Institute, Mar. At: <www.cdfai.org/PDF/The%20Newly%20Emerging%20Arctic%20Security%20Environment.pdf>.

Honderich, John. 1987. *Arctic Imperative: Is Canada Losing the North?* Toronto: University of Toronto Press.

Indian and Northern Affairs Canada. 2010. *Northern Oil and Gas Annual Report 2009*, 6 May. At: <www.ainc-inac.gc.ca/nth/og/pubs/ann/ann2009/ann2009-eng.pdf>.

Kirton, John,and Don Munton. 1992. 'The *Manhattan* Voyages, 1969–70', in Don Munton and John Kirton, eds, *Canadian Foreign Policy: Selected Cases*. Toronto: Prentice-Hall.

Lackenbauer, P. Whitney. 2008. 'The Canadian Rangers: A "Postmodern" Militia that Works', *Canadian Military Journal* 6, 4: <www.journal.forces.gc.ca/vo6/no4/index-eng.asp>.

Lajeunesse, Adam. 2008. 'The Northwest Passage in Canadian Policy: An Approach for the 21st Century', *International Journal* 63 (Autumn): 1037–52.

Lasserre, Frédéric. 2007. 'La souveraieté canadienne dans le passage du nord-ouest', *Policy Options* 28 (May): 34–41.

McCalla, Robert J. 2010a. 'Arctic Shipping: Possibilities and Challenges in the Canadian Sector', presentation at the Prairie Summit annual meeting of the Canadian Association of Geographers, Regina, 2 June.

———. 2010b. 'Adventure Cruise Shipping in Polar Regions: An Overview', in Aldo Chircop, Scott Coffen-Smout, and Moira McConnell, eds, *Ocean Yearbook*, no. 24. Boston: Brill, 359–75.

Marinelog. 2010. 'Resolve Marine Group Starts *Clipper Adventurer* Salvage', 3 Sept. At: <www.marinelog.com/DOCS/NEWSMMIX/2010sep00033.html>.

National Snow and Ice Data Center. 2007. 'The Contribution of the Cryosphere to Changes in Sea Level', *State of the Cryosphere*, 12 Mar. At: <nsidc.org/sotc/sea_level.html>.

———. 2010. 'Sea Ice', *State of the Cryosphere*, 27 Oct. At: <nsidc.org/sotc/sea_ice.html>.

Richards, S. 1999. *Glaciers: Clues to Future Climate?* USGS General Interest Publications. At: <pubs.usgs.gov/gip/glaciers/glaciers.pdf>.

Rosenberg, Matt. 2010. 'The Volcanic Mount Pinaturbo Eruption of 1991 That Cooled the Planet', 17 Nov. At: <geography.about.com/od/globalproblemsandissues/a/pinatubo.htm>.

Stewart, E.J., A. Tivy, S.E.L. Howell, J. Dawson, and D. Draper. 2010. 'Cruise Tourism and Sea Ice in Canada's Hudson Bay Region', *Arctic* 63, 1: 57–66.

Stocker, Thomas, and Gian-Kasper Plattner. 2009. 'The Physical Science Basis of Climate Change: Latest Findings To Be Assessed by WG I in AR5', United Nations Climate Change Conference, COP 15 IPCC Side Event, 'IPCC Findings and Activities and Their Relevance for the UNFCCC Process', Copenhagen, 8 Dec.

United Nations Convention of the Law of the Seas (UNCLOS). 1982. *Straits Used for International Navigation: Transit Passage*. Part III, Section 2. At: <www.un.org/Depts/los/convention_agreements/texts/unclos/part3.htm>.

———. 2007. 'The United Nations Convention on the Law of the Sea (A Historical Perspective)', *Oceans and Law of the Sea*. At: <www.un.org/Depts/los/convention_agreements/convention_historical_perspective.htm>.

Vachon, Paris W. 2010. 'New Radarsat Capabilities Improve Maritime Surveillance', 18 Oct. At: <geos2.nurc.nato.int/mrea10conf/reports/pdf/Vachon,%20New%20RADARSAT%20Capabilities%20Improve%20Maritime%20Surveillance.pdf>.

Van Aken, Hendrik M. 2007. *The Oceanic Thermohaline Circulation: An Introduction*. New York: Springer Science & Business Media.

Victoria Times Colonist. 2007. 'Harper on Arctic: Use It or Lose It', 10 July. At: <www.canada.com/topics/news/story.html?id=7ca93d97-3b26-4dd1-8d92-8568f9b7cc2a&k=73323>.

Weber, Bob. 2011. 'Treaty to Boost Arctic Search and Rescue', *The Globe and Mail*, 5 Jan., A5.

9

Looking to the Future

Change remains the common denominator in the Canadian North. By the end of the first decade of the twenty-first century, Arctic sovereignty, global warming, and the Northwest Passage had emerged as critical issues. These topics—each related to Aboriginal and environmental issues—are now front and centre, and together these five themes, along with resource development, continue to reverberate in the halls of Parliament and in the boardrooms of corporate Canada. A northern version of 'development' no longer is or can be limited to the narrow image of the North as an exotic northern Eldorado, as typified by the Klondike gold rush and the poems of Robert Service, nor can our understanding of this vast Canadian region be reduced to quaint images of Aboriginal hunters pursuing game along the land-fast ice of the Arctic Ocean. The breadth of our new image of the North includes ice as it affects both the Inuit way of life and Canada's claim to Arctic waters. The old dream of the Northwest Passage has revived, but in a new and less romantic form. The stirring chorus of the 1981 Stan Rogers song, 'The Northwest Passage', provides a measure of our earlier, romantic images of the North and, indirectly, indicates how popular support has motivated Canada's claim to these waters:

> Ah, for just one time I would take the Northwest Passage
> To find the hand of Franklin reaching for the Beaufort Sea
> Tracing one warm line through a land so wild and savage
> And make a northwest passage to the sea

While the central issues facing this region and its people will be tested, redesigned, and tested again, what distinguishes the North from other regions of Canada is its ongoing and necessary search for the fair resolution of issues separating Aboriginal and non-Aboriginal peoples. One powerful reason for this search is that, unlike in southern Canada, in the Territorial North and in northern areas of some provinces Aboriginal peoples form a majority. Only in Canada's North is traditional ecological knowledge so important in the environmental assessment process, and that fact provides but one indicator of the northern search for 'fair' resolution.

The simple image of the North as either a resource frontier or an Aboriginal homeland is no longer valid. Like myths, these two visions expressed important yet idealized realities. They organized the past and foretold the future in two different ways. While these images simplified the complex ebb and flow of past events, they missed out on the dynamics where forces of change are creating a new order. First, these dynamics focused on Aboriginal peoples obtaining comprehensive land claim agreements, gaining rights for impact benefit agreements, and securing a place in environment assessment processes. With their rate of natural increase, the northern Aboriginal population continues to outpace that of other Canadians, and their demographic dominance is assured for the foreseeable future. Second, the North is no longer a 'zombie' region that only infrequently lurches up from the dead; rather, with Arctic sovereignty, global warming, and the potential of ocean shipping through the Northwest Passage confronting us today, this region of Canada has captured the spotlight. Frances Abele (2009: 19) succinctly summarizes the current situation:

> The Canadian North is the site of impressive and sometimes daring social and political innovations. It is also, increasingly, the focus of gathering international economic and political pressure, as well as new economic opportunity. Northern development will be an increasingly important aspect of Canadian political, social and economic vitality, and it should be a priority for public discussion and debate.

The Past and Present Purpose of Northern Development

How has the past shaped events unfolding in the twenty-first century? At the midpoint of the twentieth century, the southern half of the North, the Subarctic, began to be absorbed into the world economy as a resource hinterland. Resource development took the form of industrial megaprojects with little regard for Aboriginal peoples and their traditional lands. Driven by the demand for iron ore, American steel companies made a huge investment in northern Quebec. Private capital developed mines in permanently frozen ground, built the resource town of Schefferville, and provided rail access to Sept-Îles and the St Lawrence River. This huge industrial complex needed power, and that need provided the catalyst for the construction of the Churchill Falls hydroelectric project in nearby Labrador. Yet, Aboriginal people who hunted on these lands—before the days of recognition of Aboriginal title—were ignored in the wave of industrial development and, with virtually no regulations, companies had no reason to pay attention to the environment.

Twenty-five years after this giant iron ore development, the North became a battleground between energy companies and Aboriginal peoples. The focus of this struggle was in two different regions of the North—the Mackenzie Basin and northern Quebec. Two projects, one proposed by oil and gas companies and the other by a Crown corporation, Hydro-Québec, placed the leaders of industry and government in face-to-face confrontations with Aboriginal leaders. The outcomes—the James Bay and Northern Quebec Agreement (1975) and the Mackenzie Valley Pipeline Inquiry (1974–7)—tipped the balance of power away from both private and public resource developers.

Two legal developments added to this shift in power to Aboriginal groups. First, Aboriginal title to traditional lands was recognized. Second, environmental legislation became law. The vast Crown lands of the North were no longer available for governments to sell or lease to resource companies without addressing Aboriginal title and holding environmental assessment hearings. One outcome of the recognition of Aboriginal title was comprehensive land claim agreements; another was impact and benefit agreements.[1] Aboriginal title made resource development more acceptable to Aboriginal peoples, who receive benefits in return for the use of their traditional lands. In this ever-evolving relationship, companies have also had to respond to public regulations that protect the environment. Clearly, decision-making is no longer the exclusive prerogative of those in the boardrooms of resource companies. The first vast project proposal of the new century, the Mackenzie Gas Project, has undergone a rigorous environmental review with two agencies, the National Energy Board and the Joint Review Panel. Now approved, the project has missed the boat because, through new discoveries, the supply of natural gas has driven the price of this product down, making construction of a pipeline from the Arctic coast up the Mackenzie Valley to northern Alberta unlikely for at least a decade. But could this delay actually work to the project's advantage? Here is a possible scenario: if climate change continues at its present pace, an alternative transportation route and market may emerge with LNG ships taking liquefied gas from the Mackenzie Delta gas fields to ports in China. In February 2011, Korea Gas Corporation commenced investigations into such a plan and assigned Kogas, the world's largest LNG operator, to prepare a report that includes locating a deepwater port, perhaps at Cape Bathhurst (Vanderklippe, 2011a: B1).

Unlike the Arctic, the Subarctic's economic prospects depend heavily on its forest industry. Its major market is the United States. For the past decade, prices for forest products have dipped to record low levels because of the weak US demand for softwood lumber and other forest products. While the Arctic stands on the threshold of a boom, the Subarctic is in the midst of a near collapse of its forest industry. Given the dire situation in single-industry forestry towns, the federal government announced a special aid package in 2008 and again in 2010, but such assistance has not yet shaken the forest industry out of its doldrums—what is needed is a revival of US house construction.(Laghi et al., 2008; Canada, 2010).

This review brings us back to the question, 'What is the purpose of development?' As we noted earlier (Vignette 5.8), in 1939 Peter Drucker envisaged a capitalistic world where multinational companies diverted some of their energies from profit-making to social well-being. In the real world, companies expect governments to look after the rather nebulous concept of 'social well-being'. In the Canadian context, the place of Aboriginal peoples who have special rights adds another dimension to this concept. Alan Cairns, one of Canada's leading political scientists, refers to Aboriginal peoples as *Citizens Plus*, meaning that these First Peoples have rights of other Canadians plus those gained from treaties.

The cold reality is that federal governments did not respond to the impacts of resource development until forced to act by the courts, non-governmental organizations, and public opinion. Initial jolts came from the Berger Inquiry and a Quebec court injunction that halted construction on the James Bay Project and led to the James Bay and Northern Quebec Agreement. These were the initial instruments of social change. In the post-Berger years, the federal, provincial, and territorial governments accepted the legal ramification of Aboriginal title, passed environment legislation, and then began the process of comprehensive land claim negotiations. However, the first comprehensive land claim agreement in 1984 (the Inuvialuit Final Agreement) was silent on self-government. Yet, the 1975 James Bay and Northern Quebec Agreement laid the groundwork for regional Aboriginal governments in northern Quebec and eventually for the rest of the Aboriginal world (Salisbury, 1986; Koperqualuk, 2001). Twenty years later, the federal government agreed to include self-government in comprehensive land claim negotiations (see Chapter 7). The Aboriginal community showed great flexibility and negotiating skills to create 'hybrid' institutions that expand the nature and complexity of Canadian governance. For example, in 1984, the Inuvialuit persuaded the federal negotiators to accept a dual structure—one that kept a foot on the land (Inuvialuit Game Council) and another in the marketplace (Inuvialuit Regional Corporation). Later, in 1993, Inuit negotiators convinced federal officials to include self-government at the territorial level in the Nunavut agreement. Also, Quebec Inuit, with the experience of administering local affairs in Nunavik for over 35 years, pushed the federal and Quebec governments to embrace the concept of regional government for Nunavik. Thus, within the global economy and the Canadian system of government, local forces have created hybrid agreements that better match the interests of Aboriginal peoples and, in a few cases, regional Inuit forces have persuaded governments to approve regional homelands.

Northern Vision for the Twenty-First Century

The road to this new northern reality is not an easy one. Bumps along the road are common. Yukon lost its principal mine (Faro); the dream of Denendeh vanished; and the Voisey's Bay nickel project stumbled badly before finding its feet. Beyond promoting resource development, Ottawa and its provincial and territorial counterparts have failed to formulate a joint long-term northern development strategy that could lead to the eventual diversification of the northern economy, including within remote communities. The exception is Quebec's announcement in 2011 of its $80-billion Plan Nord.

How do we find a better balance between market-driven resource projects and public-driven social development? Past grand projects and public strategy have often fallen short of their target. In the 1950s, Canada was treated to John G. Diefenbaker's 'Northern Vision' and its 'Roads to Resources' program. As an offshoot of Diefenbaker's Northern Vision, in the 1960s Richard Rohmer launched his Mid-Canada Development Corridor proposal to build a northern railway across the Subarctic. His idea—modelled after Canada's nineteenth-century completion of the CPR line that tied the country together and enabled western settlement, and perhaps after the Soviet dream of a northern version of their Trans-Siberian Railway—failed because Ottawa, which would supply the cash, realized that the days of railway-building to stimulate regional development were long past (see Chapter 3). The grand project for the twenty-first century may be the Mackenzie Gas Project. This undertaking would create an energy corridor along the Mackenzie Valley to Alberta, thus opening the western Arctic to further development. The Joint Review Panel for the Mackenzie Gas Project provided a blueprint for a new balance between this project and social/environmental matters, but these recommendations were too much for Ottawa to swallow (see Chapter 6).

Ironically, two megaprojects in the twentieth century, one proposed and one completed, caused a shift towards more public involvement by focusing on northern benefits, especially for Aboriginal peoples. The James Bay Hydroelectric Project resulted in the James Bay and Northern Quebec Agreement, and this project keeps on giving. Hydro-Québec has now signed long-term compensation agreements with the Quebec Cree and Inuit so that further development can be pursued in northern Quebec. The Mackenzie Valley Pipeline Project associated with the Berger Inquiry never got off the ground, but it led to several important outcomes: recognition of the potential social impacts on Aboriginal peoples; recognition of the need to resolve land claims prior to development; and transformation of the way environmental impact assessment takes place by shifting hearings from the boardrooms to community halls.

On a smaller scale, Canada needs a joint territorial/provincial/federal vision for remote Native communities. Should the public purse support them or abandon them? At the moment, most Native communities are struggling both economically and socially. Since the market economy cannot function well in remote Native settlements, should governments intervene in a more substantial way or simply let the marketplace have its way? Some Aboriginal northerners have moved to southern Canadian cities, but, given the very high birth rates, this out-migration has not put a dent in the numbers residing in Arctic and Subarctic centres, and even under trying circumstances many do not want to leave their northern homelands. For instance, the Cree First Nation of Kashechewan in northern Ontario, with a population of less than 2,000, chose to return to their community on the Albany River near James Bay, despite years of trouble with their water treatment system, rather than accept relocation to Timmins.

How did such communities living in 'Fourth World' conditions originate? The answer goes back to the 1950s with the beginning of the relocation program for Aboriginal peoples living on the land. Some 60 years later, life chances for these people, especially the youth, are extremely limited because of poor education and employment opportunities. As well, the option for an on-the-land existence of hunting and trapping is an uphill struggle except in Quebec's north, where such activities are heavily subsidized. These limitations, along with the inevitable loss of traditional culture, certainly are factors behind the high levels of drug and alcohol use and high suicide rates often found in such communities. Remnant Aboriginal cultures, especially languages and hunting practices, are maintained in these remote centres with mixed economies, but at what cost? Few governments, except for those of Quebec and Nunavut, have paid attention to supporting hunting and trapping activities in remote communities.[2] Given the persistence of these 'failed' communities (failed in the sense that they lack a firm economic base), has the federal government misjudged the aspirations and needs of Aboriginal populations living in remote centres? If so, do Ottawa and the territories and provinces need to revisit the consequences of the relocation strategy and devise, with Aboriginal communities, a fresh approach?

Residents of remote Aboriginal communities have benefited from comprehensive land claim agreements (payment of benefits, for example) and a few others have become commuters to mining sites, yet the main source of employment is found in the public sector, especially in the capital cities of Whitehorse, Yellowknife, and Iqaluit. Only the government of Nunavut has decentralized some public agencies to regional centres. While a step in the right direction, these efforts are insufficient, leaving remote communities beyond the orbit of economic well-being. We now need to re-examine three questions posed at the beginning of this book.

Question 1: Can the Resource Economy Support the Northern Labour Force?

In both the short and long runs, the resource economy simply cannot support the workforce in the North. In the long term, non-renewable resource operations exhaust their resources and closure follows. Clearly, the resource economy is not sustainable at the community level and places the regional economy in a boom-and-bust scenario. In the short run, only a few communities are fortunate enough to benefit from resource development. The vast majority are beyond the economic and hiring orbit of resource developments.

Even during the construction phase of megaprojects, the northern labour force plays a minor role. Although the resource industry remains the driving force behind the northern economy in terms of value of production, resource companies employ relatively few workers (except in the construction stage of megaprojects). Also, resource firms frequently hire skilled labour from the southern workforce because such labour is in short supply in the North. Finally, the skill level of the Aboriginal labour force falls below that required by resource companies for trades and management positions. This serious mismatch between the skilled labour required by resource companies (and other employers) and the qualifications of Aboriginal workers has prevented and continues to prevent proportional employment levels between the Aboriginal and non-Aboriginal labour forces. Jull (2000: 119) forecasted that this problem would haunt Nunavut (and, by extension of his argument, Native settlements across the North). Only a massive infusion of funding for public infrastructure would provide temporary relief to the massive unemployment dogging the Aboriginal labour force. But even in this instance, the shortage of skilled Aboriginal trades people to provide the muscle to undertake major infrastructure projects—such as ports and port facilities in the Far North—remains the fundamental challenge. Big resource projects, if and when they come, do not hire large numbers of unskilled local labour.

For Nunavut, this critical issue—the shortage of skilled Inuit trades people and professional workers—shows no signs of resolution. The failure of Inuit to acquire trades certification is partly due to access to trades programs, but, more significantly, relatively few Inuit have obtained a high school degree, which is necessary to enter a trades program. Similarly, a shortage of Inuit administrators and professionals has prevented the government of Nunavut from fulfilling its goal of a majority of Inuit employees in its civil service. The Inuit labour force remains some distance from being able to take full advantage of existing employment opportunities in Nunavut. To change this situation, a stronger connection between education in English and job access must occur in the Aboriginal community. But now we are on thin cultural ice—would such a connection represent assimilation, i.e., the loss of Aboriginal language and culture? The Nunavut government, by introducing the Official Languages Act and Inuit Language Protection Act (2007), is struggling to find a balance between the Inuktitut language and jobs for Inuit (primarily government jobs). Faced with strong opposition from Inuit organizations that want even stronger language legislation, Louis Tapardjuk, Nunavut's Language and Culture Minister, says, 'the . . . language law package offers the strongest possible formula for protecting and promoting the Inuit language and . . . the government is unlikely to change the two bills' (J. Bell, 2007). Language issues are not simple matters, as Vignette 9.1 indicates.

Increasing the number of Inuit in Nunavut's government is not an easy task. This goal is complicated because of the limited number of Inuit who qualify for jobs in the public sector. Even at the lowest entrance level, the public sector normally requires a high school degree. In Nunavut, for example, the Inuit have achieved their political

---

Vignette 9.1 Two Views on Inuktitut

In a hard hitting article, Colin Campbell began by stating that 'As southern culture encroaches in the North, it [saving Inuktitut] will be a tough fight.' With English so entrenched in government circles, maybe the cards are stacked against Inuktitut? Campbell observed two views on this subject and these are summarized below.

Imposing Language Laws
In a controversial move in 2006, Nunavut Premier Paul Okalik ordered that senior bureaucrats learn the language or lose their jobs, saying that 'they have to be fluent, they have to work with members and with people within Nunavut' (CBC News, 2006b). The government is also drafting two new language laws designed to help make Inuktitut Nunavut's working language by 2020 and lift employment barriers for Inuktitut speakers. 'It really does open the door for Inuit', says Johnny Kusugak, Nunavut's [now former] language commissioner. 'Inuit kids can now look up and see that there are lots of positions in the government where they can reach their goals.'

Kusugak says he's not worried the new language laws will drive away qualified bureaucrats. To reverse Inuktitut's decline, young Inuit need to see that they can hold the most important positions in society while maintaining their culture, he adds. 'When we were growing up, our parents told us that we have to learn English if we want to work in the changing world. There will be jobs and security, they said.' Now that the Inuit have their own land and their own government, he says, that view must change.

But What about English?
Not everyone backs Nunavut's plans to prop up Inuktitut. For young people to succeed as professionals (e.g., doctors or lawyers), they must have strong English skills, insists Nancy Gillis, a city councillor in Iqaluit. Basic English skills and a strong education system risk being lost in the scramble to preserve Inuktitut, she says. 'By the time children hit Grade 4, they're behind already', she says. Others say forcing bureaucrats to learn Inuktitut is also misguided. Many of the skilled managers and bureaucrats here are not Inuit, or Inuktitut speakers. The majority of Inuit today lack the professional training and post-secondary education to fill top-tier jobs. While the goal of government was to have 85 per cent of its employees drawn from the Inuit workforce by 2007, only about 47 per cent of the government jobs were held by Inuit (who make up 85 per cent of the population), and most of those jobs were lower-level positions (Légaré, 2008: 367). A critical barrier to increasing the number of Inuit employees in the Nunavut civil service is that more than half of the government positions require a college or university degree—and a fluency in English, not Inuktitut, is essential to succeed in college or university (ibid.).

Source: Adapted from Campbell (2006).

goal, thus enabling political solutions. One such effort was the government's policy of decentralizing its agencies to share public employment among its 26 communities. Another one is the proposed language law described above.

Besides language and culture, distance presents a formidable barrier to employment opportunities. Native communities, especially in the Arctic, are spread over an enormous area. Air commuting offers one solution, albeit a limited one. Air commuting also offers a cultural bridge by providing Aboriginal workers with employment at the remote mines and still allowing their families to remain in their home communities. But air commuters number in the hundreds while thousands more want jobs. Of course, not everyone wants to work at a mine site, let alone underground. Turnover is a problem. Perhaps the strategy of adding Aboriginal workers to a non-Aboriginal crew is flawed. Would an employment and retention breakthrough occur if the companies created a Cree or Inuit work environment for one complete crew? Certainly this is not a simple task, and possibly it would be a very expensive one. Yet the stakes are high. Given the cultural hardships faced by Aboriginal workers in a Canadian mine and the urgency to deal with the high unemployment in Native settlements, such an approach is worth examining. The Agnico-Eagle company, with the people in the region of Baker Lake, developed a different kind of model in their Impact and Benefits Agreement. In this agreement, instead of cash payments to communities, as in the Raglan IBA, the Baker Lake IBA focuses on preferential treatment for service contracts and employment opportunities (George, 2010). In one example, the newly completed 110-km road connecting Baker Lake with the mine provided Inuit employment while Inuit companies were awarded service contracts for the road construction.

Question 2: How Can Government Ensure That the Resource Industry Limits Its Impact on the Environment?

Government regulations and consistent monitoring of the resource industry's operations are necessary to assure that environmental impacts are kept to a minimum and are within the limits determined by the approved project proposal. But the three levels of government—federal, provincial, and territorial—have had to be pushed by non-governmental organizations and public opinion to establish proper regulatory and monitoring regimes. And this remains the case, especially when stridently business-friendly governments are in power. Governments have clarified environmental regulations and companies are following them. Understandably, companies are concerned about both the environment and their costs of production. With a 'level playing field', meaning all companies must follow the same rules, the costs of environmental protection affect all companies equally.

Bruce Mitchell (2005: ix, x) reports that Canada has made five distinctive contributions in the field of environmental impact assessment. They are:

- incorporating traditional and scientific knowledge into environmental impact statements;
- developing partnerships with communities and local groups in the design and implementation of the assessment process;

- handling issues specific to the context and distinctive physical and cultural environment of the Canadian North;
- dealing with inter-jurisdictional issues between Ottawa and the provinces;
- designing and implementing a 'package' (policy, administrative arrangements, and legal framework) for environmental impact assessment.

A sixth contribution can be added to Mitchell's list, namely, joint environmental impact assessments on lands controlled by Aboriginal groups. Stemming from comprehensive land claim agreements, their significance is twofold. First, joint environmental impact assessment ensures that local issues are fully assessed (rather than the emphasis being on 'national and regional issues' as in past Ottawa-based assessment processes). Second, sharing of power between Ottawa and the respective Aboriginal organization empowers Aboriginal people, giving them a strong voice at the decision-making table over issues of particular concern to them, such as wildlife.

Over the last 30 years, three developments have helped to reduce the threat of environmental damage caused by resource projects and, at the same time, have ensured that the land will be restored to its former natural state. One major step was more demanding environmental regulations through legislation. These regulations ensure that the reclamation of a work site takes place after a project has wound down. Another step forward was the creation of co-managed (joint) environmental impact assessment agencies in comprehensive land claim settlement areas. The third step involves a shift in corporate attitude about the environment. While the resource industry continues to put its industrial stamp on the northern landscape, its executives have accepted their social responsibilities—care for the environment that is reinforced by legally binding environmental agreements with federal, provincial, and territorial governments.

Alberta's oil sands are a major source of greenhouse gas emissions. Environment Canada reports that, while most economic sectors are expected to reduce their emissions significantly by 2020, a major increase in total emissions is predicted to come from the oil sands. This industry, unless it can drastically curb its release of greenhouse gases, is projected to nearly double its already massive emissions between 2010 and 2020, from 49 to 92 **Mt CO^{2}e** (Environment Canada, 2011a: Table 5). Yet, companies operating in Alberta's oil sands have been slow to accept their social responsibilities in caring for the environment and, equally troubling, Alberta-led environmental monitoring has not been up to the task, thus creating a false sense of acceptable impact. For over a decade, the monitoring of Alberta's oil sands has remained badly flawed, suggesting that the Alberta and federal governments turned a blind eye because of the economic importance to Alberta and Canada of the oil sands projects.

The Aboriginal presence in the oil sands region is not inconsequential, but a comprehensive-like agreement similar to the JBNQA of 1975 does not exist. According to the Alberta government (2011), an estimated 23,000 Aboriginal people live in Alberta oil sands areas, with 18 First Nations and six Métis settlements located in the region. While Aboriginal workers have gained some jobs in the oil industry and Aboriginal businesses have obtained contracts from oil companies, the legal concept of traditional lands might have sparked a grand agreement between these Aboriginal peoples and the oil companies, but this has not occurred.

The damage to the environment is not in dispute—and all industrial projects have negative environmental impacts. But what was not appreciated was the degree of damage, the geographic range of the impact, and the time horizon to restore the mined areas. In 2010, an independent study led by Dr David Schindler reported that toxic wastes far above acceptable levels for human health were discovered in the Athabasca River. When Dr Schindler met with the then federal Minister of the Environment, Jim Prentice, and made him aware of the impact on the Athabasca River, Prentice ordered an inquiry by his department. In December 2010, the inadequacy of the existing monitoring system was confirmed by an independent review authorized by Environment Canada. Alberta then stepped into the fray and appointed another independent panel to examine the existing monitoring system. The conclusion was similar—the monitoring system was not up to the task. As pointed out in Chapter 7, the Alberta government is scrambling to justify its failure to ensure an adequate environmental monitoring system, let alone its claim to have a world-class monitoring system for the oil sands. Yet, the very day in January 2011 when the formation of the Alberta environmental panel was announced, another open-pit mine (Total's Joslyn North mine located 70 km north of Fort McMurray) was approved. The Joint Review Panel for this proposal, as with previous JRPs of oil sands projects, declared that 'the project would have "no significant adverse effect" on wildlife and water quality and would meet targets on reducing waste, known as tailings, from the oil sands extraction process' (CBC News, 2011). The discrepancy between the two positions can only mean that the bar used by the JRPs was set far too low. The environment of northern Alberta continues to deteriorate, but in July 2011 the federal government made a sharp U-turn and is now insisting on a world-class monitoring system for the oil sands (Environment Canada, 2011b).

Question 3: Is There a Place for Aboriginal Corporations in the Resource Economy?

To be sure, Aboriginal development corporations have a role to play in resource projects. In fact, this model of resource development offers the potential for widespread and lasting benefits to specific Aboriginal corporations and their larger communities through direct employment and business opportunities. The shortcoming of this model is geographic inequality due to the distribution of oil and mineral resources and the absence of comprehensive land claim agreements in most provinces (see Figure 7.4).

Still, the growing strength of Aboriginal corporations in the market economy is a remarkable shift of economic power. Aboriginal leaders have chosen to participate in order to create jobs and business opportunities for their members. Modern land claim agreements have provided cash payments and a business structure that allow Aboriginal organizations the option to participate in major projects and to develop their own companies that can take advantage of various opportunities related to northern development. The first two agreements, the James Bay and Northern Quebec Agreement in 1975 and the Inuvialuit Final Agreement in 1984, created the Cree Board of Compensation[3] and the Makivik Corporation in northern Quebec and the Inuvialuit Regional Corporation in the Northwest Territories. These organizations were responsible for receiving and managing the funds flowing from the two agreements. In each instance, funds were invested in the marketplace.

The injection of Aboriginal corporations into selected parts of the northern world has altered its economic and political power structure. Recent agreements in Quebec—such as the Paix des Braves for the Cree and the Sanarrutik agreement for Inuit—are the product of the James Bay and Northern Quebec Agreement, which created the economic atmosphere making possible such agreements, i.e., the Cree/Inuit institutions that received the cash settlements and manage those funds. The situation for First Nations that do not have the benefit of such structures is more challenging. The agreement between the Manitoba Hydro and Nisichawayasihk Cree Nation over the Wuskwatim generating station represents a micro version of this partnership model for resource development (see Chapters 5 and 7).

Partnership agreements mark the future trend of resource development in the North, thereby assuring northern benefits to Aboriginal peoples. As we have seen, the Aboriginal Pipeline Group, representing most peoples of the Mackenzie Valley, secured a right to own one-third of the Mackenzie Valley natural gas pipeline, subject to providing one-third of the cost of the pipeline. At the same time, the Joint Review Panel's recommendations provide some assurances that a place for wildlife and, hence, for harvesting country food remains an option. Unexpectedly, the economics of this project soured in 2011 due to the discovery of huge deposits of natural gas in B.C. and other areas of North America, thus dashing the hopes of the Aboriginal Pipeline Group.

Nagging Old Challenges

Why are so many Aboriginal peoples in remote northern communities living in squalid conditions? This nagging matter goes back to the relocation of Aboriginal peoples to settlements, which at best represented social engineering and at worst neo-colonialism. In the 1950s, the federal government finally recognized that 'living on the land' was not the 'answer' for Indians, Métis, and Inuit. But living on the land was their way of life for centuries, and such a lifestyle was the basis of their culture. When times were good, the hunting life was the best of all worlds. When times were bad, the hunting life brought the risk of hardships that were unacceptable in a modern country and hard to bear by those on the land—extreme hunger, even starvation, and untreated illnesses. Ottawa felt the eyes of the international community and was compelled to take action. Thus, some 60 years ago, the peoples of the land were relocated to the sites of trading posts. In most instances these relocation settlements were strategically located with respect to access to hunting and trapping lands, and, most importantly for Ottawa, they provided access to store food, medical services, and basic schooling. From the federal perspective, relocation achieved several goals: store food eliminated the threat of starvation; medical attention reduced the mortality rate; and basic school was the key to participating in the larger Canadian society. Construction of administration buildings, nursing stations, schools, staff housing, and public housing provided an initial boom and employment opportunities. But beyond providing a place for Aboriginal peoples, these settlements had no economic purpose. While country food obtained through hunting, fishing, and gathering remains highly valued, community living thrust people into a different economic system and cultural world where wage income or government payments were essential to pay for the high cost of living and where social interaction was conducted in English. Many people, particularly in the smaller Native settlements, have continued to have close ties to hunting, trapping, and fishing. Unfortunately, such activities count for little in the cash economy, and cash is a necessary commodity in settlement life. At the time of the James Bay and Northern Quebec Agreement, Quebec recognized two facts: the importance of hunting and trapping to the Cree and Inuit, and the lack of sufficient jobs in the Native settlements. Quebec's solution was to provide financial assistance for hunters, thereby reducing social assistance and the loss of self-esteem and enhancing the meaning of life. Since then, only Nunavut has followed Quebec's example.

Because the market economy is not functioning in Native communities, economic and social problems have emerged with high unemployment rates, heavy reliance on welfare payments and other forms of public assistance, and dysfunctional community life. The high birth rate associated with settlement life increases the demand for more services, housing, and jobs. Yet, without an economic base (or more public investments), the economic and social problems remain a fact of life. Not surprisingly, then, suicides among young people are extremely high. Even the new Innu community of Natuashish, created at a cost of $152 million to relocate people from the earlier Davis Inlet relocation disaster, offers no escape from these social ills. Isolated Native settlements lie outside of the economic orbit of Canada's economy and, therefore, their residents are trapped. As the population of these settlements increases, so do the unemployment, housing, and social problems. Not surprisingly then, suicide rates are much higher than in southern communities. While a subsidized hunting and trapping program like Quebec's Income Security Program does not provide the whole answer, living on the land for part of the year provides a safety valve for those who do not fit well into settlement life.

As noted earlier, Kashechewan, a First Nation community situated in the Hudson Bay Lowland of northern Ontario, represents a community in crisis. In addition, its physical site is subject to flooding. Since 2004, the residents have been evacuated three times due to spring flooding, sewage backup, and health-threatening E. coli levels in the reserve's drinking water. Kashechewan is only one of many Native settlements confronted with site problems, often associated with polluted drinking water. In 2007, the options for Kashechewan members were:

• Move to another site where flooding is less of a problem.
• Relocate to Timmins.
• Stay put.

The federal government rejected the first option, citing the high cost of constructing a new community. Most band members did not want to relocate to Timmins. By means of a referendum, these Cree decided to stay put and, with federal funds, improve their local infrastructure, including the dyke system and drinking water supply (Vignette 9.2). Since then the federal government has funded a Capital Planning Study with Kashechewan First Nation identifying the needs of the community (Kashechewan First Nation, 2010). The study involved analysis of existing housing conditions, community layout, commercial and institutional needs and infrastructure capacity, and examination of alternative sites for relocation of the community, a curious proposition given the government's earlier refusal to consider relocation except to Timmins.

Figure 9.1 First Nation Community of Kashechewan, Ontario

Aerial photograph of the Cree community of Kashechewan and the Albany River. Along the river, part of
the ring dyke is visible. In the foreground are rows of residential housing; in the middle are a community
nursing station and public school.

Source: Weeneebayko Health Ahtuskaywin (2006). Reprinted by permission of Weeneebayko Health Ahtuskaywin.

Emerging Challenges

Global warming threatens the Arctic environment and may eventually melt a substantial amount of permafrost, reduce the extent and thickness of Arctic ice, and alter
the Arctic shoreline (Hassol, 2004; Fortier and Fortier, 2006). With an ice-free Arctic
Ocean, the Northwest Passage would become a commercial waterway and Canada
would have to monitor such shipping and enforce environmental violations. Added to
those responsibilities will be the additional sea floor that falls under Canada's jurisdiction through the upcoming UNCLOS submissions from Canada and others of the Arctic
Five. For the Inuit, global warming threatens their culture, which is closely associated
with sea hunting on land-fast ice. While Canadians are aware of the changing nature
of climate, sea ice, and permafrost, the Inuit have expressed strong concerns about the
possible impact on their lives. Joint university/Inuit projects, such as the International
Polar Year (IPY) Inuit Sea Ice Use and Occupancy Project led by Claudio Aporta and
Gita Laidler, may be the wave of the future by bringing together scientific and traditional knowledge about sea ice and the Inuit. Yet, the urgent need for assessment and
management of the loss of ice on Inuit culture remains unclear. As a leading Inuit
spokesperson, Sheila Watt-Cloutier (*CASR*, 2007), stated: 'As long as it's ice, nobody
cares except us, because we hunt and fish and travel on that ice. However, the minute
it starts to thaw and becomes water, then the whole world is interested.'

Vignette 9.2   Kashechewan: To Stay or To Go?

Kashechewan, a Cree community with a resident population of approximately 1,500, lies on the flood plain of the Albany River, which flows into James Bay. This leaves the community susceptible to flooding each spring. Before the 1950s, these Cree, like other Aboriginal peoples in the North, lived on the land. Encouraged by the federal government, they relocated to sites within their traditional homeland that were selected by federal officials. First, they settled at Fort Albany, but in 1957 they relocated to the present community of Kashechewan. In the late 1990s, a ring dyke some 3–4 metres in height was built around the community to prevent flooding (Figure 9.1). Apparently, the 2006 spring flood did not breach the dyke but water seeped through the dyke, perhaps augmented by river water flowing through the dyke's culvert due to a jammed butterfly valve. The purpose of the culvert is to allow spring snow melt and summer rain waters to drain into the river.

In 2006, former Chief Friday presented a proposal to locate a new site for the community, 30 km west of the existing location on the Albany River. Other relocation sites were proposed. In 2006, Alan Pope was hired by the federal government to investigate the situation at Kashechewan and to make a series of recommendations. His principal recommendation was that 'a new reserve be created for the Kashechewan First Nation on the outskirts but within the geographic boundaries of the City of Timmins'. The reaction from the Chief, band councillors, and the community at large was less enthusiastic. First, the Cree place a high value on their traditional hunting lands. Relocating to Timmins (even with the promise of trips to their traditional hunting grounds) would cut them from their historic roots. Second, the value of the 'subsistence' economy lies in the country food produced, not in the dollars derived from such activities. Third, relocating to a 'southern' town may draw its members into the larger community and such integration could strengthen the economic well-being of some individuals, but integration could weaken the First Nation. The compromise, as it turned out, was to improve existing community infrastructure and housing with the help of additional federal funds.

Sources: CBC News (2006a); Weeneebayko Health Ahtuskaywin (2006); Pope (2006); Indian and Northern Affairs Canada (2007); personal communication with Howard Chambers (2007).

Sovereignty over Arctic waters and its seabed is a delicate and partly unresolved international legal matter (Carnaghan and Goody, 2006; Pharand, 2007; Lalonde, 2007; Coates et al., 2008; Byers, 2009; Grant, 2010). Recent events associated with Arctic sovereignty are outlined in Table 9.1. (For earlier events related to Arctic sovereignty, see Table 3.4.) One unresolved legal matter involves the passage through Canada's Arctic Archipelago. The frozen waters stretching beyond 12 nautical miles (22.2 kilometres) in Canada's Arctic Archipelago could be considered international waters, but because they are frozen Canada could claim they are an extension of the land. Such a claim is strengthened by Canadian Inuit using the ice for hunting and travel (Huebert, 2003; Macnab, 2004; Griffiths, 2004; Birchall, 2006; Byers, 2009). The claim could be weakened, however, as ice retreats.

Is an Arctic shipping route a threat to Canada? Or does the Northwest Passage remain a romantic dream of a shorter route from Europe to Asia? In September 2007, the Northwest Passage was, for a short time, ice-free (see Chapter 8). Most scientists

Table 9.1 Recent Events Associated with Arctic Sovereignty

Date	Event
1985	Canada declared a 200-mile exclusive economic zone, extended its 3-mile territorial waters to 12 nautical miles, and passed the Arctic Waters Pollution Prevention Act 1985.
1988	In 1988, the Arctic Co-operation Agreement between Canada and the United States was signed. One term of the agreement stated that the US government would seek Canadian approval to traverse the Northwest Passage.
1997	Ottawa passed the Oceans Act. In doing so, Canada extended its national jurisdiction over most of the Canadian continental margin, marine resources, and seabed.
2003	Canada ratified the United Nations Convention on the Law of the Sea, which provided Ottawa 10 years to file its claim to the Arctic seabed.
2007	RADARSAT-2, Canada's second Earth observation satellite, has the capacity to track and identify ships in the Canadian Arctic.
2013	Canada's deadline to file for ownership of the Arctic seabed beyond its 200-mile economic zone. Already, Canada has established an economic zone that stretches from the country's coastline into the Arctic Ocean. Canada's claim to the Arctic seabed—potentially an area similar in size to the three Prairie provinces—depends on 'proof' that the ocean floor adjacent to Canada's Arctic continental shelf is geologically linked.

believe that the ice cover on the Arctic Ocean is retreating, and retreating rapidly. Even so, the Northwest Passage is some distance from becoming an international shipping route. As Birchall (2006: iv) puts the case:

> Canadian sovereignty over the waters of the Arctic Archipelago, and specifically the Northwest Passage, is not under threat as a result of increased international commercial shipping at this time, nor will it be in the near future. The erratic nature of archipelagic ice conditions, coupled with the imminent return of the polar winter, will dissuade any reputable shipping firm from using the Northwest Passage instead of the more reliable Panama Canal or Cape Horn routes.

Still, the retreat of sea ice has accelerated since Birchall cast doubt on the immediacy of a commercial sea route through the Northwest Passage (Figure 9.2). As we have seen, the passage has been made annually for the past decade. As well, the trend towards more open water each summer has provided another reason for circumpolar countries to map their potential sections of the seabed of the Arctic Ocean. This is the last large available territory in the world, so circumpolar nations want to prove their underwater shelves extend into the unclaimed portions of the Arctic seabed, thus allowing them to demonstrate 'ownership'. At the moment, the area around the North Pole is considered an international area, which is administered by the UN's International Seabed Authority. To a greater or lesser extent, Canada's claim to an 'extended' exclusive economic zone doubtless will be accepted by the commission set up by the United Nations under the Law of the Sea Convention—unless something entirely unforeseen changes the geopolitical seascape.

When the Canadian government, under pressure from foreign research projects in the disputed waters, announced in 2007 a more aggressive northern policy, this more

Figure 9.2   Proposed Site of the Federal Government's Arctic Harbour
on the Northern Tip of Baffin Island

Coast Guard ship *Des Groseilliers* at the land dock for the abandoned mine of Nanisivik in 2007.
Source: P. Bell (2007).

robust policy included the geological mapping of our seabed and a combination of Arctic patrol ships, a naval port at Nanisivik, radar surveillance of above-water and underwater movements, and an increase in the number of Canadian Rangers, who are part-time reservists from Inuit communities.[4] The purpose of this $7.5 billion investment is to protect and extend Canada's Arctic sovereignty.

Added to the Arctic Sovereignty commitment, in October 2011, the federal government committed $33 billion over 30 years to ship yards in Halifax and Vancouver to create a polar naval force, including a polar-class icebreaker (Chase and Marotte, 2011:A1). With sovereignty in mind, Prime Minister Stephen Harper (Canada, 2007) announced support for the International Polar Year (2007–8): 'Scientific inquiry and development are absolutely essential to Canada's defense of its North, as they enhance our knowledge of, and presence in, the region.' Support for its Arctic strategy continued in the June 2011 federal budget (Canada, 2011). Ottawa's support may extend to Canadian universities, whose involvement in the North hit a low point at the end of the twentieth century due to insufficient federal funding (Canada, 2000). Northern research is expensive and requires a long-term commitment.[5] In the present political atmosphere, Ottawa may well see fit to commit substantial long-term funding so necessary for professors and their students to engage in northern research and to establish a working relationship with people in northern communities.

Future Directions

The key question remains: What is the purpose of northern development? While the North's economy will remain focused on energy and mineral extraction for world markets, aspects of that economy require greater involvement of Aboriginal peoples and greater respect for the environment. But oil companies in particular show little sign of respect and, after pressuring the federal government to open the Beaufort Sea for oil wells, were disappointed with the report that documented the difficulty of an oil spill cleanup in Arctic waters (Vanderklippe, 2011b; Weber, 2011). For the Inuvialuit, the Arctic environment remains a precious feature of their culture and economy and a major oil spill would have devastating impacts on the sea and coastal environments.

Economic solutions do not come easily or without cost. Resource development alone cannot solve the perplexing problem of poverty and chronic housing shortages in Aboriginal communities. Ironically, both are products of the federal relocation program of the 1950s to sites with little economic promise, and prospects for stronger local economies are not bright—unless the federal government intervenes. Such intervention could take the form of a hunter/trapper subsidy; another would involve raising the role (and pay) of Canadian Rangers, a militia of local Inuit part-time soldiers who serve as one part of Canada's surveillance system and assist in search-and-rescue missions (Lackenbauer, 2008).

Practical Solutions for the Near Future

Canada needs practical social programs to complement resource development. As well as actual programs, consultations with Aboriginal stakeholders are a necessary element in developing programs that will work and for creating a spirit of partnership. From past experience, top-down initiatives have little chance of success. Such partnerships, which would include Native groups, governments, and/or private developers, could be modelled after impact and benefit agreements, which require a negotiated agreement. Two types of approaches are proposed—facilitating solutions and breakthrough solutions.

Facilitating solutions include extending education opportunities, experimenting with on-the-job training programs, expanding the geographic range of fly-in/fly-out commuting to include all communities within a certain area, revisiting the concept of a hunting subsidy within the context of the twenty-first century, and extending the joint assessment and management systems that flowed out of comprehensive land claim agreements to other areas, such as resource projects. Each approach, in its own way, provides direction for the future. For example, those Aboriginal groups that have negotiated comprehensive agreements have several advantages, including a substantial cash settlement and a dual administrative structure that allows a foot in the new world of commerce and another in the old world of hunting and trapping. In fact, this structure serves as a balance between the new ventures and the traditional lifestyle. First, it allows participation in the market economy and the traditional one. Second, its very nature allows those involved in the traditional economy associated with the environment and wildlife to confront those in the capitalist resource extraction economy to ensure that the environment is protected and that renewable resources are managed properly.

Yet, this concept of a foot in each world has a serious flaw: living on the land, romantic as it may appear, does not alone provide the cash needed to survive in a settlement-based society. A Quebec-style subsidy for hunting families could solve this economic matter. The rationale is simple: the social welfare costs of unemployed adults in northern communities offset the stay-on-the land subsidy but, perhaps more important, such a program changes unproductive workers into productive ones, thus breaking the dependency on welfare, enhancing self-worth, and mitigating the social ills that result from the dependency trap. As a family unit program, the downside is that children are unable to attend school regularly. However, in the modern techno-logical world, a variant of distance education might solve this problem.

Breakthrough solutions require visionary leadership from Ottawa and a long-term commitment by all levels of governments and industry. Recognizing that the North has limited opportunities for economic diversification compared to other regions of Canada, how can its small population and the great distance to markets work to its advantage? One way is to make more use of resources within the North; another is to harness unused resources. Could wind power development achieve several seemingly elusive goals, such as making sustainability of northern communities closer to reality?

Currently, each community and remote mine must import diesel fuel at great cost to produce electricity. In time, these costs can only become more of a burden because

Figure 9.3   Wind Energy: Pipe Dream or Clean Energy Solution?

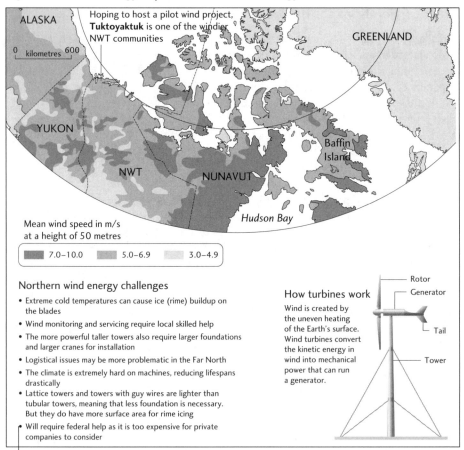

Hoping to host a pilot wind project, **Tuktoyaktuk** is one of the windier NWT communities

ALASKA

GREENLAND

YUKON

Baffin Island

NWT

NUNAVUT

Hudson Bay

Mean wind speed in m/s
at a height of 50 metres

| 7.0–10.0 | 5.0–6.9 | 3.0–4.9 |

0 kilometres 600

Northern wind energy challenges

- Extreme cold temperatures can cause ice (rime) buildup on the blades
- Wind monitoring and servicing require local skilled help
- The more powerful taller towers also require larger foundations and larger cranes for installation
- Logistical issues may be more problematic in the Far North
- The climate is extremely hard on machines, reducing lifespans drastically
- Lattice towers and towers with guy wires are lighter than tubular towers, meaning that less foundation is necessary. But they do have more surface area for rime icing
- Will require federal help as it is too expensive for private companies to consider

How turbines work

Wind is created by the uneven heating of the Earth's surface. Wind turbines convert the kinetic energy in wind into mechanical power that can run a generator.

Rotor
Generator
Tail
Tower

Source: O'Neill (2008). Reprinted with permission from *The Globe and Mail*.

of the increasing global price of oil. Wind power, as a supplementary source of energy to reduce dependency on fossil fuels, could utilize the extensive wind energy found in the North, but it requires organization—perhaps an innovative combination of private industry, governments, and Aboriginal organizations. The wind energy exists but it needs to be harnessed. Canada's Arctic coast, for instance, registers some of the highest average annual wind speeds in North America. In an Arctic environment, design and maintenance are issues, but they are resolvable. Producing electricity from wind energy has several advantages:

- It is a renewable resource.
- Training of local utility workers would be required, and they would increase the size of the local workforce.
- Wind power technology could be the basis of a new scientific industry in Canada where a wind turbine technology centre could undertake research design and product development for small northern communities and isolated industrial sites; offer advanced training in the operation and maintenance of this technology to northerners; and, with the help of northern leaders, government officials, and company officials, sell this technology to similar communities and industrial sites around the circumpolar world.
- Substituting wind energy for fossil fuel energy would help Canada reduce its greenhouse gas emissions.
- As a joint venture led by resource companies but strongly supported by governments and Aboriginal organizations, such a project could serve as a model for other joint efforts to search for breakthrough solutions to northern problems. Resource companies operating in a hinterland will find it to their advantage to go beyond their economic goal of profit-making and harness their vast power to tackle social and environmental issues of concern to northerners. Small-scale wind power generation could be one such challenge for the stakeholders in the Canadian North: resource companies, governments, and Aboriginal organizations.

Narrowing the Divide

Beyond practical solutions for the near future, a long-term national goal for the twenty-first century must involve a challenging—even daunting—test of national purpose and resolve: narrowing the economic and social divide between northern Canada and southern Canada. Given the enormous economic and social distance between these two Canadas, extraordinary efforts are needed to close the gap. Federal leadership is essential, but more than that, Canadians must recognize and accept this challenge.

Corporations, so dominant in the northern economy, certainly have a role in turning the tide, but Aboriginal peoples, particularly their youth, also must accept the challenge by obtaining the schooling and training necessary for full participation in this unfolding world. While epic projects, like the construction of the CPR to the Pacific coast in the late nineteenth century, are rare in Canada history, a northern epic project or two, involving corporate, public, and Aboriginal elements, could narrow the divide. The magnitude of such a nation-building project could easily span 50 years or more, but the key elements are a vision and a commitment.

Given the huge size of the North, one epic project for each natural region of the North is appropriate. In the case of the Subarctic, provincial hydroelectric systems of dams, generating stations, and high-voltage transmission lines already exist (Vignette 9.3). The trick is linking them into a national system. Surplus production exists in both Quebec and Manitoba while Ontario is an energy-deficit area. Of course, such a project must be cost-efficient and that is one question that an experts' panel of Aboriginal leaders, scientists, and business people would examine, doubtless keeping in mind the ever-increasing cost of fossil fuels.

Later in this century, climate warming may free the Northwest Passage of ice for long periods of time and ocean-going vessels may ply these waters. This would warrant

---

Vignette 9.3   Can a National Dream Become a Reality?

Geography has given the Subarctic vast water resources. In the twenty-first century, might a national electrical grid, linking northern hydroelectric projects in Canada's Subarctic with national markets in a partnership with local Aboriginal groups, be created by Ottawa, provincial governments, and their Crown corporations? Given the experience of Hydro-Québec, perhaps this Crown corporation could play a lead role.

Energy, especially clean energy, represents a valuable asset and could shape Canada's future. Since energy falls under provincial jurisdiction but transfers across borders are a federal responsibility, a federal–provincial conference to examine such a bold scheme would be a necessary first step. Such a grand project would require substantial federal and provincial funding, and, of course, political leadership. Then, too, the environmental and Aboriginal concerns must be addressed.

The beginnings of a potential east–west grid are already in place, and provinces with surplus energy, such as Newfoundland and Labrador, Quebec, and Manitoba, are anxious to supply energy-deficient Ontario. The cost of construction would be great, the timeline would be long, and political negotiations are necessary to allow power transfers across provincial boundaries. Yet such a bold concept would give direction and purpose to the country, and one bonus for national unity would be the solving of the bitter and long-standing dispute between Quebec and Newfoundland and Labrador over Churchill Falls and access to the Hydro-Québec high-voltage transmission system. Failure of Hydro-Québec to find a way to allow electricity from the proposed Lower Churchill Falls project to cross its territory at a fair price is driving an alternative (and arguably dubious) proposal to build an underwater transmission system to Nova Scotia, from where the power could be transmitted to markets in the United States. If the underwater transmission system comes into reality, the electrical power from Churchill Falls will likely no longer flow across Quebec but go to the proposed system.

The first stage of a national plan would involve transmitting power from the proposed Lower Churchill River dam in Labrador to energy-deficient southern Ontario, across Quebec, thus resolving somehow the current dispute between the two provinces. The second stage or leg would send the newly created power from the proposed Conawapa dam on the Nelson River to southern Ontario. The final stage could extend the high-voltage transmission line further west, connecting Saskatchewan, Alberta, and British Columbia. Following the Hydro-Québec model, surplus electricity would be sold to utilities in the United States and profits would help finance the Subarctic grid.

a second epic venture. In anticipation of an ice-free Northwest Passage, marine infrastructure could be put into place over a long period of time, the role of the Canadian Rangers could expand, and more public housing could be built. Marine infrastructure could include building a wharf for each Arctic community. By constructing one wharf per year, eventually each Arctic community would benefit in three ways: first, its loading and unloading transportation costs would be substantially reduced; second, local tourism would benefit from global crews and passengers of small vessels and sailing boats docking and spending time in the community; and third, a local fishing industry could take root. Also, given the increase in the area of open water anticipated, the fishing industry is expanding, with Arctic char, flounder, clams, crabs, and scallops among the key fisheries. With the recent resurgence of northern cod, Labrador and Nunavut fishers might have another opportunity to harvest cod. Such harvesting in northern waters, however, must learn from the collapsed fisheries of the recent past: in all likelihood, they must be community-based inshore fisheries, not the domain of factory trawlers that remain at sea for extended periods of time. The primary economic advantage comes from processing fish, which would be a great boon to Inuit communities that face extremely high unemployment rates.

Summing Up

For much of the twentieth century, northern development was seen from the narrow perspective of a resource frontier. This Western-based perspective was based on market-driven modernization theory, which purported to provide benefits to all. This assumed, of course, that all have an equal chance. Such is not the case in the North. Past failures of northern development to achieve a meaningful 'trickle down' of benefits have forced a much broader interpretation of northern development. This shift began in the late twentieth century, and a recent example of a more enlightened corporate approach is exemplified by the Canadian firm, Agnico-Eagle, which reached an IBA with the Kivaliq Inuit Association in 2006 and the following year signed a development partnership with the Nunavut government. These agreements established a formal partnership allowing for Baker Lake residents to have a meaningful place in employment and business opportunities. The Baker Lake example comes on the heels of past accomplishments that have changed the playing field for Aboriginal peoples. These accomplishments include comprehensive land claim agreements that have created Aboriginal homelands, where co-management of the environment, wildlife, and resources empowers Aboriginal partners. Over the past half-century, Aboriginal peoples have fought for their place, and as a consequence Canadian society—including its First Peoples—has begun to change. But if the twenty-first century is to become, in any sense, Canada's century, more social, political, and economic change must occur, and this change must account for what appears at present to be the overriding change in Canada's North—a warming climate, and all that this entails.

By discarding old ideas of the twentieth century and by accepting the challenges of the present and foreseeable future, Canadians could, with federal leadership, redefine the purpose of northern development and, at the same time, push the rest of Canada towards a more progressive, harmonious, and united society. Perhaps Canada needs an epic venture or two to ignite this process of national unity?

Challenge Questions

1. Change is a fact of life in all regions of Canada. What makes change in the Canadian North different?
2. 'They [Tom Thomson and the Group of Seven] believed that by heading into the bush they were leaving behind an older, European civilization and discovering a more authentic—and more distinctively Canadian—world of nature.' Did Ross King (2010: 49) capture that elusive image of Canada's North in the above statement?
3. The North has always been an important element of the national psyche, but changing it requires a most significant event. The Berger Report of the 1970s was such an event while Richard Rohmer's Mid-Canada Development Corridor fell on deaf ears. Do you think the two epic ventures described in this chapter are just idle words or potential historic game-changers?
4. Can you make a case linking the chronic public housing shortage in Aboriginal communities to the federal 'relocation strategy' of the 1950s?
5. The Nunavut government, by introducing the Official Languages Act and the Inuit Language Protection Act (2007), is struggling to find a balance between language (Inuktitut) and jobs for Inuit (primarily government jobs). What are the advantages and disadvantages of this proposed legislation?
6. The field of environmental impact assessment has evolved and companies today accept their environmental responsibilities. Is this the case with the developers of the oil sands? Explain.
7. What must Canada do to secure its claim to the international seabed that lies beyond national boundaries?
8. With very high average annual wind speeds along the Arctic coast and the cost of diesel fuel increasing rapidly, what physical factors are holding back the development of wind-generated power (as a supplement to diesel power)?
9. Given the need for Arctic security coupled with the bleak employment situation in Arctic communities, should Ottawa expand the number and functions of Canadian Rangers and adjust their pay accordingly?
10. Can you see a positive consequence for the Mackenzie Gas Project if the Northwest Passage becomes an important global seaway?

Notes

1. In the 1990s, impact and benefit agreements between Aboriginal groups and resource companies became mandatory. These agreements have the potential to increase Aboriginal employment, but low education levels, lack of job experience in the resource industry, and limited technical skills have so far prevented Aboriginal workers from realizing this potential. As well, the problem of hiring Aboriginal workers becomes even more complex on union jobs for two reasons: (1) very few Aboriginal workers hold union cards; and (2) union hiring halls are located in southern cities.
2. The annual budget for the Nunavut Hunter Support Program is generated primarily by the interest on a $30 million fund, and, to a much lesser degree, by drawing down the principal. In 1994, the government of the Northwest Territories and Nunavut Tunngavik Inc. contributed $15 million each to this fund. According to Légaré (2000), additional contributions would be necessary by 2010 to keep this program functioning. Unfortunately, the Nunavut Hunter Support Program has not received such funding and Nunavut Tunngavik has had to

restrict its support to a maximum grant of $3,000 awarded to hunter and trapper organizations, which are then responsible for organizing community harvests to benefit community members (Nunavut Tunngavik, 2010).

3. The Board of Compensation (managing funds from the 1975 JBNQA) finances economic ventures directly and manages Air Creebec, Cree Construction, Valpiro, and Cree Energy through its holding company, Creeco. Eeyou Corporation manages the funding received under the 1986 La Grande agreement and invests in community development and economic development ventures. The Cree Development Corporation set up under the Paix des Braves with Quebec (7 February 2002) is a vehicle for investment in economic ventures using the funding from this agreement (see www.gcc.ca/cra/economicdevelopment.php).

4. Established in 1947, Canadian Rangers are responsible for protecting Canada's sovereignty by reporting unusual activities or sightings, collecting local data of significance to the Canadian Forces, and conducting surveillance or sovereignty patrols as required. In the Arctic, there are approximately 1,500 Rangers, most of whom are Inuit. In 2010, over 4,250 Canadian Rangers belonged to 'a subcomponent' of the Canadian Forces, with the number expected to reach 5,000 in 2012 (Department of National Defence, 2011). The Rangers reside in 169 northern communities in the NWT, Yukon, and Nunavut.

5. Dr Ross Mackay, Department of Geography, University of British Columbia, spent his professional life undertaking geomorphic and permafrost field research in the Arctic. Many years ago, I had the pleasure of taking a physical geography class from Dr Mackay, who punctuated his lectures with riveting examples of his Arctic research. A summary of his remarkable career is available at <cgrg.geog.uvic.ca/jrm2.htm>.

References and Selected Reading

Abele, Frances. 2009. 'Northern Development: Past, Present and Future', in Frances Abele, Thomas J. Courchene, F. Leslie Seidle, and France St-Hilaire, eds, *Northern Exposure: Peoples, Powers and Prospects in Canada's North*. Montreal: Institute for Research on Public Policy, 19–65.

———, Katherine A. Graham, and Allan M. Maslove. 1999. 'Negotiating Canada: Changes in Aboriginal Policy over the Last Thirty Years', in Leslie A. Pal, ed., *How Ottawa Spends 1999–2000*. Toronto: Oxford University Press, 251–92.

Alberta. 2011. 'Aboriginal People', *Alberta Oil Sands*. At: <www.oilsands.alberta.ca/aboriginalpeople.html>.

Bell, Jim. 2007. 'Tapardjuk: GN Won't Change Language Laws', Nunatsiaq News, 14 Dec. At: <nunatsiaqnews.com/archives/2007/712/71214/news/nunavut/71214_780.html>.

Bell, Patricia. 2007. 'Harper Announces Northern Deep-Sea Port, Training Site', CBC News. At: <www.cbc.ca/canada/story/2007/08/10/port-north.html>.

Berger, Thomas R. 1977. *Northern Frontier, Northern Homeland: The Report of the Mackenzie Valley Pipeline Inquiry*. Ottawa: Department of Supply and Services.

Byers, Michael. 2009. *Who Owns the Arctic?* Vancouver: Douglas & McIntyre.

Birchall, S.J. 2006. 'Canadian Sovereignty: Climate Change and Politics in the Arctic', *Arctic* 59, 2: iii–iv.

Cairns, Alan C. 2000. *Citizens Plus: Aboriginal Peoples and the Canadian State*. Vancouver: University of British Columbia Press.

Campbell, Colin. 2006. 'Saving Inuktitut: Nunavut Has Passed Legislation Intended to Keep the Inuit Language from Fading', *Maclean's*. 20 Sept. At: <www.macleans.ca/canada/national/article.jsp?content=20060925_133597_133597>.

Canada. 2000. *From Crisis to Opportunity: Rebuilding Canada's Role in Northern Research*. Final Report to the Natural Sciences and Engineering Research Council of Canada and the Social Sciences and Humanities Research Council from the Task Force on Northern Research. Ottawa: NSERC and SSHRC. At: <www.nserc.gc.ca/Pub/crisis.pdf>.

———. 2007. 'Prime Minister Harper Bolsters Arctic Sovereignty with Science and Infrastructure Announcements', *International Polar Year 2007–2008*. At: <www.ipy-api.gc.ca/me/nr/05-10-07_e.html>.

———. 2010. 'Investments in Forest Industry Transformation Program', *Action Plan*. At: <www.actionplan.gc.ca/initiatives/eng/index.asp?mode=2&initiativeID=181>.

———. 2011. Budget 2011, 6 June. At: <www.budget.gc.ca/2011/home-accueil-eng.html>.

Canadian American Strategic Review (CASR). 2007. 'Carving Up the Arctic Seabed—Two Options: The "Meridian" Method or the "Sector" Solution', Oct. At: <www.sfu.ca/casr/id-arctic-empires-4.htm>.

Carnaghan, Matthew, and Allison Goody. 2006. *Canadian Arctic Sovereignty*. Ottawa: Library of Parliament. At: <www.parl.gc.ca/information/library/PRBpubs/prb0561-e.htm>.

CBC News. 2006a. 'Kashechewan: Water Crisis in Northern Ontario'. At: <www.cbc.ca/news/background/aboriginals/kashechewan.html>.

———. 2006b. 'Learn Inuktitut or Iqqanaijaaqajjaagunniiqtutit, Mandarins Told', 7 June. At: <www.cbc.ca/news/canada/north/story/2006/06/07/nor-bilignual-senior.html>.

———. 2011. 'Oilsands Mine Approval Condemned', 28 Jan. At: <www.cbc.ca/news/business/story/2011/01/28/total-oilsands-mine-approval.html >.

Chambers, Howard G. 2007. Personal communications, Safe Water Session at the Aboriginal Policy Research Consortium (International) meeting, London, Ont., 6 Nov.

Chase, Steven, and Bertrand Marotte. 2011. 'Halifax, Vancouver win $33-billion in shipbuilding', *The Globe and Mail*, 20 October: A1.

Coates, Kenneth, P. Whitney Lackenbauer, William R. Morrison, and Greg Poelzer. 2008. *Arctic Front*. Toronto: Thomas Allen.

Dawson, Jackie, Emma J. Stewart, Patrick T. Maher, and D. Scott Slocombe. 2008. 'Climate Change, Complexity and Cruising in Canada's Arctic: A Nunavut Case Study', in Robert B. Anderson and Robert M. Bone, eds, *Natural Resources and Aboriginal People in Canada: Readings, Cases and Commentary*. Toronto: Captus Press.

Department of National Defence, 2011. 'Canadian Rangers', 27 June. At: <www.army.dnd.ca/land-terre/cr-rc/index-eng.asp>.

Environment Canada, 2011a. *Canada's Emissions Trends,* July. At: <www.ec.gc.ca/Publications/E197D5E7-1AE3-4A06-B4FC-CB74EAAAA60F%5CCanadasEmissionsTrends.pdf>.

———. 2011b. 'An Integrated Monitoring Plan for the Oil Sands', 21 July. At: <www.ec.gc.ca/default.asp?lang=En&n=56D4043B-1&news=7AC1E7E2-81E0-43A7-BE2B-4D3833FD97CE>.

Fortier, M., and L. Fortier. 2006. 'Canada's Arctic, Vast, Unexplored and in Demand: Canadian-led International Research in the Changing Coastal Canadian Arctic', *Journal of Ocean Technology* 1, 1: 1–8.

George, Jane. 2010. 'A Nunavut Milestone: Meadowbank's First Gold Bar', *Nunatsiaq Online*, 1 Mar. At: <www.nunatsiaqonline.ca/stories/article/98767_a_nunavut_milestone_meadowbanks_first_gold_bar/>.

Grant, Shelagh D. 2010. *Polar Imperative*. Vancouver, Douglas & McIntyre.

Griffiths, F. 2004. 'Pathetic Fallacy: That Canada's Arctic Sovereignty Is on Thinning Ice', *Canadian Foreign Policy* 11, 3: 1–16.

Hassol, Susan Joy. 2004. 'Impacts of a Warming Arctic: Arctic Climate Impact Assessment'. At: <amap.no/acia>.

Harrison, C. 2006. 'Industry Perspectives on Barriers, Hurdles, and Irritants Preventing Development of Frontier Energy in Canada's Arctic Islands', *Arctic* 59, 2: 238–42.

Hicks, Jack, and Graham White. 2000. 'Nunavut: Inuit Self-Determination through a Land Claim and Public Government?', in Dahl et al. (2000: 30–117).

Huebert, R. 2003. 'The Shipping News: How Canada's Arctic Sovereignty Is on Thinning Ice', *International Journal* 58, 3: 295–308.

Indian and Northern Affairs Canada. 2007. 'Canada's New Government Signs Agreement with Kashechewan First Nation to Redevelop Community'. At: <www.ainc-inac.gc.ca/nr/prs/m-a2007/2-2915-eng.asp>.

Jull, Peter. 2000. 'A Different Place', in Dahl et al. (2000: 118–36).

Kashechewan First Nation. 2010. *Community Planning Study*. At: <www.kashechewan.firstnation
.ca/node/29>.

King, Ross. 2010. 'What Tom Thomson Saw', *The Walrus* 7, 9: 44–51.

Koperqualuk, Lisa, ed. 2001. 'Premiere Issue: Our Land, Our Future', *Nunavik* 1: 1–17.

Lackenbauer, P. Whitney. 2008. 'The Canadian Rangers: A Postmodern "Militia" That Works', *Canadian
Military Journal*, 14 Aug. At: <www.journal.forces.gc.ca/vo6/no4/north-nord-03-eng.asp>.

Laghi, Brian, Karen Howlett, and Rhéal Séguin. 2008. 'Charest, McGuinty to Push PM on Economy:
Harper Government to Unveil $1-Billion in Aid for One-Industry Towns', *The Globe and Mail*,
10 Jan. At: <www.theglobeandmail.com/servlet/story/RTGAM.20080110.wpremiers10/BNStory/
National/home>.

Lalonde, Suzanne. 2007. 'Le passage Nord-Quest: zone canadienne ou internationale?', *Le Multilatéral*:
1–13.

Légaré, André. 2008. 'Canada's Experiment with Aboriginal Self-Determination in Nunavut: From
Vision to Illusion', *International Journal on Minority and Group Rights* 15: 335–67.

Macnab, R. 2004. 'Canada's Arctic Waterways: Future Shipping Crossroads?', *Meridian*
(Fall/Winter): 1–5.

Mitchell, Bruce. 2005. 'Foreword', in Kevin S. Hanna, ed., *Environmental Impact Assessment: Practice
and Participation*. Toronto: Oxford University Press.

National Post. 2002. 'Nunavut's Bill 101', 4 Feb., A15.

National Snow and Ice Data Center (NSIDC). 2007. 'Arctic Sea Ice News Fall 2007'. At: <nsidc.org/
news/press/2007_seaiceminimum/20070810_index.html>.

Nunavik Commission. 2001. *Amiqqaaluta—Let Us Share: Mapping the Road toward a Government for
Nunavik*. Quebec: Nunavik Commission.

Nunavut Tunngavik Inc. 2010. 'Nunavut Harvest Support Program'. At: <www.tunngavik.com/
documents/beneficiaryProgramForms/NHSP%20Community%20Harvest%20Program%20
Description%20ENG.pdf>.

O'Neill, Katherine. 2008. 'Territories Hope Arctic Winds Pack Power', *The Globe and Mail*, 3 Jan. At:
<www.theglobeandmail.com/servlet/story/RTGAM.20080103.wind03/BNStory/National/home>.

Pharand, Donat. 2007. 'The Arctic Waters and the Northwest Passage: A Final Revisit', *Ocean
Development and International Law* 38, 1 and 2: 3–69.

Pope, Alan. 2006. *Report on the First Nation of Kashechewan and Its People*. Ottawa: Indian and
Northern Affairs Canada. At: <www.ainc-inac.gc.ca/nr/prs/s-d2006/kfnp_e.html>.

Salisbury, Richard F. 1986. *A Homeland for the Cree: Regional Development in James Bay, 1971–1981*.
Montreal and Kingston: McGill-Queen's Press.

Vanderklippe, Nathan. 2011a. 'South Korean Firm Eyes Arctic Natural Gas', *The Globe and Mail*, 20
Apr., B1, B6.

———. 2011b. 'Oil Industry Outlines Cleanup Strategy for Arctic Spill', *The Globe and Mail*, 9 June,
B1.

Weber, Bob. 2011. 'Arctic Oil Spill Cleanup Impossible One Day in Five, Energy Board Report
Finds', *The Globe and Mail*, 1 Aug., B1.

Weeneebayko Health Ahtuskaywin. 2006. 'Kashechewan'. At: <www.wha.on.ca/a_kash.html>.

Appendix I

The Method of Calculating Nordicty

While the division of the North into the Arctic and Subarctic demonstrates the existence of two physical environments, nordicity permits us to measure differences between places and to create human/physical regions. But what exactly does nordicity measure? A Canadian geographer, Louis-Edmond Hamelin, created this term to quantify 'northernness' into a single value measured as polar units. In sum, nordicity is a quantitative measure based on 10 variables found at a particular northern place. Nordicity is based on both physical elements, such as annual cold, permafrost, and ice cover of water bodies, and human elements such as population size, accessibility by land, sea, and air, and degree of economic activity. These 10 physical and human elements seek to represent all facets of the North. Hamelin set numeric values (polar units) for each variable. At a particular place, the sum of polar values for the 10 variables results in a single numeric value. This value represents the nordicity of that particular place. The North Pole, for example, has a nordicity of 1,000 polar units. Isachsen, the northernmost weather station in Canada, has 925 polar units while Vancouver has only 35. Hamelin determined that the boundary between northern and southern Canada followed a line representing 200 polar units. Hamelin is describing the Canadian North from a southern perspective. Northerners may not have the same mental map of Canada. For them, southern Canada is a distant and different place while the terms Middle North, Far North, and Extreme North have no meaning for them.

The method of calculating the number of polar units for a place is summarized below. For example, the number of polar units assigned to varying degrees of latitude ranges from zero units for latitudes of 45°N or less to 100 units at the North Pole (90°N).

List of 10 Variables and Their Polar Units		
Variable		**Polar Units**
1. Latitude	90°	100
	80°	77
	50°	33
	45°	0
2. Summer Heat	0 days above 5.6°C	100
	60 days above 5.6°C	70
	100 days above 5.6°C	30
	>150 days above 5.6°C	0

Variable			Polar Units
3. Annual Cold		6,650 degree days below 0°	100
		4,700 degree days below 0°	75
		1,950 degree days below 0°	30
		550 degree days below 0°	0
4. Types of Ice	Frozen ground	Continuous permafrost 457 m thick	100
		Continuous permafrost <457 m thick	80
		Discontinuous permafrost	60
		Ground frozen for less than 1 month	0
	Floating ice	Permanent pack ice	100
		Pack ice for 6 months	36
		Pack ice <1 month	0
	Glaciers	Ice sheet >1,523 m thick	100
		Ice cap 304 m thick	60
		Snow cover <2.5 cm	0
5. Annual Precipitation		100 mm	100
		300 mm	60
		500 mm	0
6. Natural Vegetation		Rocky desert	100
		50% tundra	90
		Open woodland	40
		Dense forest	0
7. Accessibility	Land or sea	No service	100
		For two months	60
		Up to four months by both land and sea	15
		Continuous service by either land or sea	0
8. Accessibility	Air	Charter only	100
		Weekly regular service	25
		Daily regular service	0
9. Population	Settlement size	None	100
		About 100	85
		About 1,000	60
		>5,000	0
	Population density	Uninhabited	100
		1 person per km^2	50
		4 persons per km^2	0
10. Economic Activity		No production	100
		Exploration	80
		20 hunters/trappers	75
		Interregional centre	0

Source: Hamelin (1979: ch. 1).

Appendix II

Population by Northern Census Divisions

Northern Census Divisions	2001	2006	% Change
Yukon	28,674	30,372	5.9
Fort Smith, NWT	28,824	32,272	12.0
Inuvik, NWT	8,536	9,192	7.7
Baffin, Nunavut	14,372	15,765	9.7
Keewatin, Nunavut	7,557	8,348	10.5
Kitikmeot, Nunavut	4,816	5,361	11.3
Territorial North	**92,779**	**101,310**	**9.2**
British Columbia	**198,289**	**196,027**	**–1.1**
Bulkley-Nechako	40,856	38,243	–6.4
Fraser-Fort George	95,317	92,264	–3.2
Peace River	55,080	58,264	5.8
Stikine	1,316	1,109	–15.7
Northern Rockies	5,720	6,147	7.5
Alberta	**100,479**	**112,362**	**11.8**
16	42,971	53,080	25.5
17	57,508	59,282	3.1
Saskatchewan	**32,029**	**33,919**	**5.9**
18	32,029	33,919	5.9
Manitoba	**66,622**	**68,279**	**2.5**
21	22,556	21,606	–4.2
22	35,077	38,421	9.5
23	8,989	8,252	–8.5

Northern Census Divisions	2001	2006	% Change
Ontario	**416,475**	**413,446**	**–0.7**
Cochrane	85,246	82,503	–3.2
Algoma	118,567	117,461	–0.9
Thunder Bay	150,860	149,063	–1.2
Kenora	61,802	64,419	4.2
Quebec	**515,126**	**499,503**	**–3.0**
Rouyn-Noranda	39,621	39,924	0.8
Abitibi-Ouest	21,984	20,792	–5.5
Abitibi	24,613	24,275	–1.4
La Vallée-de-l'Or	42,375	41,896	–1.1
La Tuque	15,862	15,448	–2.6
Le Domaine-du-Roy	32,839	31,956	–2.7
Maria-Chapdelaine	26,900	25,767	–4.2
Le Saguenay-et-son-Fjord	166,780	163,717	–1.8
La Haute-Côte-Nord	12,894	12,303	–4.6
Manicouagan	33,620	33,052	–1.7
Sept-Rivières-Caniapiscau	38,931	38,661	–0.7
Minganie-Basse-Côte-Nord	12,321	11,895	–3.5
Nord-du-Québec	38,575	39,817	3.2
Newfoundland & Labrador	**47,735**	**44,668**	**–6.4**
9	20,091	18,084	–10.0
10	25,230	23,950	–5.1
11	2,414	2,634	9.1
Provincial North	**1,376,755**	**1,368,204**	**–0.6**
Territorial North	**92,779**	**101,310**	**9.2**
Canadian North	**1,469,534**	**1,469,514**	**0.0**

Glossary

Aboriginal peoples: The original inhabitants of North America. Canada's Constitution Act, 1982 defines Aboriginal peoples as including 'the Indian, Inuit and Métis peoples'.

Aboriginal title: A legal right to land; the claim depends on documenting current and traditional occupancy and use of the land. See *Calder* and *Delgamuukw*.

Air mass: A large body of air that has similar horizontal temperature and moisture characteristics.

Albedo: Proportion of solar radiation reflected by the Earth's surface back into the atmosphere. The more radiation reflected away from the Earth by light-coloured or white surface, the higher the albedo.

Arctic Archipelago: The group of islands in the Arctic Ocean representing a total land area of 1.3 million km². This Canadian archipelago consists of nearly 37,000 islands, many of which are tiny. Some of the islands are very large—Baffin Island is the fifth largest in the world, and other large islands include Victoria, Ellesmere, Banks, Devon, Axel Heiberg, Melville, and Prince of Wales.

Arctic Circle: This imaginary line, like latitudes, extends in an east/west direction at 66° 33'N; at the time of the summer solstice (21 June), the Sun at this location on the Earth's surface does not set below the horizon for the entire day; similarly, at the winter solstice (21 December), the sun does not rise above the horizon for 24 hours.

Arctic Clause: Article 234 of UNCLOS, which allows Canada to enforce its Arctic Waters Pollution Prevention Act in its exclusive economic zone (200 nautical miles from shore).

Arctic Council: Group of eight nations (the Arctic Five plus Finland, Iceland, and Sweden) that plays an advisory role on Arctic issues, especially related to the environment, but not those related to boundaries or resource development. Representatives of Aboriginal organizations have observer status at Council meetings.

Arctic Five: The five countries (Canada, Denmark, Norway, Russia, and the United States) with coasts bordering the Arctic Ocean and that therefore can make sovereign claims to the Arctic seabed under UNCLOS. Denmark's claim is through its possession of Greenland.

Arctic ice pack: The Arctic ice cap, consisting of ice that covers much of the Arctic Ocean throughout the year, although in recent years this ice pack has reduced in size during the late summer. See *pack ice*.

Beaufort Gyre circulation system: The principal ocean currents in the Beaufort Sea have a clockwise direction (anticyclone) in the winter months and counterclockwise direction (cyclone) in the summer, at which time, cold and relatively fresh water and ice flow into Baffin Bay.

Berger Report: The 1977 report of the Mackenzie Valley Pipeline Inquiry headed by Justice Thomas Berger. The report emphasized environmental issues and the potential clash between resource projects and the Aboriginal way of life. Berger recommended that no pipeline be built through the northern Yukon and that a pipeline through the Mackenzie Valley should be delayed for 10 years.

Biodiversity: The variety of life forms on Earth or that inhabit a particular ecosystem or geographic region.

Biomagnification: The increase in concentration of toxic substances in organisms as contaminants are passed up food chains from lower forms, such as plankton and fish, to higher forms, such as whales and seals.

Bitumen: A thick, black, viscous oil consisting of naturally occurring hydrocarbons mixed with other substances such as sand and clay.

Bitumen-Royalty-in-Kind program: An Alberta program, initiated in 2009, to ensure a bitumen supply for Alberta's existing and proposed upgraders by having royalty payments in bitumen rather than cash.

Boom-and-bust cycle: A business cycle often associated with an economy based heavily on a single commodity. During the boom cycle, demand and prices for the commodity increase quickly and the region experiences a boom; during the bust cycle, demand and prices fall even faster, leaving the region's economy in tatters.

Borders: Demarcations between two political units (countries, provinces, municipalities) as applied to land.

Boundary: Demarcation of political jurisdiction between two countries as applied to a body of water.

Calder: *Calder v. Attorney General of British Columbia*, a 1973 Supreme Court of Canada decision that determined 'Indian title' is a legal but undefined right, based on Aboriginal peoples' historic 'occupation, possession and use' of traditional territories.

Canadian Environmental Assessment Act: Legal foundation for EIA under federal jurisdiction. In 1995, the Canadian Environmental Assessment Act replaced the Environmental Assessment Review Process.

Canadian Environmental Assessment Agency: Agency that oversees federal EIA and implementation of the Canadian Environmental Assessment Act. Formed in 1994, this agency replaced the Federal Environmental Assessment Review Office.

Circumpolar world: Those lands and peoples found in the higher latitudes of the northern hemisphere; also the countries bordering on the Arctic Ocean whose environments consist of Arctic and Subarctic biomes. In a more narrow sense, it refers to the Arctic and its original inhabitants.

Clean energy policy: US policy announced by President Obama on 15 June 2010. The policy is linked to US efforts to reduce greenhouse gas emissions. Later in 2010, the question of the US importing oil from Alberta's oil sands became a political issue that saw the term 'dirty oil' become a rallying point against approval of the extension of the Keystone pipeline to Gulf state heavy oil refineries.

Climate change: Changes measured over a period of time by significant variations in global temperatures and precipitation due to natural factors or human activity. Climate is normally thought of as 'stable', but since the Earth was formed many climate changes have taken place. Today, the concern is that human-caused climate change could be irreversible.

Colonialism: The imposition of one society/country on another society in another part of the world; associated with cultural, economic, and political domination.

Core/periphery model: A theoretical model that provides a geographic framework for interpreting the North, which is seen as a resource hinterland dominated and exploited by an industrial core (southern Canada, as well as wider North American and global economic forces). In reality, the flow of wealth from the North is somewhat offset by the injection of public funds through transfer payments.

Country food: Food obtained by hunting, fishing, and gathering. While country food is no longer the chief source of food for many Native northerners, it remains important both nutritionally and culturally, particularly those in more remote communities.

Cree Hunters and Trappers Income Security Program: A subsidized provincial program for Quebec Cree hunters and trappers that allows them to stay on the land for part of the year harvesting furs and country food for Cree communities.

Cryosolic soils: Thin soils formed in the continuous permafrost zone. These soils have active or thawed layers less than one metre thick.

Delgamuukw: *Delgamuukw v. British Columbia*, a case that reached the Supreme Court of Canada, which in its 1997 decision defined how Aboriginal title may be proved and outlined the justification test for infringements of Aboriginal title.

Demographic transition theory: A sequence of demographic changes in which populations progressively move over time from high birth and death rates to low birth and death rates.

Dependency: A corollary of dominance; a situation where a region or people must rely on other regions or people for their economic well-being. Dependency can also mean that external capital and technology play a paramount role in the regional economy and that local politicians have little power relative to higher levels of government.

'Dirty oil': Oil extracted from Alberta's oil sands; a term used by opponents of the oil sands projects, which are heavy polluters of the air, water, and land and use enormous amounts of water and natural gas in the extraction and upgrading processes. See 'ethical oil'.

Drumlins: Elongated hills composed of glacial deposits formed by massive subglacial flooding; the long axis of a drumlin parallels the direction of glacier flow.

Economic cycle: The business cycle, which consists of periods of economic expansion and contraction. This cycle is irregular and thus cannot be predicted.

Ecumene: Portion of the land that is permanently inhabited by humans.

Environmental Assessment Review Process: The first EIA process affecting federal lands, created in 1973 by the federal government.

Epidemiological transition: A sequence of health changes that progressively move over time from high infant mortality to low infant mortality and death coming at higher ages. Control of infectious diseases accounts for the decline in infant mortality rates while control of degenerative diseases has resulted in longer lifespans.

Erratics: Boulders moved by glacial ice and deposited in another place.

Eskers: Ridges of sand and gravel, sometimes many kilometres long, deposited by streams beneath or within a glacier.

'Ethical oil': A slogan used by supporters of oil sands development and adopted by both the federal and Alberta governments to counter the term 'dirty oil'. The term 'ethical oil' first appeared in the Canadian media in September 2010 with the release of Ezra Levant's book, *Ethical Oil: The Case for Canada's Oil Sands*.

Ethnocentrism: The belief in the superiority of one's own culture or people and, thus, judging other cultures or groups using criteria specific to one's own group.

Ethnosphere: Term coined by Wade Davis to encompass the cultures and languages of past and present human populations; the global social web of life.

Exclusive Economic Zone: The territorial sea that extends 200 nautical miles from shore as measured from its baseline, i.e., at low tide.

Within this area, coastal nations have sole rights over all natural resources.

Federal Environmental Assessment Review Office: Federal agency created in 1973 that later evolved into the Canadian Environmental Assessment Agency. Its responsibility was to implement Environmental Assessment Review Process.

Gelifluction: The movement of thawed soil downslope forming a series of distinct lobes; occurs in permafrost area. See *solifluction*.

Geopolitics: A branch of political geography focusing on the interplay of geography, national power, and international relations; often takes the political form of national strategies to maximize favourable resolutions to territorial disputes.

Glacial till: Unsorted and unstratified material deposited by ice sheets; also known as drift.

Glaciofluvial: Deposits or landforms produced by glacial meltwater streams.

Glaciolacustrine: Ancient lake bottoms, formed from the draining of glacial lakes.

Gleysolic soils: Water-saturated soils formed in marshy areas of the Subarctic.

Global circulation system: The general movement of air in the atmosphere and of water in the oceans from tropical areas to polar areas.

Global warming: The increasing temperature of the Earth due to the burning of fossil fuels, which adds greenhouse gases to the atmosphere.

Greenhouse effect: The effect on the Earth's temperature by certain atmospheric gases known as greenhouse gases that trap energy from the sun. These gases consist of water vapour, carbon dioxide, nitrous oxide, and methane.

Heavy oil: Crude oil with a higher density than light crude oil so that it does not flow easily; product of upgraders.

Holocene Epoch: The most recent geological epoch that began some 10,000 years ago, marking the beginning of an interglacial phase and warmer climates than those experienced in the Late Wisconsin glacial period.

Homeland: A region where the inhabitants have both a strong attachment and a commitment to its social and political institutions; a sense of place.

Horizontal drilling: A type of well drilling that deviates from its vertical drill hole and travels horizontally through a producing layer of natural gas or oil.

Hydraulic fracturing: Pumping a fluid into the well at very high pressure to create cracks in the reservoir rock and thus release natural gas or oil trapped in the reservoir rock.

Indian: A legal term in Canada for a person whose name is on the band list of any Indian community in Canada or on the central registry list in Ottawa. The main Indian linguistic groups in the Canadian North are Algonkian (e.g., Cree, Innu) and Athapaskan (e.g., Dene).

International Seabed Authority: Autonomous international organization established under the 1982 Law of the Sea Convention (UNCLOS) that organizes and controls activities of the ocean floor lying in international waters (known as 'the area').

Inuit: Aboriginal people whose homeland is the Arctic and who, traditionally, were nomadic hunters, especially of sea mammals such as seal, whale, and walrus; they comprise about 85 per cent of the population of Nunavut and also live in Arctic Quebec, northern Labrador, and the western Arctic.

Inuit Circumpolar Council: Non-governmental agency founded in 1977 that represents approximately 150,000 Inuit of Alaska, Canada, Greenland, and Russia (Chukotka).

Isostatic uplift: Rebound in the Earth's crust to a state of balance (isostasy) after massive ice sheets that depressed the crust have melted.

Joint Review Panel: Under the Canadian Environmental Assessment Act, a panel formed by the federal Minister of the Environment to conduct socio-economic assessment of an industrial project. In 2004, the minister appointed a JRP to examine the Mackenzie Gas Project. The panel's report, published in late 2009, expressed serious concern about the social and environmental impacts of the project if its recommendations were not met.

Land-fast ice: Newly formed sea ice attached to shore but extending for some distance into the sea; in the winter, land-fast ice merges with the Arctic ice cap while in the late summer, this ice becomes detached from shore and floats into open water as ice floes. Also known as fast ice. Unlike the Arctic ice pack, land-fast ice is attached to the land.

Late Wisconsin glacial period: The last glacial phase in North America, also known as the Late Cenozoic Ice Age, when huge ice sheets covered practically all of present-day Canada and northern portions of what is now the United States. The Late Wisconsin glaciation spanned a period from 25,000 BP to 10,000 BP.

Leads: Open water in the Arctic ice cap created by cracks resulting from its movement. The cracks cause smaller areas of solid ice with water between them; these leads can open and close quickly.

Life expectancy: The average number of additional years a person would live if current mortality trends were to continue; most commonly cited as life expectancy at birth.

Little Ice Age: A period of cool temperatures that lasted some 400 years from around 1450 to 1850.

Malthusian population trap: Concept introduced by the English economist Thomas Malthus (1766–1834) that populations, if unchecked, tend to increase at a geometric rate while subsistence increases at an arithmetic rate. The trap occurs when a country or region has a higher rate of population increase than its rate of economic growth.

Maunder Minimum: The period roughly spanning 1645 to 1715 when sunspots became exceedingly rare, which may have been a causal factor in the Little Ice Age.

Megaprojects: Large-scale industrial undertakings, which, because of their enormous size, dominate the local and regional economy during the construction phase. Construction costs usually exceed $1 billion and the start-up phase can extend for several years.

Métis: The offspring of fur traders and Indian (primarily but not exclusively Cree and Ojibwa) women who developed a separate culture and history focused on the buffalo hunt and centred around the Red River colony in present-day

Manitoba. In 1982, the Métis gained official recognition as one of the three Aboriginal peoples of Canada, and the definition of Métis has been expanded by some groups and scholars to include all Canadians of mixed Indian–European heritage who pursue a common local or regional culture and lifestyle.

Modernization theory: Popular social scientific theory in the twentieth century that attempts to explain human progress over time from less advanced (i.e., non-Western) to more advanced (i.e., Western) societies, stressing those social elements that advance or hinder social advancement.

Mt CO²e: Megatonnes of CO^2 equivalent; unit used to report the amount of greenhouse gas emissions or reductions.

Multi-year ice: Ice that does not melt for one year or more; much thicker and harder than new ice.

Muskeg: A Cree term for bogland covered by sphagnum moss; found primarily in the Subarctic.

National Energy Board: An independent federal agency that regulates several parts of Canada's energy industry. Its purpose is to regulate pipelines and energy development and trade in the Canadian public interest.

Native peoples: Those Canadians of Indian, Inuit, and Métis ancestry; an older term describing Aboriginal peoples.

Natural increase: The surplus (or deficit) of births over deaths in a population over a given time period.

New ice: Sea water that freezes and then thaws within one year; also referred to as seasonal or young ice. See *multi-year ice*; *land-fast ice*; *pack ice*.

Non-status Indians: Those Canadian Indians who by birth, marriage, or choice have no legal status, under the Indian Act, to benefit from reserve lands and special federal programs.

Nordicity: Concept created by Louis-Edmond Hamelin to measure the degree of 'northernness' of a place. Nordicity provides a quantitative definition of the southern boundary of the North and is based on 10 variables, such as latitude,

degree of isolation, and annual cold, that are supposed to represent all facets of the North.

NORDREG: Canada's Northern Canada Vessel Traffic Services Zone Regulations. Mandatory since 2010, NORDREG requires vessels to report and receive permission to enter Canadian ice-covered waters.

Nunataks: Unglaciated mountain peaks that stood above the ice sheets.

Nunavik: Inuit-dominated territory in Arctic Quebec, soon to become a semi-autonomous region within the province.

Nunavut: Territory in Canada's eastern Arctic, formally established in 1999, that was hived off from the present Northwest Territories; means 'our land' in Inuktitut, and represents a political expression of the vision of an indigenous homeland.

Pack ice: Floating ice of varying age, size, and thickness that makes up the Arctic ice cap.

Patterned ground: Stones and pebbles arranged by frost action in a geometric pattern, e.g., circles or polygons. It is widespread in Arctic environments where frost action is the dominant geomorphic force.

Pingos: Ice-cored hills found in permafrost areas. Pingos range in height from a few metres to 50 metres, and expand in size as water seeps into the core and freezes. Most pingos are found in the Mackenzie Delta.

Plan Nord: A grand vision for Quebec's North announced by Premier Jean Charest in July 2011. Designed to foster sustained development, the plan calls for private companies and Crown corporations to invest over $80 billion over the next 25 years in new mines and forest enterprises as well as substantial provincial expenditures in transportation infrastructure and community development.

Pleistocene Epoch: A geological epoch associated with the last ice age; included at least four major ice advances, including the Wisconsin; began some 1.6 million years ago and ended some 10,000 years ago when the Holocene Epoch began.

Podzolic soils: Acidic soils found in the boreal forest.

Polynya (north water): Large area of open water in the Arctic Ocean surrounded by sea ice.

Population density: A measure of the number of people in a given area. In Canada, for instance, density is measured by population per km², and for a community to be considered urban the density must be at least 400 people per km².

Push–pull migration theory: A model explaining the movement of people from economically depressed, conflict-ridden, or environmentally damaged areas or countries to regions or countries with more economic opportunity, less conflict, or a more supportive environment. Factors in the home area push migrants from that location and factors in the receiving area pull or attract migrants to that location.

Sense of place: The intense feeling of belonging and loyalty to a region.

Slave Geological Province: An extremely rich mineral area in the Northwest Territories portion of the Canadian Shield.

Solifluction: The slow downslope movement of waterlogged soil. See *gelifluction*.

Status Indians: Canadian Indians who have 'status' and are registered under the Indian Act. They have a right to use reserve lands held by their band and access to federal funding for programs such as housing and education. Those status Indians whose ancestors signed a treaty also have treaty rights.

Steam-assisted gravity drainage: An oil recovery technology for extracting bitumen by a pair of wells drilled into the oil sands, one a few metres above the other. Low-pressure steam is continuously injected into the upper wellbore to heat the bitumen and reduce its viscosity, causing the heated bitumen to drain into the lower wellbore, where it is pumped out.

Subarctic: A natural region of North American distinguished by its natural vegetation, the boreal forest; also the traditional homeland of northern Indians.

Subsistence economy: An economic system of relatively simple technology in which people produce most or all of the goods to satisfy their own and their family's needs by hunting,

gathering, and/or subsistence farming; little or no exchange occurs outside of the immediate or extended family.

Tailings ponds: Huge ponds containing a slurry-like water/oil mixture laced with toxic chemicals, a by-product of bitumen extraction. The process of sedimentation separates the fine particles suspended in the water, causing the particles to settle at the bottom of the pond. In 2010 Suncor developed technology involving polymers to hasten the sedimentation process, which otherwise might not have been completed in this century.

Tar sands: Original name for deposits of bitumen, a viscous mixture of oil, sand, and clay, such as the deposits in northern Alberta; now more commonly called oil sands.

Traditional ecological knowledge: Knowledge about a local environment or ecosystem based on personal observation and experience, especially as it has been passed on from one generation to another living in a specific area.

Transpolar Drift: The movement of water and ice from the shores of Siberia across the North Pole, where it joins the East Greenland current that flows into the North Atlantic Ocean.

Tribal groups: Groups of Indians united by language and customs and belonging to the same band.

United Nations Convention on the Law of the Sea (UNCLOS): An international agreement, first established in 1982, that defines the rights and responsibilities of nations in their use of the world's oceans, including the ownership of coastal waters and seabeds; most countries have ratified this agreement with one notable exception: the United States.

Upgrader Alley: The industrial heartland of Alberta, extending from Edmonton to Fort Saskatchewan where most bitumen upgraders outside of Fort McMurray are located.

Upgrader plants: Operations that convert bitumen into synthetic crude oil, which then can be transported by pipelines to refineries where the synthetic crude is refined into various commercial products. Bitumen that has not been processed by an upgrader is thinned by adding lighter oil. This product, known as dilbit, can flow through pipelines to refineries.

Index